Sustainable Development Goals Series

World leaders adopted Sustainable Development Goals (SDGs) as part of the 2030 Agenda for Sustainable Development. Providing in-depth knowledge, this series fosters comprehensive research on the global targets to end poverty, fight inequality and injustice and tackle climate change.

Sustainability of Future Earth is currently a major concern for the global community ans has been a central theme for a number of major global initiatives viz. Health and Well-being in Changing Urban Environment, Sendai Framework for Disaster Risk Reduction 2015–2030, COP21, Habitat III and Future Earth Initiative. Perceiving the dire need for Sustainable Development, the United Nations and world leaders formulated the SDG targets as a comprehensive framework based on the success of the Millennium Development Goals (MDGs). The goals call for action by all countries, poor, rich and middle-income, to promote prosperity while protecting the planet earth and its life support system. For sustainability to be achieved, it is important to have inputs from all sectors, societies and stakeholders. Therefore, this series on the Sustainable Development Goals aims to provide a comprehensive platform to the scientific, teaching and research communities working on various global issues in the field of geography, earth sciences, environmental science, social sciences and human geosciences, in order to contribute knowledge towards the current 17 Sustainable Development Goals.

Volumes in the Series are organized by the relevant goal, and guided by an expert international panel of advisors. Contributions are welcome from scientists, policy makers and researchers working in the field of any of the following goals:

No poverty
Zero Hunger
Good Health and Well-Being
Quality Education
Gender Equality
Clean Water and Sanitation
Affordable and Clean Energy
Decent Work and Economic Growth
Industry, Innovation and Infrastructure
Reduced Inequalities
Sustainable Cities and Communities
Responsible Consumption and Production
Climate Action
Life Below Water
Life on Land
Peace, Justice and Strong Institutions
Partnerships for the Goals

The theory, techniques and methods applied in the contributions will be benchmarks and guide researchers on the knowledge and understanding needed for future generations. The series welcomes case studies and good practices from diverse regions, and enhances the understanding at local and regional levels in order to contribute towards global sustainability.

More information about this series at http://www.springer.com/series/15486

Ali Asgary
Editor

Resettlement Challenges for Displaced Populations and Refugees

Springer

Editor
Ali Asgary
Disaster & Emergency Management
School of Administrative Studies
Faculty of Liberal Arts and Professional Studies
York University
Toronto, ON, Canada

ISSN 2523-3084 ISSN 2523-3092 (electronic)
Sustainable Development Goals Series
ISBN 978-3-030-06445-7 ISBN 978-3-319-92498-4 (eBook)
https://doi.org/10.1007/978-3-319-92498-4

This Springer imprint is published by the registered company Springer Nature Switzerland AG
The registered company address is: Gewerbestrasse 11, 6330 Cham, Switzerland

This book is dedicated to late Graham Saunders. Through decades, Graham provided a significant contribution to knowledge about resettlement challenges particularly around housing and sheltering for refugees and disaster displaced populations. He helped us better understand the complex links between recovery, resettlement, temporary sheltering and permanent housing reconstruction. As a frequent collaborator of the i-Rec (Information and Research for Reconstruction) community, Graham provided valuable support to i-Rec conferences, student competition, and the international online debates.

In one of the i-Rec international online debates about the benefits and drawbacks of temporary housing, Graham rightly concluded that "the question that should be asked is not whether temporary housing hinders the recovery process, it is why support for recovery and reconstruction is not the default response, instead of the provision of temporary housing, and what are the institutional impediments within affected governments, donors and humanitarian agencies to make this the norm" (see his complete contribution here: https://oddebates.com/first-debate/).

Graham's ideas transformed the way we see the role of humanitarian and development agencies in post-disaster and conflict contexts, demonstrating that the resettlement challenges are not technical but social and political ones. He had clever and original ideas to challenge young professionals and students interested in post-disaster and post-conflict recovery and reconstruction. To a group of architecture students he once said "I challenge you to provide a 50$

solution to sheltering, because....this is the budget we sometimes have." He considered his duty to test new ideas and to challenge solutions based on misconceptions about humanitarian aid. He was always looking for empirical evidence and wanted policy to be based on knowledge obtained through rigorous research and thinking.

A senior and respected leader at IFRC, Graham was generous—sharing time and resources with scholars and students (students and young professionals from developing countries benefitted from scholarships he gave through i-Rec events). We will miss his clever ideas, his sense of duty, and his engagement in reducing vulnerabilities and finding appropriate housing solutions to reduce people's suffering after disasters and crisis.

Preface

Annually a large number of people are displaced and uprooted from their homes, communities, and countries. These people are resettled in various places in diverse conditions and under different roles and regulations. Depending on the cause (e.g., disaster, conflict, or war) and context (e.g., developing or developed countries), displacement could be short, medium, or long term. According to the UNHCR, more than 65.6 million people were forcibly displaced in 2016. About 34.30% (22.5 million) of them are refugees. The rest are displaced due to natural and technological disaster events or as the results of climate change impacts.

Resettlement of displaced populations is a challenging task for local, national, and international community. Provision of adequate and standard support for these people requires significant resources, planning, and operations at various levels. Many of the refugees and displaced populations may never be able to go back and rebuild and live in their original houses, communities, and businesses. Therefore, it is important that the recovery and reconstruction professionals, researchers, and policy makers examine how refugees and displaced populations can rebuild their life in new locations and recover from conflicts and disasters that have impacted their lives, livelihoods, and communities. As the title of this book *Resettlement Challenges of Refugees and Displaced Populations* suggests, there is an urgent need to better understand the multidimensionality and complexity of population displacement and the role that reconstruction and recovery knowledge and practice can play in this regard.

There is a great connection between forced population displacement and sustainable development. On the one hand, it can be argued that population displacement is rooted in unsustainable social, political, and economic processes, and on the other hand population displacement, if not managed properly, could exacerbate unsustainability. While, vulnerability, poverty, hunger, insecurity, inequality, lack of access to health, education, sanitation, and infrastructure push people to displace, displacement, though it may provide short-term security, it may not guarantee a better life for refugees and displaced populations. In other words, it can be argued that an important measurable outcome of Sustainable Development Goals (SDGs) can be seen in the total number of displaced populations and refugees. It is expected that as international community attempts to achieve the SDGs, we should see less displaced population and refugees. Similarly, it is expected that by adhering to SDGs, refugees and displaced populations live in settlements that have better qualities of life compared to previous years.

This book is a selection of papers that were presented in the 8th i-Rec Conference that was held at York University on June 1–2, 2017. i-Rec is an international network focused on the study of reconstruction and recovery after conflicts and disasters[1]. i-Rec deals with information exchange between its members in order to contribute with knowledge related to building activities in situations of crisis, particularly disasters in developing countries. It creates links between researchers, specialists, and policy makers in the field of post-disaster reconstruction, particularly in the areas of architecture, engineering and construction, urban planning, humanitarian aid, international development, and social sciences. i-Rec organizes an international conference every 2 years: Montreal (Canada), 2002; Coventry (UK), 2004; Florence (Italy), 2006; Christchurch (New Zealand), 2008; Ahmedabad (India), 2010; Ascona (Switzerland) 2013; and London (UK), 2015. One of the key goals of the 2017 i-Rec conference was to understand how disaster recovery and reconstruction knowledge and practice can contribute to the recovery and reconstruction of refugee and displaced populations. In particular, the conference aimed to push the boundaries of knowledge on recovery and reconstruction and develop new areas of inquiry that help to solve the complex problems and challenges of resettlement of forced displacement.

This book contains a total of 16 papers divided into two main parts with and introductory paper. The first paper which serves as an introduction to the book is titled *Sustainability, Design Futuring, and the Process of Shelter and Settlements*. This paper, written by Liz Brogden, provides an insightful review of shelter literature as it relates to the sustainability and the "sustainable development" paradigm. The first part contains the papers that focus on various aspects and challenges of resettlement of refugees. Three papers are dealing with the resettlement challenges of Syria refugees which make up a significant portion of the global refugees in recent years in different contexts and from different perspectives. These papers look at resettlement of refugees in Canada and neighboring countries such as Lebanon and Jordan. These are followed by four more papers examining resettlement issues and challenges of refugees in Italy, Uganda, Myanmar, and Kenya.

The second part consists of eight papers that are related to resettlement challenges of people affected and displaced by major disasters in the recent years. Two papers are dedicated to the 2003 Bam earthquake and two other cover different aspects of temporary and transitional shelters and settlements after the 2011 earthquake, tsunami, and nuclear disasters in Japan. Another paper examines the recovery and reconstruction process after the 2015 Earthquakes in Nepal using a build back better framework. This part also includes some special topics, particularly related to the effects of disasters on displaced workers and decision-making methods for selecting sustainable temporary shelters.

Toronto, ON, Canada — Ali Asgary

[1] i-Rec facebook group: www.facebook.com/groups/iReconstruction/
i-Rec website: www.grif.umontreal.ca/i-Rec.htm

Acknowledgments

We would like to acknowledge the supports provided by York University's Advanced Disaster, Emergency and Rapid Response Simulation (ADERSIM funded by Ontario Research Fund and NSERC), Disaster and Emergency Management Programs, and Centre for Refugee Studies. We also would like to thank i-Rec board members for giving the York University team to organize the 8th i-Rec Conference and publish its outcomes in this book.

Special thanks to the scientific committee members including *Professor David Alexander* (UCL Institute for Risk and Disaster Reduction, University College London, London, United Kingdom), *Dr. Hakan Arslan* (Düzce University, Düzce, Turkey), *Dr. Camillo Boano* (The Bartlett Development Planning Unit, University College London, London, United Kingdom), *Professor Roberto Bologna* (College of Architecture, University of Florence, Florence, Italy), *Dr. Lee Bosher* (School of Civil and Building Engineering Loughborough University, United Kingdom), *Dr. Jennifer Duyne Barenstein* (Centre for Research on Architecture, Society and the Built Environment (CASE), ETH Zurich, Faculty of Architecture, Switzerland), *Dr. Estella Carpi, Postdoctoral Fellow* (The Bartlett Development Planning Unit, University College London, London, United Kingdom), *Professor. Colin Davidson* (School of Architecture, University of Montreal, Montreal, Canada), *Professor David Etkin* (Disaster & Emergency Management, York University, Toronto, Canada), *Professor Alireza Fallahi* (Faculty of Architecture and Urban Planning, Shahid Beheshti University, Tehran-Iran), *Dr. Mahmood Fayazi* (School of Architecture, University of Montreal, Montreal, Canada), *Dr. Rohit Jigyasu* (Research Centre for Disaster Mitigation of Urban Cultural Heritage, Ritsumeikan University, Kyoto, Japan), *Dr. Cassidy Johnson* (Senior Lecturer, The Bartlett Development Planning Unit, University College London, London, United Kingdom), *Dr. Sneha Krishnan,* (London School of Hygiene & Tropical Medicine, London, U.K.), *Prof. Gonzalo Lizarralde* (School of Architecture, University of Montreal, Montreal, Canada), *Professor. Susan McGrath* (School of Social Work & Centre for Refugee Studies, York University, Toronto, Canada), *Dr. Niru Nirupama* (Disaster & Emergency Management, York University, Toronto, Canada), *Dr. Mojtaba Rafieian* (Department of Urban & Regional Planning, Tarbiat Modares University, Tehran, Iran), *Dr. Adriano O Solis* (School of Administrative Studies, York University, Toronto, Canada), *Dr. Jason von Meding* (School of Architecture and Built Environment,

University of Newcastle, Newcastle, Australia), *Professor Jianhong Wu* (*Canada* Research Chair in Industrial & Applied Mathematics, Department of Mathematics and Statistics, York University, Toronto, Canada).

Finally, many thanks to Dr. Jenaro Nosedal-Sanchez, Alison Kingelin, Richard Karsseboom, Zhara Yazdizadeh, and Shichao Zhang who helped with organizing the conference.

Contents

Part II Resettlement Challenges of Disaster Displaced Populations

Contributors

Oluwadunsin Ajulo University of Newcastle, Newcastle, Australia

S. M. Amin Hosseini Universitat Politècnica de Catalunya (UPC), Barcelona, Spain

April Aryal University of Auckland, Auckland, New Zealand

Roberto Bologna DIDA Department of Architecture, Università degli Studi di Firenze, Florence, Italy

Lisa Bornstein The Canadian Disaster Resilience and Sustainable Reconstruction Research Alliance (Oeuvre durable), McGill University, Montréal, Canada

Liz Brogden Queensland University of Technology, Brisbane, Australia

Gisella Calcagno DIDA Department of Architecture, Università degli Studi di Firenze, Florence, Italy

Georgia Cardosi The Canadian Disaster Resilience and Sustainable Reconstruction Research Alliance (Oeuvre durable), The Fayolle-Magil Construction Research Chair in Architecture, the Built Environment and Sustainability, Université de Montréal, Montréal, Canada

Sandra Carrasco University of Melbourne, Melbourne, Australia

Alice Chang-Richards University of Auckland, Auckland, New Zealand

Albert de la Fuente Universitat Politècnica de Catalunya (UPC), Barcelona, Spain

Mahmood Fayazi The Institute of Disaster Managment and Reconstruction, Polytechnic University of Hong Kong and Sichuan University, Chengdu, China

Giuseppe Forino University of Newcastle, Newcastle, Australia

Thayaparan Gajendran University of Newcastle, Newcastle, Australia

Hedyeh Gamini Shahid Beheshti University, Tehran, Iran

Thomas Johnson University of Newcastle, Newcastle, Australia

Catherine Kenny York University, Toronto, Canada

Hamid Amouzad Khalili University of Melbourne, Melbourne, Australia

Susan N. Kibue Jomo Kenyatta University of Agriculture and Technology, Nairobi, Kenya

Faten Kikano The Canadian Disaster Resilience and Sustainable Reconstruction Research Alliance (Oeuvre Durable), Université de Montréal, Montreal, Canada

Kenji Koshiyama Faculty of Societal Safety Sciences, Kansai University, Osaka, Japan

Gonzalo Lizarralde The Royal Society of Canada, Professeur École d'architecture, Université de Montréal, Titulaire de la Chaire Fayolle-Magil Construction en architecture, bâtiment et durabilité, Directeur du Groupe de recherche IF (grif) et OEuvre durable, Montreal, Canada

Chinh Luu University of Newcastle, Newcastle, Australia

Aaida Mamuji York University, Toronto, Canada

David O'Brien University of Melbourne, Melbourne, Australia

Oriol Pons Universitat Politècnica de Catalunya (UPC), Barcelona, Spain

Erica Seville Resilient Organisations, Christchurch, New Zealand

Patrick Tang University of Newcastle, Newcastle, Australia

Maria Gabriella Trovato American University of Beirut, Beirut, Lebanon

Jason von Meding University of Newcastle, Newcastle, Australia

Bernard Walker University of Canterbury, Christchurch, New Zealand

Suzanne Wilkinson University of Auckland, Auckland, New Zealand

Sustainability, Design Futuring, and the Process of Shelter and Settlements

Liz Brogden

Abstract

The need for adequate shelter and resettlement solutions for the growing number of displaced people worldwide is a problem of such magnitude that adequate solutions currently elude us. Humanitarian organisations are calling for greater engagement from experts who are able to assist with this crisis, including built environment professionals such as architects. With vague career pathways and a lack of exposure to this crucial area of practice in architects' formative years, further education and postgraduate studies targeting humanitarian practice have only recently begun to emerge. This research reviews current literature in the areas of shelter and settlements, illustrating that the architectural profession has a responsibility to contribute towards positive resettlement outcomes for displaced populations. This paper also discusses a particular area of theory known as 'Design Futuring' by Tony Fry, exploring how it can provide a theoretical framework to support engagement from architects in both research and education targeting this area of practice. The paper identifies key areas of conceptual overlap including the emphasis on sustainability, shelter and design processes in favour of product thinking, and the reframing of a more collaborative and transdisciplinary approach to architectural practice. It is anticipated that this research can inform approaches to architectural education and research focusing on shelter and settlements design, representing a new area of theory-based inquiry.

Keywords

Shelter and settlements · Displacement · Sustainable design · Design Futuring . Humanitarian Architecture

1 Introduction

In a global context that is characterised by uncertainty about our collective future, the impacts of climate change, violent conflict, and access to energy and resources are culminating in mounting geopolitical tensions. Not only does this unrest fuel political conditions which exacerbate the problem of population displacement, it simultaneously inhibits collective action in the reduction of carbon emissions. The result of which is *more* extreme weather events, *more* disasters, and *more* displacement. In effect, the problem of population displacement is a self-perpetuating one. We are currently witnessing record levels of forced displacement. Over 65 million people were displaced in 2015 as a result of persecution,

L. Brogden (✉)
Queensland University of Technology,
Brisbane, Australia
e-mail: liz.brogden@qut.edu.au

A. Asgary (ed.), *Resettlement Challenges for Displaced Populations and Refugees*,
Sustainable Development Goals Series, https://doi.org/10.1007/978-3-319-92498-4_1

conflict, violence, or human rights violations (UNHCR 2015a). This is paralleled by the increased magnitude and frequency of disaster events worldwide (UNISDR 2015). Humanitarian aid systems are stretched beyond their response capacity. The High Commissioner for Refugees has observed a paradigm change and, "an unchecked slide into an era in which the scale of global forced displacement as well as the response required is now clearly dwarfing anything seen before" (Guterres, UNHCR 2015b).

For displaced populations, first-line humanitarian aid and the provision of shelter set in motion processes of recovery and reconstruction; the critical first step of broader human settlement and development patterns. Such patterns are both tangible and intangible, reflected in the physical built environment but also through the observed social impact. Intangible effects are evident in shelter's capacity to support ordinary day-to-day life, to enable livelihoods and cultural practices, and to strengthen community networks. The long-term ramifications of first-response aid can be significant, and early decisions impact upon the success or failure of resettlement projects. Complexities associated with how aid funding is allocated, collaboration between agencies, and the urgency of high-risk situations can result in scenarios where differing first-response strategies are implemented simultaneously. These do not necessarily contribute towards a cohesive, sustainable vision for a relief and recovery project.

The resettlement of displaced populations through repair, reconstruction, or relocation inevitably involves built environment expertise. Architects are uniquely qualified to design buildings with social impact in mind, but despite this, there is very limited engagement from the profession of architecture in shelter and settlement projects (Harris 2011). Architects are rarely equipped, particularly in post-disaster contexts as shelter projects call for different approaches to design than those required under "normal" circumstances (Aquilino 2011; Charlesworth 2014b). The nature of a post-disaster context highlights a gap in architectural knowledge with architects typically trained to assume a top-down position of control over a project. Humanitarian and development projects are increasingly calling for more participatory, community-led approaches in which the role of the architect is necessarily that of a "humble facilitator". Charlesworth (2014a) asserts that the lack of architects involved in design for a disaster response stems from architectural education, and the absence of training in design and problem-solving relevant to post-disaster contexts. Further to this, conventional design education conditions architects to take a design approach that is characterised by making personal marks that are rarely appropriate in the aftermath of an emergency (Sanderson, cited in Charlesworth 2014b).

1.1 Shelter and Settlements

The Sphere Project (2015) handbook or *Humanitarian Charter and Minimum Standards in Humanitarian Response* outlines standards for shelter based on the premise that "everyone has the right to adequate housing" (The Sphere Project 2015, p. 243). Even though the document outlines the basic elements of 'adequate' housing, the most recent State of the Humanitarian System report by ALNAP (2015) found that the consensus amongst practitioners was that there was no easy answer when it came to adequately meeting shelter needs (ALNAP 2015, p. 57). The problem of providing adequate shelter has been discussed for decades, as was highlighted by Davis (1978) in his early publication *Shelter after Disaster*. It was observed in a review by McGraw (1978) at the time that "the facts and conclusions presented have been known for a long time – but sadly are not heeded" (1978, p. 270).

The fact that the global shelter requirements for displaced populations exceed the response capacity of the humanitarian system (Davis, 2011), can be illustrated in part as a financial deficit, with assistance requirements reaching a record $17 billion in 2014 and only 50% of this amount raised (Youssef, 2015). But this resource shortfall is not only financial; it is also knowledge-based. An observed lack of consensus and understanding surrounding humanitarian action prompted the 2011 Humanitarian Emergency Response Review (HERR), which noted a lack in training, experience, and skills across the sector. Regarding shelter, the report states that providing adequate shelter is "one of the most intractable problems in international humanitarian response" (DFID 2011). The absence of expertise in the sector has been a long-standing issue, with concerns expressed by Saunders (2004) regarding, "the limited involvement within the relief sector of professionals with backgrounds and expertise in planning, design, planning and construction" (2004, p. 166). Corsellis has also highlighted that "approximately 85% of stakeholder staff regionally have no technical training whatsoever" (cited in Davis 2011, p. 203).

In order to address this, the humanitarian system is calling for greater collaboration between aid organisations and a more diverse range of actors, including the private sector. Zyck and Kent (2014) identify that new types of private sector engagement and partnerships between aid agencies, businesses, and corporations in recent years have great potential. In addition to this, they observe that this represents "a significant challenge to the humanitarian sector as traditionally conceived" (2014, p. 1).

1.2 An Emergent Specialisation of Architecture

The call for private sector and specialist engagement by humanitarian agencies is paralleled by growing interest from architects and other built environment professionals seeking involvement in the shelter sector. This is in turn reinforced by moves to recognise alternative 'humanitarian' career pathways within the profession of architecture (Charlesworth 2014b). Prior to the 2004 Indian Ocean tsunami, architects as design practitioners were rarely involved in the provision of humanitarian aid (Aquilino 2011; Charlesworth 2014a; Harris 2011), and in cases where architects were involved, it was typically in a technical or logistical capacity and not necessarily that of a designer (Charlesworth 2014b). In recent years, the value of engaging architects in shelter and reconstruction projects has been increasingly recognised as critical to long-term recovery outcomes as argued by numerous scholars such as (Aquilino 2011; Charlesworth, 2014b; Harris 2011), but there is also extensive debate surrounding the nature of this involvement and the role architects should fulfil.

The term "humanitarian architecture" has been used in recent years to describe the specialist form of architectural practice defined by Charlesworth (2014b) as "using *design skills* to assist vulnerable communities, particularly after the crises of social conflict, war and natural disaster" (2014b, pp. 268, emphasis added). Sometimes referred to as socially engaged architecture, or public interest architecture, these modes of architectural practice are growing to encompass a broad range of project types. These typologies extend beyond the scope of disaster response, including favela (slum) upgrade projects, marginalised and disadvantaged community projects, and refugee camp shelter for conflict-generated internally displaced people (IDP).

Many shelter sector practitioners who are trained as architects do not readily relate to the term "humanitarian architect" and resist the idea that social consciousness in architecture should be considered anything out of the ordinary, or deemed to be "alternative". Charlesworth (2014a) states, "That this should be a special category within the field of building speaks volumes about the condition of the design profession" (2014a, p. viii). While the objectives of design for crisis situations align closely with thinking around social architecture or design activist agendas, there is as yet no universally agreed upon descriptor for architects seeking to pursue a humanitarian style of practice. Professions such as law and medicine acknowledge particular standards and ethics specific to humanitarian contexts. These professions have embraced modes of practice that have been adapted to be relevant within systems which may be "nonexistent, rudimentary, or dysfunctional because of poor state policies and/or inadequate resource allocation, facing collapse, or has already collapsed" (Medecins Sans Frontieres 2016). Despite the fact that shelter is identified as a fundamental need in Article 25 of the Universal Declaration of Human Rights (UN General Assembly 1948), the profession of architecture is only recently engaging with the idea of a specialist humanitarian form of practice and has yet to integrate it within professional training or the main curriculum of architectural education. Regardless, "humanitarian architecture" as referring to a subtype of architectural practice appears to be gaining momentum as a term in architectural culture and discourse.

1.2.1 New Courses

Further educational qualifications for architects seeking to engage in shelter and settlement projects have only recently begun to emerge. Since 2011, the first Master's courses and modules focusing on post-disaster shelter started through Copenhagen/Lund Universities in collaboration with the IFRC. At the same time, the Centre for Development and Emergency Practice (CENDEP) began a postgraduate certificate in "Shelter after Disaster" at Oxford Brookes University, along with a "Master of International Cooperation Sustainable Emergency Architecture" established at the International University of Catalonia (UIC). Following from this, the Royal Melbourne Institute of Technology (RMIT) commenced their "Master of Disaster, Design and Development" (MoDDD) in 2016.

These courses follow on from undergraduate degrees, with exposure to humanitarian shelter and settlements projects rarely embedded in the main curricula of architectural education. Wagemann and Ramage (2013) argue that given the nature of our changing environment, the profession needs to address how the design of our built environment should respond. Further to this, the value of lessons learned through disaster relief can be applied beyond extreme crisis situations, such as any project which may involve a "lack of resources and land, urgency and speed of construction, prefabrication and use of local materials, transportation and logistics" (Wagemann and Ramage 2013, p. 133).

2 Purpose of the Research

Shelter and settlement projects frequently call for an urgent response to an acute context that is characterised by extreme unpredictability and highly condensed development timeframes. Olshansky et al. (2012) refer to the effect as a condition of "compressed time", particularly evident in post-disaster settings. Tauber (2014) observes that the distinctiveness of a post-disaster working context has yet to be examined and that the particular skills required to create people-oriented housing are not clearly understood. Additionally, shelter and settlement projects confront practitioners with conditions of trauma and grief throughout the resettlement process.

If architects are to engage in this field of practice, they must respond with design skills and new knowledge that is different to what is utilised under ordinary circumstances. In light of this observed gap in knowledge, this research aims to suggest a theoretical framework that can inform approaches to design in humanitarian contexts and assist with issues of displacement and resettlement.

2.1 Methods

This research paper outlines a contextual review of literature relating to shelter and settlements, exploring areas of conceptual overlap existing within a particular area of design theory known as 'Design Futuring' developed by Tony Fry. Both areas of knowledge are discussed, exploring the extent of conceptual alignment and making the argument for the value of greater engagement from design-focused practice and research into shelter and settlements.

With limited precedence of design research into shelter and displacement it is anticipated that this theory-based inquiry will contribute towards emergent design education programmes, research, and practice in the area of displacement, shelter, and resettlement.

3 Sustainable Futures and the Shelter Process

Fry's ideas of Design Futuring centre primarily on the notion of sustainability. He pushes past weak and fragmented understandings of "green" approaches to development, and proposes that design has an ethical imperative to assist in creating a future in which we have a hope of survival. The theory tackles issues of climate change and resource depletion head-on, arguing for a correction and redirection of some of our most fundamentally held paradigms. Where such an extreme ontology might not be readily adopted in ordinary day-to-day design practice (although Fry would argue this is the very place where such shifts must occur), it certainly lends itself to the extraordinary circumstances of disaster and crisis situations that cause population displacement. These are the very limits of human survival that Design Futuring seeks to circumvent, and such scenarios present opportunities to redirect inbuilt vulnerabilities and build in future resilience.

The following review provides a concise summary and overview of definitions relating to three identified areas of interest within shelter sector literature and the book "Design Futuring" (Fry 2009). These include (1) sustainability, (2) process-thinking in preference to a product-only focus, and (3) reframing approaches to architectural and design practice.

3.1 Sustainability

The concept of sustainability holds far greater potential than widespread application of the buzzword would suggest. It is a notion freely embedded within both mainstream professional and academic spheres, commonly stated in policy documents and vision statements as if simply incorporating the term will attract a green tick of "eco-approval". When framed according to conventional thinking, sustainability goals can run contrary to the agenda of an ever-expanding market economy. Our collective tendency to deny responsibility for our impact on the planet and its inhabitants, according to Fry, makes us "dangerous beings in our failure to recognise that we are unknowingly throwing problems into the future while failing to deal with problems of the present" (Fry 2009, p. 187).

This section explores the idea of sustainability as it relates to shelter and settlements as well as how it is defined by Fry (2009). This paper does not provide an in-depth exploration of theory relating to sustainability; it introduces the theme as it relates to UN activities, the shelter sector, and explores a new ontological view of sustainability that can inform approaches to practice, research, and education targeting shelter and settlements design.

3.1.1 The "Sustainable Development" Paradigm

Early references to Sustainable Development (SD) appeared in the 1987 Brundtland Report, where it was defined as "development that meets the needs of the present without compromising the ability of future generations to meet their own needs" (World Commission on Environment and Development 1987). Since this report, SD has grown to become central to both environmental and development discourse. The Brundtland Report was commissioned in recognition of the need for a common vision for a future that is not under threat. Also titled *Our Common Future*, the report responded to perceived failures arising from both poverty and the short-sighted way in which prosperity is frequently pursued (World Commission on Environment and Development 1987). The report alludes to failed civilisations of the past, highlighting that where pressures were previously more localised, we are now facing broader, global "interlocking crises" of population growth, available resources, food security, damaged ecosystems, energy, industry impacts, and the morphology of human settlements. Critically, the report recognised the interconnectedness of all these areas and the need for a comprehensive strategy.

Since the Brundtland Report, the 1992 UN Conference on Environment and Development (also known as the "Earth Summit" or "Rio Conference") saw the publication of both the Earth Charter and Agenda 21 outlining an action plan for SD in the twenty-first century. In 2000, the Millennium Development Goals (MDGs) were agreed upon at the Millennium Summit at the UN in New York, summarised as eight international development goals to be reached by 2015. Since then, the MDGs have been built upon and are now reflected in 17 *current* Sustainable Development Goals (SDGs) for 2030. These were agreed upon by 193 world leaders in attendance at the UN Sustainable Development Summit (United Nations 2015).

Despite efforts of the UN and partner countries, unsustainable development, rising carbon emissions, and the subsequent increase in disaster events see the need for plans that specifically target disaster risk reduction. The Hyogo Framework for Action 2005–2015 has now been succeeded by the Sendai Framework for Disaster Risk Reduction 2015–2030, which reiterates the Hyogo Framework's commitment to disaster risk reduction and resilience-building initiatives in light of the urgent requirement for more sustainable modes of development. Importantly, the Sendai Framework highlights that disasters "impede progress towards sustainable development" and, inversely, that sustainable development slows the impacts of climate change and associated disaster events (UN 2015 pp. 9–10).

Of particular relevance to the resettlement of displaced populations, is the recent update to the 1982 "Shelter after Disaster" guidelines by Saunders et al. (2015), which introduces new topics relating to sustainability. The publication now includes criteria for successful shelter after disaster and acknowledges that technical competence in key fields such as architecture and planning is required, as well as shelter designs that are environmentally sustainable, adaptable, and climatically and culturally appropriate (Saunders et al. 2015).

3.1.2 A Sustainable Future

> In our endeavour to sustain ourselves in the short term we collectively act in destructive ways toward the things we and all other beings fundamentally depend upon. (Fry 2009, p. 22)

There is considerable debate surrounding the term "sustainability", at least in part due to its frequent use without a clear definition or sense of any quantifiable outcome. Fry (2009) critiques the sustainable development paradigm as outlined in the Brundtland Report (1987). He argues that this ties sustainability directly with economic growth, implying that this view of future security is gained through capitalist logic alone. Fry questions the view that simply ensuring the availability of natural resources will guarantee a certain quality of life for future generations, arguing that there needs to be more awareness of the interconnected dependences between species and various forms of life. A development-based understanding of sustainability reinforces a sole focus on human life to the exclusion of other biological systems (Fry 2009). A number of Fry's concerns have been addressed since, through the Sustainable Development Goals (SDG) which highlight areas of focus such as SDG15 "Life on Land" and SDG14 "Life Below Water", yet the UN still frames them as development goals. Griggs et al. (2013) also argue that in order to bring about sustainable change, the conventional "three pillars" model of sustainable development which emphasises economic, social and environmental factors need to be reframed as a more nested concept in which everything is connected to earth as our life-support system.

3.1.3 Design and Sustainability

Fry presents the concept of sustainability – quite literally – as *the ability to sustain*. This is more than an exercise in semantics; it is indicative of a deliberate attempt to emphasise our agency in bringing about sustainable outcomes. In "Design Futuring" the term is cited in a hyphenated form: "sustain-ability" in an effort "to suggest a more materially grounded objective and agency" (Fry 2009, pp. 6–7). This framing of sustainability sets itself apart from commonly held and frequently vague understandings of the term, also illustrating how the concept can be understood from a design philosophy perspective. Of particular significance is the incorporation of the notion of *time* in this theory. Sustainability as defined by Fry is "a means to secure and maintain a qualitative condition of being over time"; describing a process and not only an endpoint goal. The definition offered is absolute: a state in which "all that supports and extends being exceeds everything that negates it" (Fry 2009, p. 43). Fry proposes that this understanding of sustainability is future-enabling and that through sustainable change, we are actively *future-making*.

The opposite of sustainability – unsustainability – is described by Fry as a form of *defuturing*: "A state in which the short-term priorities of our sustainment culminate as a destructive force on the very things we depend upon in the long term" (Fry 2009, p. 22). Fry demands we acknowledge that through our acts of making we inevitably also "unmake"; clearing land, consuming resources, and disrupting ecologies. He argues that our inherent destructiveness is unavoidable but that we can no longer conceal this fact as we go about creating the world in which we live. In order to ensure accountability for the social and environmental impacts of human actions, Fry states that we must first visualise the nature and quantity of what will be destroyed. In this way, we might be able to justify the necessity and attempt to offset any destruction. This call to halt our current patterns of unsustainability is made to everybody and not only designers. As elaborated upon in his subsequent publication "Becoming Human by Design", Fry (2013) proposes that all humans are essentially creative by nature and that our drive to *make* sees us all designing the world we inhabit and the futures we create. This world, in turn, designs our interactions and shapes the way we humans are. Rather than framing design from a disciplinary sense, Fry propels the idea of design-as-verb across the actions of all people, claiming that "design is one of the main operative agents of the social, cultural and economic functioning and dysfunctioning of humanity's made world" (Fry 2009, p. 25). Where sustainability is elsewhere presented in terms of "sustainable development", Fry argues that we must identify collectively as designers. He proposes that we all need to be makers of the sustainable future we wish to inhabit and ask the question, "is what we are making world-making or negating?" (Fry 2009, p. 25).

Sustainability in design provides a robust theoretical basis for architecture and design practice that targets resettlement projects. It is a readily adopted theme across the humanitarian and shelter sectors and also within the profession of architecture. Furthermore, it complements resilience-focused schemas, as both resilience and sustainability agendas strive for similar objectives that relate to survival. In order to build resilience, sustainable strategies call for innovation and engagement from creative thinkers and designers. This requires that we begin to consider how every product we create can contribute towards a future in which communities cannot only survive but thrive during changing times.

3.2 Process Versus Product Thinking

Having established that sustainability requires consideration of how short-term decisions impact upon future conditions, the problem of shelter and settlements for displaced populations represents a formidable sustainability challenge. In order to develop or adapt human settlements in response to sudden disruption and displacement, the urgent requirement for solutions does not allow for extensive consideration of long-term impacts. One way to approach the problem of compressed development timeframes is to describe shelter as a process and not just as a single end product. This thinking is also evident where many agencies refer to shelter aid in terms of "shelter and settlements", incorporating the broader context that exists around singular shelter units. This acknowledges that shelter is more than a simple roof over people's heads but that it sits within a multitude of complex systems associated with the human settlements they form.

3.2.1 Shelter as Process

There has been a determined effort in the shelter sector to move away from product thinking in which shelter is conceived as a discrete unit, in favour of viewing shelter as a process. This counters traditional notions of staged shelter which Saunders et al. (2015) describe as problematic. Conceptualising shelter strategies as a process supports a view in which sheltering cannot be subdivided into defined products or stages but that "the reality is a continual, seamless sheltering process" (Saunders et al. 2015, p. 40).

Misconceptions regarding the nature and value of *design* parallel the same oversimplified "product-driven" attitudes that are pervasive in shelter and settlements discourse. In the same way that shelter projects can catalyse either recovery or a kind of second disaster (Harris 2011), "the enormous power of design for good and bad has to be brought out of the shadows" (Fry 2009, pp. viii).

Turner famously highlighted in 1972 that housing is not only a noun but a verb. He emphasised that the verb "to house" describes the process or activity of housing, and further, that housing is not just what it is, but what it does (Turner and Fichter 1972 p. 151). This observation also applies to a synonym of housing: shelter. It was Davis (1978) who first made this connection regarding post-disaster shelter, stating that "shelter must be considered as a process, not as an object" (1978, p. 33). Amongst many, similar process-oriented perceptions of shelter have been presented by Johnson et al. (2006) whose systems-based analysis focused not on the individual shelter elements but "the important relationships between them, and between them and their environment" (2006, p. 369). This also aligns with Kennedy et al. (2008) who assert that particularly after a disaster, "shelter must be considered as a series of actions for fulfilling certain needs other than as objects" (2008, p. 25). The conceptualisation of shelter as a process is also emphasised in recent guidance materials, including "Transitional Shelter Guidelines" (Shelter Centre, 2012), the recent update to "Shelter after Disaster" (Saunders et al. 2015), and "Recovery from Disaster" (Davis and Alexander 2016).

The idea of shelter as a verb – or process – can be further elaborated upon through architectural theory. "Shelter" is a fundamental, frequently abstract concept in architectural discourse, with highly condensed, multiple meanings relating to the essential functions and nature of architecture itself. Through the use and understanding of words such as shelter, dwelling, and building, each is conceived as both processes and products, verbs and nouns. A shelter is a structure or building, yet also *to shelter* is an act that incorporates rich layers of tradition, culture, and religion. Heidegger (1971) questioned what it is to dwell, suggesting that building (as both verb and noun) *belongs* to dwelling or simply that the act of building *is* dwelling. He illustrates this fact using the Old English and High German meanings of the word for building, *bauen*, which means to dwell; and neighbour, *nachbar*, near dweller (Heidegger 1971, p. 1). More fundamentally, according to Heideggan thinking, "dwelling is the manner in which mortals are on earth" (Heidegger 1971, p. 2), revealing the density of meaning attributed to such terms in architectural theory as processes, products, and even existence itself.

3.2.2 Transitional Shelter

Burnell and Sanderson (2011) have observed that "To say that the provision of shelter after disaster by humanitarian aid actors is costly, complicated and fraught with problems would probably be one of the greatest understatements in humanitarian aid" (Burnell and Sanderson 2011, p. 189). There is a great deal of debate surrounding appropriate post-disaster shelter strategies, and despite a substantial amount of research emerging in recent years, as observed by Batchelor (2011), "opinions are still divided over the most appropriate shelter response" (2011, p. 61). The main challenge has been identified by Pelling (2012) in that decision-makers are required to "integrate and balance the needs of the emergency with long-term requirements of sustainability" (2012, p. 19).

Much of the debate surrounding shelter response centres on the use of transitional shelter. Critics of the approach argue that they too often become unintentional permanent solutions or that they absorb large amounts of financial and human aid resources for the sake of a short-term solution (Burnell and Sanderson 2011). Clermont et al. (2011) argue that transitional shelters are frequently a waste of money, counter-developmental, and more suiting of NGO timeframes and marketing needs rather than those of disaster survivors. Such arguments are common but may not be referring to the broader concept of transitional shelter as a strategy or *process* but to isolate examples of ineffective *products*. As observed by Pelling (2012), such failed shelters often imply standardisation, resulting in "repetition of a 'universal' unit that rarely respond to the specifics of climate, topography, local customs and local forms of living" (2012, p. 10).

Despite a lack of consensus, transitional shelters have become the preferred response of choice for nearly all large aid agencies (Burnell and Sanderson 2011). Following a disaster event, they are one solution that is fast and able to last the required average of 3–5 years while reconstruction occurs (IFRC 2013). Tom Corsellis, author of the Oxfam/Shelter Centre Transitional Settlement guidelines, claims that other than incremental transitional shelter, there are "insufficient options able to support communities' shelter requirements whilst permanent reconstruction takes place" (cited in Davis 2011*)*. Saunders et al. (2015) highlight the necessity to understand both the value and limitations of a transitional shelter approach based on the context of the disaster setting. They acknowledge that where time is necessary for complex reconstruction, transitional shelter is beneficial as it allows for planning and risk assessments that ensure future resilience (Saunders et al. 2015). Essentially, transitional shelter is a solution that bridges time between emergency relief and permanent reconstruction and recovery.

3.2.3 Designing in Time

> I've started working 'upstream'. Almost before the beginning, we need to craft and design the problems that we act on with care and with attention. The way in which we shape these problems will determine the outcomes. Trying to back out of what we perceive as a good solution and hit the right problem from reverse is not a particularly good strategy!Marie Aquilino, cited in (Vitiello 2017).

Fry (2009) proposes that design needs to be directional and conducted in consideration of both time and orientation as its primary focus, with form being secondary to this. This understanding deviates from the frequently held perception that design is nothing more than an "aesthetically elevated object" (Fry 2009, p. 29). It injects the idea that it is possible not only to design with the future in mind (in time) but to utilise design as a redirective tool in shaping a different future.

The process of designing in time involves examining possible negative or positive outcomes and designing against undesirable possibilities. It is not "an ability to see into the future" (Fry 2009, p. 58) but is described as a methodological framework in which scenario building is employed to "design from the future to the present". Fry refers to this idea as "futuring", involving the identification of impact events which will influence the conditions of a "continually modified present". These impacts are then traced, and subsequent design decisions trigger wilful change (Fry 2009, p. 147). It is envisioning a scenario that involves a moving "present" *and* a moving future and projecting designs into that space to bring about sustainable outcomes.

Fry intends to reframe an understanding of design as more than an objectified function (a product). This parallels efforts from experts in the shelter sector seeking to reframe shelter as a process and not a product. Fry's explanation of four aspects of design activities can also be used to inform an understanding of what constitutes shelter and settlement design. Fry (2009) defines design activities as:

1. Forming a sense of what needs to be designed
2. The act of designing
3. The nature of the design object
4. The consequences of its actions in the world, both immediately and over a considerable expanse of time

The object or product is presented here as only one aspect of "design". Similarly, shelter is more than the nature of the singular unit produced. The shelter process extends back to (1) first-response emergency shelter, risk assessment, and context analysis; (2) project brief development, design methodologies such as participatory planning, community-led/owner driven reconstruction, or prefabricated strategies; (3) informing the shelter strategy employed in response to these stages;

and, finally, (4) the ways in which the shelter enables recovery and builds upon future disaster resilience.

Fry establishes that the full power of design can only be realised once it is understood ontologically, in that designed objects have agency and go on designing the nature of our world over time, thus shaping future realities. In this sense, he argues that the designed world, in turn, designs us as even our interaction with the world is designed (whether intentionally or not). Fry (2009) states, "It is not just that we are born into a designed world but that our interaction with this world is also designed - our built environment, forms of work, modes of transport, manufactured products, media, infrastructure systems and myriad other things are all designed in relation to use" (2009, p. 25). In essence, he is saying that everything designed goes on designing, including us.

The need for more sustainable ways of designing calls for a paradigm shift and a move away from the market-driven mindset of perpetual economic growth. Designing in time allows for a new set of priorities and measurements of benefit to emerge. Conceptualising longer durations for processes enables quality to become the primary objective, rather than the quick profits of quantity. This way, the feasibility of significant sustainability ventures can be framed accordingly, but this requires a shift in thinking on many levels, including in architectural education and practice.

3.3 Reframing Architectural Practice

Philip Johnson famously stated that all architecture is shelter, referring to an understanding of shelter in terms of providing the basic needs for human survival. In crisis settings, however, meeting basic needs does not necessarily translate to employing basic approaches. It must be acknowledged that the conditions under which architects design in conventional practice and those seen in humanitarian and crisis contexts are radically different. The provision of shelter requires greatly intensified efforts and alternative skill sets suited to conditions where there may be no societal or economic structure, limited social capital, and broken social networks and in an environment that is characterised by insecurity and fear (Henrotay 2008).

The nature of the role of architects engaged in shelter sector work has been explored by various scholars (Boano and Garcia 2011; Burnell and Sanderson 2011; Charlesworth 2014a; Lizarralde 2015; Tauber 2014), and this section of the paper builds upon this work. It explores how Design Futuring theory by Fry (2009) informs a theoretical understanding of sustainable shelter processes that are enabled by transdisciplinary practice and collaborative, bottom-up design methodologies.

3.3.1 Transcending Disciplines

The argument for more diverse forms of humanitarian action was made in a recent report released by the ODI and HPG. Bennett et al. (2016) argue for "a new model of humanitarian action, one that requires letting go of the current paradigm" (2016, p. 8). They propose that the current model of action needs to become "a diverse, devolved and decentralised model that genuinely recognises and embraces the contributions of new donors" (Bennett et al. 2016, p. 8). The report acknowledges the tensions which exist in reconciling the "exceptionalism of humanitarian assistance", with the strategic and operational requirement of opening up the humanitarian system to a more diverse set of actors. These actors often differ in such fundamental ways as their interpretation of what it means to be a humanitarian (Bennett et al. 2016, p. 8).

Coupled with the condition of an historically inaccessible humanitarian system is the architectural discipline's tendency to self-isolate within a self-referential "black box" of cultural prestige (Banham 1990). The profession's autonomy is referred to by Till (2009) as "deluded detachment" (p. 7) in that it refuses to engage with contingencies of the world beyond. The idea of design's disconnection from the world is explored by Fry, who maintains that in order to ensure a sustainable future, design must acknowledge its *relationality.* Relationality requires that we ask the question "what will that which has been designed design?" and acknowledges the "dynamic complexity of interconnected and multidirectional causes" resulting from designed objects (Fry 2009, pp. 34, 31). This multifarious framing of design as relational (or "contingent" as Till would say) requires that design engages with new knowledge beyond that which is conventionally observed within disciplinary boundaries. Rather than an exclusionary approach to practice, Fry advocates bridging of disciplines to form a sustainability-focused meta-discipline, facilitating exchange and dialogue between different forms of knowledge (Fry 2009).

3.3.2 Collaborative Design Practice

Harris (2011) states that incorporating expertise in the shelter sector should not be confused with imposing top-down solutions. Instead, it needs to enable capacity. Sadly, this is not always the case – with the number of inappropriate prefabricated shelter designs bearing witness to this truth. It has been observed by Sanderson (2010), that unless people are engaged to help solve their own problems, architects are often the last people needed in disaster reconstruction. This view is also maintained by Aquilino who asserts, "Places are not made by disciplines. They're made by people" (cited in Vitiello 2017). This idea was also a recurrent sentiment throughout the 15 interviews conducted by Charlesworth (2014a) in "Humanitarian Architecture". As stated by Patama Roonrakwit, "it's better to think and work in another way. The architectural knowledge and skills I learnt are important, but they never taught me that the design process should be done by an architect in partnership with, and as a servant of, the owner of the place" (cited in Charlesworth 2014a, p. 68).

It is widely acknowledged that architects seeking to pursue careers in the shelter sector must embrace collaborative design strategies that go beyond standard notions of interdisciplinary engagement. They must also work directly with the clients and communities being served. This represents a shift away from more conventional styles of practice in which the architect as authority bestows a top-down solution upon a community. Architects need to visualise their collaborative efforts laterally, both across disciplinary boundaries and professions, while also considering the client/community as able to offer valuable insights, skills, and resources to both the shelter and design process. Terms such as "bottom-up", "participatory", "community-led", or "community-driven", have emerged to describe these inverted models of practice that deviate from more conventional design processes.

Aquilino (cited in Vitiello 2017) argues that these alternative modes of design practice are more likely to give a voice to those who are ordinarily silenced or excluded. These include people such as women, those with disabilities, or ageing members of the community. These approaches are also more likely to enable local cultural practices to influence architectural outcomes, which Aquilano states should be "inseparable allies" (cited in Vitiello 2017). Boano and Garcia (2011) elaborate extensively upon the idea that shelter processes commonly perpetuate colonialist ideas. They argue that the architect's place is to enable cultural agency, assuming the role of "translator" for the people.

Fry (2009) advocates for socio-cultural plurality in which we collectively abandon Eurocentric or Western inclinations towards power and embrace "commonality in difference". He argues that it is through diversity and difference that the human race has a hope of survival, drawing parallels with biodiversity and the inherent vulnerability of monocultures which carry a high risk of extinction. Questioning the merit of the "development" paradigm that emerged after World War II, and the relentless drive for globalisation, Fry (2009) highlights that the result has proven to be largely unsustainable. He discusses the suppression and blocking of sustainable characteristics of non-Western cultures, often having emerged as an appropriate response to a particular environment. Fry argues that our current circumstances call for the building of new cultures in which old knowledge is used to design with the common good in mind and our collective need for more sustainable solutions.

Fry's definition of sustainability is presented as an appropriate basis from which to approach architectural practice, research, and education that targets shelter and settlements design. It outlines how more sophisticated modes of design are required of the architectural profession if they are to produce sustainable solutions to the very complex and multifaceted problem of displaced populations. Critically, it highlights that the profession of architecture must embrace transdisciplinary and collaborative practice in order to engage successfully as shelter sector practitioners in crisis contexts.

4 Conclusion

This paper builds upon existing theory relating to shelter and settlements. With a shortage of solutions and expertise, the humanitarian sector is calling upon professionals and the private sector to engage with humanitarian agencies and projects, including those in the shelter sector. The rise of interest in "humanitarian architecture" in recent years is seeing the emergence of a specialised form of architectural practice targeting this field.

Practice in this area requires that architects "unlearn" many of the design skills imparted during conventional architectural education and in practice, and to incorporate new knowledge that is appropriate to the humanitarian sector. Further study for architects seeking to develop skills that are relevant to crisis contexts is almost entirely limited to post-graduate study, with very little evidence of resettlement, and relief projects' incorporation into the main curriculum of architectural education.

This paper provides a review of relevant shelter sector publications as they relate to themes emerging from Fry's "Design Futuring". Sustainability and the "sustainable development" paradigm were described and related to Fry's definition of "sustain-ability" as future-making, thus asserting that all people in some way assume the role of "designer". Critically, this view of sustainability requires that we begin to frame design as directional in time, offering a design-focused theoretical basis from which to discuss the concept of a shelter process. The significance of transitional shelter as a means of bridging time was discussed in support of this conceptual overlap. Finally, this research outlined arguments for reframing of the role of architects in design practice, both in shelter response for displaced populations and more broadly in sustainable practice. This involved a summary of various arguments for more collaborative and transdisciplinary approaches to design practice in the profession and the shelter sector.

Boano and Garcia (2011) observe that built environment experts are steeped in "default modes of operation" resulting in "unsustainable solutions that focus solely on physical deliverables" (2011, p. 293). Further to this, current modes of architectural education marginalise humanitarian forms of practice as "alternative", and many architects in the field observe that their education did not prepare them in any way for humanitarian practice (Charlesworth 2014a). The theoretical framework presented here is, fundamentally, sustainability-focused, implying that sustainability must be at the centre of all decision-making from architects seeking to practice as designers in crisis contexts.

> The value of what one knows and does may have to be fundamentally altered. This means a great deal of knowledge that has historically been acquired as the corpus of the discipline may need to be discarded and replaced in order for any real ability of the remade professional to drive affirmative change.(Fry 2009, p. 11)

Fry (2009) argues that in order to redirect the course of the future, vague notions of 'attitudinal change' are insufficient, and over-inflate the idea of how much power and ability one individual can hold. This review reveals that for architects to engage as humanitarian practitioners, new design methods and knowledge must be embraced by the profession. Only through this, can architects as "remade professionals" enact affirmative change through the design of shelter and settlement solutions for displaced populations.

References

ALNAP (2015). *The state of the humanitarian system, ALNAP study*. Retrieved from London: http://www.alnap.org/resource/21036.aspx

Aquilino, M. J. (Ed.). (2011). *Beyond shelter: Architecture for crisis*. London: Thames & Hudson.

Banham, R. (1990). A black box: The secret profession of architecture. *New Statesman & Society, 22*(12 Oct), 22–25.

Batchelor, V. (2011). *Tarpaulins, transitional shelters or permanent houses: How does the shelter assistance provided affect the recovery of communties after disaster? (Master of Arts in Development and Emergency Practice)*. Oxford: Oxford Brookes University.

Bennett, C., Foley, M., & Pantuliano, S. (2016). *Time to let go: Remaking humanitarian action for the modern era*. Retrieved from

Boano, C., & Garcia, M. (2011). Lost in translation? The challenges of an equitable post-disaster reconstruction process: Lessons from Chile. *Environmental Hazards: Human and Policy Dimensions, 10*(3–4), 293–309.

Burnell, J., & Sanderson, D. (2011). Whose reality counts?: Shelter after disaster. *Environmental Hazards: Human and Policy Dimensions, 10*(3–4), 189–192.

Charlesworth, E. (2014a). *Humanitarian architecture : 15 stories of architects working after disaster*. New York: Routledge.

Charlesworth, E. (2014b). The rise of humanitarian architecture. *Architectural Research Quarterly, 18*(03), 267–271.

Clermont, C., Sanderson, D., Sharma, A., & Spraos, H. (2011). *Urban disasters – Lessons from Haiti: Study of member agencies' responses to the earthquake in Port au Prince, Haiti, January 2010*. Retrieved from

Davis, I. (1978). *Shelter after disaster*. Oxford, England: Oxford Polytechnic Press.

Davis, I. (2011). What have we learned from 40 years' experience of disaster shelter? *Environmental Hazards: Human and Policy Dimensions, 10*(3–4), 193–212.

Davis, I., & Alexander, D. (2016). *Recovery from disaster*. Milton Park, Abingdon, Oxon: Routledge.

DFID (2011). *Humanitarian emergency response review*. Retrieved from UK: https://www.gov.uk/government/uploads/system/uploads/attachment_data/file/67579/HERR.pdf

Fry, T. (2009). *Design futuring: Sustainability, ethics and new practice*. London: Bloomsbury Publishing.

Fry, T. (2013). *Becoming human by design*. London: Bloomsbury.

Griggs, D., Stafford-Smith, M., Owen, G., Rockström, J., Öhman, M. C., Shyamsundar, P., . . . Noble, I. (2013). Sustainable development goals for people and planet – Comment. *Nature's International Weekly Journal of Science, 495*(21 March 2013), 305–307.

Harris, V. L. (2011). The architecture of risk. In M. J. Aquilino (Ed.), *Beyond shelter: Architecture for crisis* (pp. 12–23). London: Thames & Hudson.

Heidegger, M. (1971). *Building dwelling thinking poetry, language, thought*. New York: Harper Colophon Books.

Henrotay, C. (2008). *A contribution to an integrated and more sustainable design approach for the material support of shelter after disaster*. Belgium: Vrije Universiteit Brussel.

IFRC (2013). *Post-disaster shelter: Ten designs*. Retrieved from Geneva, Switzerland: http://www.sheltercasestudies.org/files/tshelter-8designs/10designs2013/2013-10-28-Post-disaster-shelter-ten-designs-IFRC-lores.pdf

Johnson, C., Lizarralde, G., & Davidson, C. H. (2006). A systems view of temporary housing projects in post-disaster reconstruction. *Construction Management and Economics, 24*(4), 367–378.

Kennedy, J., Ashmore, J., Babister, E., & Kelman, I. (2008). The meaning of 'Build Back Better': Evidence from Post-Tsunami Aceh and Sri Lanka. *Journal of Contingencies & Crisis Management, 16*(1), 24–36.

Lizarralde, G. (2015). *The invisible houses*. New York: Routledge.

McGraw, J. W. (1978). Shelter after disaster – Book review. *Mass Emergencies, 3*(4), 270.

Medecins Sans Frontieres (2016). *Humanitarian medicine and ethics*. Retrieved from http://www.doctorswithoutborders.org/humanitarian-medicine-and-ethics

Olshansky, R. B., Hopkins, L. D., & Johnson, L. A. (2012). Disaster and recovery: Processes compressed in time. *Natural Hazards Review, 13*(3), 173–178.

Pelling, M. (2012). *The vulnerability of cities: Natural disasters and social resilience*. London: Earthscan Publications Ltd..

Sanderson, D. (2010). Architects are often the last people needed in disaster reconstruction. *The Guardian*. 3 Mar 2010. Retrieved from http://www.theguardian.com/commentisfree/2010/mar/03/architects-disaster-reconstruction-haiti-chile

Saunders, G. (2004). Dilemmas and challenges for the shelter sector: Lessons learned from the sphere revision process. *Disasters, 28*(2), 160–175.

Saunders, G., Davis, I., Thomson, P., & Krimgold, F. (2015). *Shelter after disaster* (2nd ed.). Geneva, Switzerland: UN-OCHA, IFRC.

Shelter Centre (2012). *Transitional shelter guidelines*. Retrieved from

Tauber, G. (2014). *Architects and post-disaster housing: A comparative study in South India*. Verlag, Bielefeld: transcript.

The Sphere Project (2015). *The sphere project: Humanitarian charter and minimum standards in humanitarian response*. Retrieved from http://www.sphereproject.org/about

Till, J. (2009). *Architecture depends*. Cambridge, Massachusetts: The MIT Press.

Turner, J., & Fichter, R. (1972). *Freedom to build: Dweller control of the housing process*. New York: Macmillan.

UN (2015). *Sendai framework for disaster risk reduction 2015–2030*. Retrieved from

UN General Assembly (1948). *Universal declaration of human rights, 217 A (III)*. Retrieved from http://www.un.org/en/documents/udhr/index.shtml

UNHCR (2015a). *Global trends: Forced displacement in 2015*. Retrieved from http://www.unhcr.org/576408cd7.pdf

UNHCR (2015b). *World at war: Global trends, forced displacement in* 2014. Retrieved from https://www.uno-fluechtlingshilfe.de/fileadmin/redaktion/PDF/UNHCR/Global_Trends_2014.pdf

UNISDR (2015). *2015 disasters in numbers*. Retrieved from http://www.unisdr.org/files/47804_2015disastertrendsinfographic.pdf

United Nations (2015). Sustainable development goals: 17 goals to transform our world. Retrieved from http://www.un.org/sustainabledevelopment/summit

Vitiello, R. (2017). Local legends: Post-crisis reconstruction expert Marie Aquilino talks conflict and common ground in the backstreets of Paris. Retrieved from http://www.local-legends.org/people/2017/4/4/8vjeda05hg6zva1kotgc7krtph62v2

Wagemann, E., & Ramage, M. (2013). Relief for the Curriculum. Architecture Education and Disaster Recovery. *Scroope: The Cambridge Architecture Journal*, (22), 129–133. https://doi.org/10.17863/CAM.6437

World Commission on Environment and Development (1987). *Report of the world commission on environment and development: Our common future, The Brundtland Report*. Retrieved from http://www.un-documents.net/our-common-future.pdf

Youssef, H. (2015). Ambassador Hesham Youssef: 6 propositions for a paradigm shift to change the system. Retrieved from http://blog.worldhumanitariansummit.org/entries/ambassador-hesham-youssef-6-propositions-for-a-paradigm-shift-to-change-the-system

Zyck, S. A., & Kent, R. (2014). *Humanitarian crises, emergency preparedness and response: The role of business and the private sector*. Retrieved from London, UK.

Part I

Resettlement Challenges of Refugees

Resettling Syrian Refugees in Canada: Challenges Faced by Nongovernmental Service Providers

Catherine Kenny and Aaida Mamuji

Abstract

Between November 2015 and the end of February 2016, the Government of Canada resettled over 25,000 Syrian refugees in Canada in response to the Syrian crisis. The Syrian refugee resettlement initiative galvanized the entire country, resulting in collaboration between all levels of government, religious and community groups, private citizens, and nongovernmental organizations (NGOs). Canada's resettlement program relies heavily on NGOs to provide resettlement services to refugees. This paper focuses on one type of Canadian NGO: Resettlement Assistance Program service provider organizations, or RAP SPOs, who provide government-funded resettlement services to refugees during their first weeks in Canada. During this initiative, RAP SPOs experienced unique challenges while planning for and providing services to Syrian refugees, largely due to the initiative's large scale and quick pace. Beyond the scale and pace of this initiative, however, structural deficits caused by changes to resettlement policies over the past decades have impacted RAP SPOs and the entire nongovernmental resettlement sector. This paper pays particular attention to social policy trends toward contractualism and neoliberal restructuring of the resettlement sector. It aims to illuminate the impact of these policy trends on the resettlement sector by focusing on challenges RAP SPOs experienced throughout this initiative, including budgetary constraints, inadequate employee training, and difficulty with volunteer management. The RAP SPO experience highlights the importance of ensuring that community service providers have both the time and resources to effectively implement essential humanitarian assistance and resettlement services in order for incoming refugees to thrive throughout the resettlement and integration process.

Keywords

Refugee resettlement · Nongovernmental organizations · Neoliberalism · Public policy

1 Canada's Resettlement Regime and the Syrian Refugee Resettlement Initiative

The ongoing Syrian crisis has produced the worst refugee crisis since World War II (Coen 2015). In mid-September 2015, when a young Syrian boy—whose family had ties to Canada—drowned

C. Kenny (✉) · A. Mamuji (✉)
York University, Toronto, Canada
e-mail: amamuji@yorku.ca

A. Asgary (ed.), *Resettlement Challenges for Displaced Populations and Refugees*, Sustainable Development Goals Series, https://doi.org/10.1007/978-3-319-92498-4_2

while he and his family attempted to flee the Syrian crisis, many Canadians demanded the government take humanitarian action (Garcea 2016). As a result, Canada's role in responding to the crisis became a prominent election issue during Canada's 2015 federal election (Carlier 2016; Garcea 2016). Justin Trudeau's Liberals committed to resettling 25,000 refugees by the end of 2015 (a deadline which was later extended to the end of February 2016) and stressed the need for Canada's refugee system to return to its humanitarian origins (Abu-Laban 2015; Carlier 2016; Liberal Party of Canada 2015; Soennecken 2014).[1]

Similar to Canada's efforts to resettle 60,000 refugees from Cambodia, Laos, and Vietnam in 1979 and 1980, the Syrian refugee resettlement initiative galvanized the entire country; collaborative partnerships developed between all levels of government, various religious and community groups, private citizens, and nongovernmental organizations (NGOs) (Alboim 2016; Beiser 2003; Alberta Association of Immigrant Serving Organizations [AAISA] 2016). The United Nations' Sustainable Development Goals highlight global partnerships and solidarity as a means to address complex social issues (United Nations 2015). The local and national partnerships that developed during Canada's resettlement initiative highlight the role that local relationships can have in promoting global solidarity, as Canadians came together to ensure that thousands of Syrian refugees arrived in Canada safely and were supported throughout the resettlement process.

[1]Trudeau's commitment contrasted with the securitization of Canadian immigration and refugee policy throughout the 2000s (Soennecken 2014). Stephen Harper's Conservatives, which led the country from 2006 until the 2015 federal election, cut funding for refugee healthcare and introduced stricter processing timelines and limited appeals mechanisms for refugee claimants from countries deemed to be "safe" (Abu-Laban 2015; Barnes 2013; Soennecken 2014). Although Harper's 2015 election campaign included a commitment to resettle 10,000 Syrian refugees by the following year, Canadians elected a new government, and Trudeau was sworn in as prime minister of Canada on November 4, 2015 (Carlier 2016).

NGOs played a particularly important role in this initiative, acting as a central point of contact for the government, arriving refugees, and private citizens interested in helping throughout the resettlement process. As has been the norm in Canada since the mid-1970s, NGOs provided Syrian newcomers with resettlement services under the auspices of the federal government (IRCC 2017b; Lippert 1998). Given the central role that resettlement agencies play in successful integration of refugees, this chapter explores the challenges faced by these nongovernmental service providers by looking specifically at the case of Syrian newcomer arrival in Canada.

Newcomer integration is inherently a national and local project (Organisation for Economic Co-operation and Development 2006). In many countries with resettlement programs, governments offer contracts to NGOs to provide refugees with resettlement and settlement services due to NGOs' local scale, flexibility, quick ability to respond to emerging issues, and their knowledge of the communities in which refugees are arriving and settling (Chekki 2006; Elliott and Yusuf 2014; Winkler 1981). In Canada, these government-funded resettlement agencies are known as Resettlement Assistance Program service provider organizations (RAP SPOs) and are the unit of focus in this chapter. Sources of social capital for refugees, resettlement agencies like RAP SPOs facilitate refugee connections to local resources and liaise between refugees and members of the receiving community (Elliott and Yusuf 2014; Ives et al. 2010; Nawyn 2006). They therefore serve as central connections for refugees in both the short- and long-term upon arriving in a new country.

Canada has one of the largest resettlement programs in the world and has a complex infrastructure in place to receive refugees (Labman 2011). Refugees arrive in Canada under one of three streams:

1. As government-assisted refugees (GARs), who receive financial support from the federal government for up to 1 year and whose resettlement services are provided by RAP SPOs

2. As privately sponsored refugees (PSRs), who receive financial, emotional, and social support from established Canadian citizens
3. Under the Blended Visa Office-Referred (BVOR) program, where the federal government and established citizens jointly provide support to refugees for up to 1 year (IRCC 2016a, c, d)

While the 25,000 Syrians who arrived between November 2015 and February 2016 consisted of refugees in each of these categories, our interest in RAP SPOs, and their relationship with the government in particular, necessitates a better understanding of policies and services that pertain to government-assisted refugees, the first category described above.

RAP SPOs deliver resettlement services to government-assisted refugees through the Resettlement Assistance Program (RAP). In addition to income support for up to 1 year, RAP provides incoming GARs with "immediate and essential services, usually needed by the refugees in the first 4–6 weeks in Canada," including port-of-entry services, translation and interpretation, temporary accommodations provided by RAP SPOs, assistance locating and moving to permanent accommodations, general and financial orientations, links to mandatory federal and provincial programs, and assessment and referral to other settlement services (Government of Canada 2016).

Twenty-seven RAP SPOs throughout Canada held government contracts to provide these services to the GARs arriving through this initiative. Of the 25,000 total refugees who arrived in Canada between November 4, 2015, and February 29, 2016, 15,000 were GARs who largely arrived in a compressed timeframe between the end of December and end of February (IRCC 2017b, d). This four-month timeframe resulted in double the amount of GARs that RAP SPOs resettled throughout all of 2014 (IRCC 2016b). However, given that the engagement between the Government of Canada and resettlement agencies only began in November 2015, shortly after Justin Trudeau was sworn in as the prime minister of Canada, RAP SPOs only had 1 month to prepare before the bulk of GARs began arriving in their communities in mid-December. RAP SPO contributions were therefore provided on an extremely short timeline and far exceeded normal organizational capacities, causing unique challenges for these organizations.

This chapter examines the government-RAP SPO relationship both before and while Syrian refugees arrived in Canada. It illuminates the challenges faced by these resettlement service provider organizations given the short-term policy planning and implementation experienced during the Syrian resettlement initiative and amid long-term structural deficits caused by changes to resettlement policy over the past decades. Particular attention is paid to social policy trends toward contractualism and neoliberal restructuring, in which private sector characteristics increasingly characterize public service provision (Hood 1990, as cited in Evans et al. 2005).

2 Understanding the Government-RAP SPO Relationship

Familiarity with New Public Management (NPM) and neoliberal restructuring of the resettlement sector provides the context necessary to understand how government policies shape on-the-ground realities of RAP SPOs. NPM literature highlights the increasing use of contracting services by the government for the provision of services to the public, while literature on neoliberal restructuring of the resettlement sector offers insight into how such policy shifts impact the day-to-day operations of resettlement agencies.

2.1 New Public Management

New Public Management emerged in response to the rising notion in the 1980s that bureaucracy was inefficient (Aucoin 1995; Barzelay 2001; Lane 2000). NPM represented a shift in governance and the emergence of a new managerialism, where management practices of the private sector were increasingly adopted in the public sector (Aucoin 1995; Lane 2000). Peter Aucoin (1995) notes that citizens became increasingly viewed as "clients" or "customers" of government services. As government managers became required to accomplish more with fewer financial resources, governments began adopting private sector practices to deliver more efficient services to citizens. Under NPM, the work of government became increasingly about "the management of contracts, [and] how they are awarded, monitored and policed" (Lane 2000, p. 9). Contracted service providers therefore became the source of expertise in particular areas, while government played an increasingly administrative role through contract provision and monitoring, and ensuring that service providers were held accountable through various reporting mechanisms (Lane 2000).

This rise of contracting services favors short-term contracts, contrasting with the traditional public sector use of long-term contracts. Long-term contracting results in "the evolution of expertise" and "agency independence" and facilitates the development of trust between the government and service provider organizations (Lane 2000, p. 134). Trust is particularly important in the human services sector, where outputs cannot immediately be measured quantitatively and where services are often long-term in nature (Lane 2000). From the perspective of NPM, however, long-term relationships result "in the monitoring problem," where government cannot simply assume that the service provider will live up to all contractual expectations (Lane 2000, p. 136). While professional groups often resist short-term contracting, bureaucrats must work to create trust between such groups and government in order to ensure continued service provision and compliance with the contracting regime.

The provision of resettlement services to refugees in Canada exhibits many characteristics of NPM. As previously mentioned, NGOs have been prominently involved in providing resettlement and settlement services in Canada since the mid-1970s, before which bureaucrats of Citizenship and Immigration Canada (CIC)—now known as Immigration, Refugees and Citizenship Canada (IRCC)—delivered resettlement services (Lippert 1998). Between the mid-1970s and mid-1990s, resettlement services were offered somewhat jointly between NGOs and government officials. Canada's refugee resettlement program changed dramatically in 1998 with the establishment of the Resettlement Assistance Program (RAP). RAP "saw a change in the service delivery method, with a shift from direct provision of services by CIC [now IRCC] to the use of third-party contractors [now known as RAP SPOs] who provided necessary programming" (IRCC 2011). With the introduction of RAP SPOs into refugee resettlement, the government acknowledged its own shifting role as that of "administrator of income support" and "program/contract management," with NGOs as the site of service provision (IRCC 2007).

2.2 Neoliberal Restructuring of Resettlement Sector

Neoliberal restructuring refers to the incorporation of market-based regulation into social policy, emphasizing management and performance measurement, a focus on fiscal constraint, and an increase in competition within the public sector, among other things (Hood 1990, as cited in Evans et al. 2005, p. 77). Research demonstrates that neoliberal restructuring has strongly impacted the entire nongovernmental industry in many ways, from day-to-day operations and programming abilities to the relationships between government funders and these organizations (Evans et al. 2005; Lippert 1998; Richmond and Shields 2004; Richmond and Shields 2005). Impacts of neoliberal restructuring on NGOs include compromised autonomy and advocacy, continuous reporting to government funders, increasing administrative activities, strained organizational capacity, and increasing employment instability within the sector (Christensen and Ebrahim 2006; Ebrahim 2003; Evans et al. 2005; Ontario Council of Agencies Serving Immigrants [OCASI] 2016).

Despite the government's reliance on NGOs for service provision, as a whole government provides less funding to the nonprofit sector in the advanced liberal era (Evans et al. 2005; Richmond and Shields 2004; Richmond and Shields 2005). With less funding overall, Evans et al. (2005) note that often "nonprofit fiscal stress [appears to be] 'managed' by laying off core staff and restraining wages and benefits – often the only area where there is flexibility in nonprofit budgets" (p. 82). For remaining employees, a loss of human resources within their organization increases individual job responsibilities, stress, and feelings of dissatisfaction, as they attempt to meet the same—or increasing—levels of client needs with decreasing resources (Evans et al. 2005; OCASI 2016).

Understanding NPM and neoliberal restructuring serves as a useful foundation upon which to examine the relationship between the Government of Canada and Resettlement Assistance Program service provider organizations (RAP SPOs) during the Syrian resettlement initiative. Overall, precarious funding and precarious employment are large concerns for employees within the immigrant- and refugee-serving sector (Le-Doux and Stephens 1992; Meinhard 2016), and the Syrian refugee resettlement initiative was evidently no exception (see OCASI 2016 for more detail on the 2015 budget cuts to Ontario's settlement sector). The following section discusses RAP SPOs' planning efforts for this initiative and the impacts that these policy shifts had on implementation.

3 Methodology

The subsequent discussion on planning efforts is based on an analysis of 21 semi-structured interviews, conducted between July and October 2016, with a total of 24 individuals who were involved in planning for and/or providing resettlement services during the Syrian refugee resettlement initiative. Among those who participated include 15 employees of 14 RAP SPOs located throughout Canada. When quoted or referenced in this paper, these participants will be referred to as RAP#, where every participant has been given a unique identifying number (e.g., RAP1–RAP15).

The remaining interviews were with four employees of IRCC, who will be referred to as public officials in this paper, four employees of organizations within the resettlement sector that did not provide RAP services, and one employee of an international NGO that was involved in the early stages of this initiative.

4 Planning for the Syrian Refugee Resettlement Initiative

When the Government of Canada officially announced the Syrian resettlement initiative on November 24, 2015, many RAP SPOs were ready and willing to play a central role in the resettlement effort (Garcea 2016; Government of Canada 2015). Some municipal governments, NGOs (including some RAP SPOs), and community groups had already been organizing themselves and/or asking government to address the Syrian conflict since September 2015 and therefore were further ahead in efforts to mobilize community support (Garcea 2016).

Official RAP SPO preparation for this initiative began in November 2015 with regular teleconferences with Immigration, Refugees and Citizenship Canada (IRCC). These discussions addressed the following issues: housing capacity for temporary and permanent accommodations for refugees, RAP SPO capacity internally (primarily in terms of human resources), each community's settlement-sector capacity (such as language assessment and training capacity and wait lists for other services), and how welcoming a community was toward newcomers generally. RAP SPOs' public collaboration with government for this initiative began when IRCC held a working forum on the initiative on November 28 and 29, 2015, in Toronto, Ontario, which brought together the resettlement and settlement sectors, as well as civil society organizations, such as the Canadian Red Cross, Canada's chapter of the United Nations High Commissioner for Refugees, the Canadian Council for Refugees, Lifeline Syria, and more (IRCC 2017c). The teleconferences and working forum were used primarily for information exchange and open discussions about how this initiative might unfold operationally.

However, the information the government provided to RAP SPOs and other stakeholders at this stage was often vague and did not assist in planning efforts in a concrete way. For instance, early in the planning process, teleconferences often did not provide information that satisfactorily answered many operational and logistical questions RAP SPO employees had, resulting in some employees feeling left "a little bit in the dark" (RAP07). As well, participants highlighted confusion over how much (or if) the RAP process would change, as well as where Syrians would be initially arriving and residing in Canada.

5 Impact of Neoliberalism on Policy Planning and Implementation

Overall, RAP SPO participants of this study applauded the government for undertaking this initiative, as well as communities across Canada for offering their time and resources to assist in resettling Syrian refugees. Participants highlighted many successful aspects of the initiative, including increased government funding flexibility for direct service provision to refugees, increased operational funding for the RAP, and increased communication and openness between RAP SPOs and government. This openness by government was a contrast from the conservative government that preceded Trudeau, as noted by RAP15, who stated there was "really an openness in sharing [between RAP SPOs and the government] that I don't think we've seen in probably a decade."

Despite an increase in communication throughout the planning and early implementation stages of this initiative, many operational challenges remained. These operational challenges and their connection to neoliberal trends include planning with short-term and uncertain funding, inadequate training and added employee stress, and difficulty with volunteer management.

5.1 Planning with Short-Term and Uncertain Funding

Planning and implementing the Syrian refugee resettlement initiative required funding that went above and beyond IRCC's normal operating costs and the budget devoted to RAP. In mid-December 2015, just after the government officially announced the initiative, IRCC increased RAP SPOs' budgets by 25% to enable them to hire additional staff, secure temporary accommodations for refugees, and fulfill other preparatory tasks prior to the arrival of Syrian refugees. While this funding flexibility was a welcome change from the status quo of rigid funding structures, the funding was neither timely nor large enough to "ramp up" RAP adequately for the majority of participants. Many participants noted that the first part of top-up funding came just before or as refugees began arriving in late December 2015 or early January 2016, a timeline which adversely impacted RAP SPOs' ability to plan beyond the weeks immediately after refugee arrival.[2] For instance, RAP13 noted, "I would say if we had to go back and redo the whole thing that I would have asked for our government to give us more lead time, and then put the funds in place way, way before the actual work starts."

Moreover, many participants expressed frustration that the top-up money was only until the end of the fiscal year and would therefore end on March 31, 2016, and that funding levels for the settlement phase—longer-term services offered to refugees after initial resettlement services—remained relatively unclear at that point. Participants highlighted that resettlement is part

[2]As previously noted, although the first refugees in this initiative arrived as early as November 4, 2015, most RAP SPO participants did not receive GARs until mid- to late-December.

of what they termed a "settlement continuum" wherein successful refugee integration depends on funding for long-term settlement service as well. For organizations that assist with immigrant and refugee integration, such as RAP SPOs, employees often worry about the impact of precarious funding on the long-term impacts of their services, noting that a "'disconnect' [exists] between practices of government and their expressed desire to offer more-holistic support services" (Meinhard et al. 2016, p. 289). For example, RAP15 noted:

> [I]n the initial phase, we were trying to plan where [we were] going to need more assistance. What sort of supports do people need? But our ability to do that was really constrained by the funding. So when the funding came, [we were] mostly focused on advocacy to the government around the need to not approach the resettlement as RAP first, then settlement [...], that they needed to go hand-in-hand.

This quotation demonstrates the inherent connection between effective settlement services and strategic, long-term funding, a characteristic which contradicts the short-term contract regime of NPM.

5.2 Inadequate Training and Added Employee Stress

Similarly, participants often highlighted that funding for human resources was inadequate in amount and timeliness, which strained organizational capacity and increased employee stress. Previous research on NGOs that assist with immigrant and refugee integration has noted that employees believe that effective NGO-government relationships depend on "having an appropriate amount of time, enough personnel, and financial resources" to effectively offer their services (Meinhard et al. 2016, p. 289). Being inadequately resourced is not a new experience for these employees (OCASI 2016; Richmond and Shields 2004). However, given the scale of the Syrian resettlement initiative, many participants mentioned that staff worked overtime and sometimes became burnt out. Being understaffed meant that "staff also put in huge numbers of volunteer hours during that time [because] everyone was aware of where the volunteers were needed" and employees wanted to ensure the needs of incoming refugees were being met (RAP01). These characteristics align with past research, which notes that immigrant-serving agencies "are doing more for less," as the norm becomes an "increased demand, without an increase in staff to match these needs" (Richmond and Shields 2004, p. 11).

For many RAP SPO participants, when financial resources did arrive, needing to hire staff very quickly was likewise stressful. Often, RAP SPOs received funding for human resources just before refugees arrived, and so they "could just absolutely not afford to provide the normal training" to these new employees (RAP02). For some, having undertrained new staff led to stress and additional responsibilities for senior employees. One resettlement agency representative noted:

> On our own staff, [it was] very stressful because we had senior staff with a lot of experience trying to support newly hired staff with much less experience. [...] So, it definitely put a big strain on the core [RAP] workers because they had to orient and coordinate and support a lot of new people who didn't have experience.

This funding timeline demonstrates the Government of Canada's inability to adequately prepare RAP SPOs for the bulk of arrivals, as new employees were thrown into an "overwhelming" environment, where they "right away [had] a ton of responsibility and tons of tasks to fulfill" despite having inadequate or rushed training (RAP02). Furthermore, for many participants, hiring people for short-term, three- and four-month appointments for this project and then dismissing them due to funding limitations was very stressful. RAP13 stated:

> [I]t was difficult because you put people who are just learning how to do their job and they were just getting good at it, and then 3 months after you hire them, now you just let them go. And that is what happens. It was really very... – we were put in a very difficult position.

This quotation highlights the personal impacts that the entrenchment of short-term contract work into an industry can have; established employees can be placed in a "difficult position" when forced to lose a competent employee amid organizational strain. However, RAP04 noted that "know[ing] when to roll up and roll back" is something that their agency is familiar with, as funding variance in contracts over the years is a routine experience for many NGOs under the current government-provided contract system.

5.3 Difficulty with Volunteer Management

Beyond the internal capacity issues that RAP SPOs experienced, RAP SPO participants highlighted community engagement as a primary pressure point in this initiative because internal capacity often did not match public offerings of goodwill. As discussed, community engagement is central to successful refugee resettlement, and so at a government forum outlining the initiative on December 1, 2015, the government encouraged all Canadians to become involved in assisting refugees to resettle and integrate (IRCC 2017c). This forum was designed to encourage "a coordinated and collaborative national effort," while "highlight[ing] the practical efforts [already] under way across the public, community and private sectors and among individual Canadians" (IRCC 2017c). With this encouragement to become involved, RAP SPOs became one potential access point for Canadians to interact and volunteer with refugees.

However, despite the importance of refugees gaining social support, community engagement programs have lost funding in recent years and NGOs in general have a harder time retaining a core group of volunteers than in the past (Richmond and Shields 2004). Over the last 10–15 years, the Government of Canada has funded less intensive community engagement programs for new immigrants and refugees (public official, personal communication; IRCC 2012). In 1990, the Government of Canada announced that the pilot Host Program would become a permanent program in the refugee and immigrant settlement sector (Lippert 1998). The Host Program was designed to "assist in resettlement and the development of social networks" (Yu et al. 2007, p. 25) by having "[v]olunteers familiar with Canadian ways help newcomers learn about available services and how to use them, practice English and French, get contacts in their field of work and participate in the community" (IRCC 2004). These relationships could be short term in nature, but were generally longer term and one on one (IRCC 2004). This program was facilitated by service provider organizations, which were given funding to support the relationship between newcomer clients and host volunteers (IRCC 2004).

In 2008, as IRCC (then Citizenship and Immigration Canada) "modernized" its approach to settlement, the Host Program and other settlement programs were amalgamated into one program (IRCC 2012). Here, modernization through amalgamation demonstrates government's continuing attempts to make settlement leaner and cheaper to support.[3] The change has made it more difficult for established Canadians to volunteer and get involved in the Syrian refugee resettlement initiative. As a result of the defunding of the Host Program, there was no consistent, national, and established program for RAP SPOs and other settlement agencies to use in order to harness the support of Canadians during the Syrian resettlement initiative.

Despite the importance of involving established Canadians in the resettlement process, many RAP SPO participants were unprepared to handle the high surges of volunteer support they received. For instance, RAP15 noted that with the federal government's call for Canadians to get involved in the initiative, "the [resettlement]

[3] To replace the Host Program, the government introduced the Community Connections program. Important distinctions exist between the two. According to one public official, Community Connections is focused more on group activities, as opposed to one-on-one or family-to-family relationships; the Community Connections program is not an effective program for providing extended support to volunteers in the same way that the Host Program could (personal communication). While the Host Program provided organizations with funding to identify, train and manage volunteers, the Community Connections program is simply one small part of a larger settlement program.

sector bore a lot of that pressure because we were getting calls" from people interested in volunteering, "saying, 'When are you going to send me those GARs?' or 'I want a GAR family.'" Although some agencies did bring in volunteer coordinators as quickly as possible, if no volunteer engagement system was in place, managing those volunteers became difficult. This organizational disconnect ultimately resulted in some volunteers adversely impacting RAP SPOs' service provision abilities. One participant, whose organization struggled with volunteer management, noted the paradox of public support:

> [The temporary accommodations] became the place to go and interact with Syrian refugees. [...] [T]here was just this energy, which, as I said, was very positive but was also totally consuming of staff time and energy. [...] We didn't think to have two or three staff on call at any time. [...] Because there were people just pulling and pulling and pulling staff time.

This experience illustrates that public support and volunteer offers often exacerbated the pressure that RAP SPOs' internal capacities were already experiencing throughout this initiative due to the high volume of work they were funded to complete for increasing numbers of refugee arrivals.

Official lessons learned reports on the short-term resettlement outcomes of Syrians in the Resettlement Assistance Program (RAP) confirm both the successes and challenges that RAP SPO experienced during this initiative. These are discussed in the following section.

6 Lessons Learned Through Resettlement Service Provision

In the months following the Syrian refugee resettlement initiative, government and resettlement sector organizations evaluated the initiative and analyzed the lessons learned through providing resettlement services to Syrian refugees (Access Alliance 2017; AAISA 2016; IRCC 2016e, 2017a; Manitoba Association of Newcomer Serving Organizations [MANSO] 2016; Saskatchewan Association of Immigrant Settlement and Integration Agencies [SAISIA] 2016). Reports highlight the robust information-sharing between stakeholders; the importance of partnerships, collaboration, and public support; the development of trusting relationships; and operational flexibility that emerged throughout this initiative, partially as necessary responses to the short timeline in which the initiative was planned and implemented (AAISA 2016; Access Alliance 2017; MANSO 2016; SAISIA 2016; IRCC 2017a). These reports also confirm the challenges outlined in this paper: a lack of timely and adequate funding, the subsequent challenge of hiring and training people quickly and adequately, and difficulties with managing volunteers (AAISA 2016; Access Alliance 2017; MANSO 2016; SAISIA 2016).

In December 2016, IRCC released a report outlining a Rapid Impact Evaluation (RIE) that they conducted of the resettlement initiative, providing unique insight into possible connections between the challenges of this initiative and early resettlement outcomes of Syrian refugees (IRCC 2016e). This assessment of refugees' resettlement experiences confirms that some RAP SPO services suffered as a result of the pace and volume of arrivals throughout this initiative. Challenges in the delivery of RAP included difficulty finding permanent housing and a lack of consistency in the standards of RAP delivery (IRCC 2016e). Given the high volume of arrivals, "RAP SPOs were not able to provide orientation to Canada to the same standard provided to other refugees" (IRCC 2016e). RAP SPOs often offered RAP orientation services in temporary accommodations and in larger groups to accommodate for the high volume of arrivals, rather than in their own RAP SPO facilities as they normally would. For some refugees, receiving orientations in large groups and where families were living was distracting and made it hard to focus on the information being presented (IRCC 2016e).

This lower level of orientation may have contributed to IRCC's finding that privately sponsored refugees (PSRs)—those who received financial and emotional support from private citizens, rather than RAP SPOs—"reported receiving more help to resettle compared to GARs" and were more likely than GARs "to indicate that their immediate needs were met" (IRCC 2016e). GARs were less likely to report that they had received adequate assistance or information on how to find a doctor on their own; how to buy clothes, furniture, and other much-needed items; how to look for permanent accommodations; and how to shop for food (IRCC 2016e). Overall, GARs were less positive about all of the resettlement services they received than PSRs (IRCC 2016e).[4] These differences between GARs and PSRs in satisfaction and access to resettlement and longer-term settlement services highlight the potential impacts of the challenges RAP SPOs faced that have been outlined in this paper and provide a strong case for addressing the consequences of short-term, contractual funding.

7 Conclusion

Canada's Syrian refugee resettlement initiative saw millions of Canadians working together to assist in the resettlement and integration of over 25,000 Syrian refugees, including approximately 15,000 GARs, between November 2015 and February 2016. This paper focused on the experiences of one group of those who assisted, the Resettlement Assistance Program service provider organizations (RAP SPOs), who were responsible for administering the initial resettlement services to government-assisted refugees. This initiative led to many positive experiences between RAP SPOs and government officials but also highlighted challenges in the resettlement sector that can be linked back to policy shifts in light of New Public Management and neoliberal restructuring. Challenges faced by RAP SPOs throughout this resettlement initiative were made more difficult by these policy shifts, which have historically been shown to create organizational strain. These include planning with short-term and uncertain funding, inadequate training and added employee stress, and difficulty with volunteer management.

Overall, IRCC acknowledges that one major shortfall of the Resettlement Assistance Program throughout this initiative was inadequate "end-to-end planning," in which "resettlement and settlement considerations are fully integrated into the planning phase" of the initiative (IRCC 2016e). Specifically, stakeholders, including the government, acknowledged after the fact that "not enough attention was given to the settlement and integration portion of the initiative" before the initiative began (IRCC 2016e). The Canadian experience highlights the importance of host governments ensuring that community service providers have both the time and resources to effectively implement essential humanitarian assistance and resettlement services in order for incoming refugees to thrive and succeed in their countries.

[4]While these findings may appear to demonstrate an inadequate level of resettlement support, it is important to note that as of December 2016, approximately 9–12 months after arrival in Canada, 77% of both GARs and PSRs studied by IRCC indicated they were either happy or very happy with their lives in Canada (IRCC 2016e). Although GARs reported lower levels of satisfaction with the resettlement services they received, the majority of both GARs and PSRs indicated that their "immediate and essential needs were mostly or completely met soon after their arrival in Canada" (IRCC 2016e).

References

Abu-Laban, Y. (2015). Transforming citizenship: Power, policy and identity. *Canadian Ethnic Studies, 47*(1), 1–10.

Access Alliance (2017). *Refugee health and settlement: Lessons learned from the Syrian response.* http://accessalliance.ca/wp-content/uploads/2017/06/SyrianResettlement_ResearchHighlightsReport_2017.pdf. Accessed 25 Oct 2017.

Alberta Association of Immigrant Serving Agencies (AAISA) (2016). *Alberta settlement and integration two-day participatory event report.* http://aaisa.ca/wp-content/uploads/2017/03/aaisa_ab_settlement-and-integration-consultation-report-1.pdf. Accessed 3 Apr 2017.

Alboim, N. (2016, May 18). Lessons learned from the Indochinese and Syrian refugee movements. Policy Options Politiques: Le Forum au Service de Bien Public. http://policyoptions.irpp.org/fr/magazines/

mai-2016/lessons-learned-from-the-indochinese-and-syrian-refugee-movements. Accessed 2 Feb 2017.

Aucoin, P. (1995). *The new public management: Canada in comparative perspective*. Ottawa, ON: Institute for Research on Public Policy.

Barnes, S. (2013). *The real cost of cutting the Interim Federal Health Program*. Toronto, ON: Wellesley Institute.

Barzelay, M. (2001). *The new public management: Improving research and policy dialogue*. Berkeley, CA: University of California Press.

Beiser, M. (2003). Sponsorship and resettlement success. *Journal of International Migration and Integration, 4*(2), 203–215.

Carlier, M. (2016). Explaining differences in the Canadian and American response to the Syrian refugee crisis. *Virginia Policy Review, 9*(2), 56–82.

Chekki, D. A. (2006). *Immigrant and refugee serving organizations in a Canadian city: An exploratory study. PCERII Working Paper Series*. Edmonton, AB: Prairie Centre of Excellence for Research on Immigration and Integration.

Christensen, R., & Ebrahim, A. (2006). How does accountability affect mission? The case of a nonprofit serving immigrants and refugees. *Nonprofit Management & Leadership, 17*(2), 195–209.

Coen, A. (2015). R2P, global governance, and the Syrian refugee crisis. *The International Journal of Human Rights, 19*(8), 1044–1058.

Ebrahim, A. (2003). Accountability in practice: Mechanisms for NGOs. *World Development, 31*(5), 813–829.

Elliott, S., & Yusuf, I. (2014). 'Yes we can, but together': Social capital and refugee resettlement. *Kōtuitui: New Zealand Journal of Social Sciences Online, 9*(2), 101–110.

Evans, B., Richmond, T., & Shields, J. (2005). Structuring neoliberal governance: The nonprofit sector, emerging new modes of control and marketisation of service delivery. *Policy and Society, 24*(1), 73–97.

Garcea, J. (2016). The resettlement of Syrian refugees: The positions and roles of the Federation of Canadian Municipalities and its members. *Canadian Ethnic Studies, 48*(3), 149–173.

Government of Canada (2015, November 24). Canada offers leadership on the Syrian refugee crisis. *News Release*. https://www.canada.ca/en/immigration-refugees-citizenship/news/2015/11/canada-offers-leadership-on-the-syrian-refugee-crisis.html. Accessed 22 Aug 2017.

Government of Canada (2016, January 21). Backgrounder: Resettlement and settlement services for government-assisted refugees. *Backgrounders*. http://news.gc.ca/web/article-en.do?nid=1028969. Accessed 20 Apr 2017.

Hood, C. (1990). Beyond the Public Bureaucracy State? *Public Administration in the 1990's*. London: London School of Economics and Political Science.

IRCC (2004, November 1). Evaluation of HOST. *Audits and Evaluations*. http://www.cic.gc.ca/english/resources/evaluation/host/intro.asp. Accessed 1 Apr 2017.

IRCC (2007). Evaluation of the Resettlement Assistance Program (RAP). *Audits and Evaluations*. http://www.cic.gc.ca/english/resources/evaluation/rap-summary.asp. Accessed Apr 2017.

IRCC (2011, November 4). Evaluation of Government Assisted Refugees (GAR) and Resettlement Assistance Program (RAP). *Audits and Evaluations*. http://www.cic.gc.ca/english/resources/evaluation/gar-rap/section1.asp. Accessed 1 Apr 2017.

IRCC (2012, January 13). Evaluation of the host program. *Audits and Evaluations*. http://www.cic.gc.ca/english/resources/evaluation/host-program/section3.asp. Accessed 1 Apr 2017.

IRCC (2016a, June 16). Government-assisted refugees program. *Refugees*. http://www.cic.gc.ca/english/refugees/outside/gar/index.asp. Accessed 11 Feb 2017.

IRCC (2016b, July 7). Evaluation of the resettlement programs (GAR, PSR, BVOR and RAP). *Audits and evaluations*. http://www.cic.gc.ca/english/resources/evaluation/resettlement.asp. Accessed 20 Aug 2017.

IRCC (2016c, August 23). Private sponsorship of refugees – Information for refugees. *Refugees*. http://www.cic.gc.ca/english/refugees/outside/private.asp. Accessed 11 Feb 2017.

IRCC (2016d, August 30). Blended visa office-referred program – Sponsoring refugees. *Refugees*. http://www.cic.gc.ca/english/refugees/sponsor/vor.asp. Accessed 11 Feb 2017.

IRCC (2016e, December 21). Rapid impact evaluation of the Syrian refugee initiative. *Audits and evaluations*. http://www.cic.gc.ca/english/resources/evaluation/rapid-syria.asp. Accessed 20 Aug 2017.

IRCC (2017a, April 20). Syrian refugee resettlement initiative – Looking to the future. *Refugees*. http://www.cic.gc.ca/english/refugees/welcome/syria-future.asp#lessons. Accessed 20 Aug 2017.

IRCC (2017b, May 4). #WelcomeRefugees: Canada resettled Syrian refugees. *Refugees*. http://www.cic.gc.ca/english/refugees/welcome/index.asp. Accessed 20 Aug 2017.

IRCC (2017c, June 12). *Syrian refugees horizontal initiative*. Departmental Performance Reports. http://www.cic.gc.ca/english/resources/publications/dpr/2016/04.asp. Accessed 20 Aug 2017.

IRCC (2017d, July 26). Canada's Syrian commitments. Refugees. http://www.cic.gc.ca/english/refugees/welcome/commitment.asp. Accessed 20 Aug 2017.

Ives, N., Witmer Sinha, J., & Cnaan, R. (2010). Who is welcoming the stranger? Exploring faith-based service provision to refugees in Philadelphia. *Journal of Religion & Spirituality in Social Work: Social Thought, 29*, 71–89.

Labman, S. (2011). Queue the rhetoric: Refugees, resettlement and reform. *University of New Brunswick Law Journal, 62*, 55–63.

Lane, J. E. (2000). *New public management*. New York, NY: Routledge.

Le-Doux, C., & Stephens, K. S. (1992). Refugee and immigrant social service delivery: Critical management issues. *Journal of Multicultural Social Work, 2*(1), 31–46.

Liberal Party of Canada (2015, September 5). Liberals commit to leadership on Syrian refugee crisis. *Liberal Party of Canada*. https://www.liberal.ca/liberals-commit-to-leadership-on-syrian-refugee-crisis. Accessed 11 Feb 2017.

Lippert, R. (1998). Rationalities and refugee resettlement. *Economy and Society, 27*(4), 380–406.

Manitoba Association of Newcomer Serving Organizations (MANSO) (2016). *Lessons learned: Settlement and Integration Consultation*. http://mansomanitoba.ca/wp-content/uploads/2016/10/MANSO-Lessons-Learned-Report.pdf. Accessed 20 Aug 2017.

Meinhard, A., Lo, L., & Hyman, I. (2016). Cross-sector partnership in the provision of services to new immigrants in Canada: Characteristics, relevance and constraints. *Human Services Organizations: Management, Leadership & Governance, 40*(3), 281–296.

Nawyn, S. J. (2006). Faith, ethnicity, and culture in refugee resettlement. *American Behavioral Scientist, 49*(11), 1509–1527.

Ontario Council of Agencies Serving Immigrants (OCASI) (2016). *Telling our stories from the frontline: Adverse institutional impacts of cuts to immigrant settlement funding in Ontario*. Toronto, ON: OCASI. http://www.ocasi.org/sites/default/files/telling-our-stories-from-the-frontline_1.pdf. Accessed 17 Feb 2017.

Organisation for Economic Co-operation and Development (OECD). (2006). *From immigration to integration: Local solutions to a global challenge*. Paris: OECD Publishing.

Richmond, T. & Shields, J. (2004). *Third sector restructuring and the new contracting regime: The case of immigrant serving agencies in Ontario*. Ryerson University Centre for Voluntary Studies Working Paper Series.

Richmond, T., & Shields, J. (2005). NGO-government relations and immigrant services: Contradictions and challenges. *Journal of International Migration and Integration, 6*(3/4), 513–526.

Saskatchewan Association of Immigrant Settlement and Integration Agencies (SAISIA) (2016). *Lessons learned from the Syrian refugee project: Engaging the Saskatchewan settlement system for better outcomes for all newcomers*. http://saisia.ca/images/Lessons-Learned-Report---Saskatchewan-Syrian-Refugee-Resettlement-Initiative.pdf. Accessed 20 Aug 2017.

Soennecken, D. (2014). Shifting up and back: The European turn in Canadian refugee policy. *Comparative Migration Studies, 2*(1), 101–122.

United Nations (2015). *Transforming our world: The 2030 agenda for sustainable development*. https://sustainabledevelopment.un.org/content/documents/21252030%20Agenda%20for%20Sustainable%20Development%20web.pdf. Accessed 25 Oct 2017.

Winkler, E. (1981). Voluntary agencies and government policy. *The International Migration Review, 15*(1/2), 95–98.

Yu, S., Ouellet, E., & Warmington, A. (2007). Refugee integration in Canada: A survey of empirical evidence and existing services. *Refuge, 24*(2), 17–34.

Settlement Policies for Syrian Refugees in Lebanon and Jordan: An Analysis of the Benefits and Drawbacks of Organized Camps

Faten Kikano and Gonzalo Lizarralde

Abstract

Six years after the eruption of the Syrian conflict, Lebanon and Jordan adopted divergent policies in hosting over two million Syrian refugees. Whereas Jordan has allowed organized camps, Lebanon adopted a non-encampment policy, leading refugees to spread informally in almost 2000 locations, with only a few camps authorized. This study examines the reasons that led to the adoption of these policies in both countries and their effects on the refugee and host populations. It also analyzes the conditions of refugees in organized camps in both contexts through two comparative case studies: first, in Zaatari, a camp in Jordan with almost 80,000 refugees and, second, in the Union of Relief and Development Associations (URDA) camp in Lebanon, which hosts about 2000 refugees. Data was collected during fieldworks in Jordan in 2015 and in Lebanon in 2015, 2016, and 2017 through field observation and discussions with 41 actors, including camp managers and refugees. The results show that most refugees in Zaatari live in harsh conditions in a context of inequality and restricted freedom. Refugees in the URDA camp live in more favorable conditions and receive increased security, aid, and free services. Whereas previous assessments have established that urban refugees in Jordan enjoy better conditions than those in camps, and urban refugees in Lebanon often live in poor conditions, in constant fear, and in insecurity, empirical findings showed here challenge the paradigm that strongly criticizes encampment. They suggest that, in specific conditions, organized camps can represent a better solution than refugee settlements in urban areas.

Keywords

Syrian refugees · Urban refugees · Organized camps

1 Introduction

Since 2011, the Syrian conflict has caused the displacement of millions of people who now represent the largest refugee population in the

F. Kikano (✉)
The Canadian Disaster Resilience and Sustainable Reconstruction Research Alliance (Oeuvre Durable), Université de Montréal, Montréal, Canada
e-mail: faten.kikano@umontreal.ca

G. Lizarralde
The Royal Society of Canada, Professeur École d'architecture, Université de Montréal, Titulaire de la Chaire Fayolle-Magil Construction en architecture, bâtiment et durabilité, Directeur du Groupe de recherche IF (grif) et OEuvre durable, Montreal, Canada
e-mail: gonzalo.lizarralde@umontreal.ca

A. Asgary (ed.), *Resettlement Challenges for Displaced Populations and Refugees*, Sustainable Development Goals Series, https://doi.org/10.1007/978-3-319-92498-4_3

world (United Nations 2016). Lebanon and Jordan, two neighboring countries to Syria, are hosting, respectively, over 1.1 million and 650,000 Syrian refugees, but unofficial sources estimate that 1.5 million Syrians reside in Lebanon (Humanitarian Aid and Civil Protection 2017). Refugees represent almost 30% of Lebanon's population, which has thus become the country with the highest refugee population per capita, and 10% of Jordan's population (UNHCR 2016). However, due to differences in the political and economic contexts, and in past experiences with refugees and camp management, the two countries have adopted divergent settlement policies (Turner 2015). Whereas Jordan has allowed organized camps (OCs) for some 100,000 Syrian refugees in areas close to the border with Syria, Lebanon has adopted a non-encampment policy, which has led refugees to spread out informally in almost 2000 locations, with only a few camps being authorized on private land.

This paper seeks firstly to understand the reasons that led to the adoption of the refugee settlement policies, by exploring the political and economic contexts and the institutional status of Syrian refugees in Lebanon and in Jordan. Secondly, it examines the situation of Syrian refugees in urban areas in both countries. Thirdly, it explores conditions in organized camps in both contexts, through two comparative case studies: in Zaatari, a camp in Jordan that hosts almost 80,000 refugees, and in the Union of Relief and Development Associations (URDA) camp in Lebanon, where almost 2000 refugees reside. The results show that, behind simplified clichés and basic representations, complex political and economic factors lead to the policies adopted in settling refugees, usually benefitting the host country. Findings also challenge the recent scholarly approach that strongly criticizes encampment. They suggest that, in specific contexts and conditions, organized camps can represent a better solution for both refugee and host population than settling refugees in urban areas.

2 Non-encampment Policy for Syrian Refugees in Lebanon

Lebanon's first response to the refugee influx was to adopt an open-border and non-encampment policy, a position praised by the international community and by humanitarian agencies (Loveless 2013; Onishi 2013). Some scholars have described it as the "policy with no policy," since it was implemented without any prior planning (El Mufti 2014). Yet, several political and economic factors elucidate the Lebanese Government's (LG) position.

2.1 Political Factors

In most countries in the Middle East, the image of camps is largely politicized and linked to the almost 70-year-long presence of Palestinian refugees (Doraï 2006). While many argue that in some cases refugees in camps are easier to monitor and control than urban refugees (Jacobsen 1996), the Lebanese experience with camps was deplorable and largely contradicted this principle. In fact, weakened by years of internal conflicts and war, the Lebanese state has lacked (and still lacks) capacity to control Palestinian camps, which have evolved from tented settlements to extraterritorial sites of radicalization and armed resistance (Hanafi and Long 2010). Moreover, the Hezbollah and other minority groups have security concerns and worry that the refugee population become a permanent one, causing a drastic sectarian imbalance in the country (Turner 2015). Finally, the fear of militarization of camps with the radicalization of Syrian refugees and the potential permanence of their presence have led to a rejection of formal camps by the LG (Janmyr 2016).

2.2 Economic Factors

During the years of the Lebanese war, many agreements were signed between Lebanon and Syria. One of them, the Agreement for Cooperation and Economic and Social Coordination, signed in 1993, provides freedom for persons between the two countries to reside, work, employ, and engage in economic activity in accordance with the laws and regulations in force in each country (Syrian Lebanese Higher Council 1993). While precise figures do not exist, estimates for the early 2000s claim that Syrians constituted between 20% and 40% of Lebanon's labor force (Chalcraft 2009). Chalcraft (2007) argues that a historically informed analysis of the Lebanese economy demonstrates that Lebanon's refusal to build formal camps is designed to serve Lebanese capital and business owners. It allows expanding the labor supply, lowering wages, and increasing workers' precariousness in both communities. Hence, the absence of formal camps has security rationale, and it is in relation to the ongoing experience of Palestinian camps, but it also has economic reasons, Syrian workers having, for decades, formed the bulk of the unskilled and informal workforce in Lebanon (Chalcraft 2007; Turner 2015).

3 Encampment Policy for Syrian Refugees' in Jordan

The Jordanian Government (JG) has first been praised for its open-border policy and humanitarian stance toward Syrian refugees. Syrians were encouraged to seek shelter in Jordan by several factors such as close cultural and geographical ties and the political stability of the Hashemite Kingdom (Achilli 2015). However, in 2012, 1 year after the beginning of the conflict in Syria, in response to the increasing numbers of refugees, most Syrians were hosted in camps organized by the Jordanian Government and the UNHCR. This policy was, as it is the case in Lebanon, instigated by political and economic reasons.

3.1 Political Factors

While in Lebanon the experience of Palestinian camps was one of settlements becoming armed extraterritorial sites (Hanafi and Long 2010), in Jordan, which has a stronger and more centralized state than Lebanon, Palestinian camps are successfully managed and well surveyed by the government (Peteet 2005; Turner 2015). Hence, Jordan's encampment policy must be understood as an incentive to manage and control the Syrians' displacement within the country.

3.2 Economic Reasons

While many argue that camps are a space of political containment and control of displaced populations (Agier 2008; Agamben 1998; Bernardot 2008), Turner (2015) rightfully defines the camp as a tool through which states spatially segregate refugees of certain socioeconomic classes to keep them from being on the job market. This is the case in Jordan where the poorest refugees who might be expected to exert the strongest pressure on wages are consigned to camps. Only Syrians with sufficient access to capital and social connections are able, through a costly bailout system and with the support of a Jordanian sponsor, to move into Jordanian host communities (Turner 2015).

4 Refugees' Institutional Status in Lebanon and Jordan

Neither Lebanon nor Jordan has signed the 1951 Convention on Refugees (Organisation des Nations Unies 1951). Both countries signed a Memorandum of Understanding (MoU) with the UNHCR (UNHCR 2014, 2011), according to which Lebanon and Jordan are countries of transit where refugees are allowed to stay, respectively, for a period of 1 year and 6 months (Janmyr 2016). Jordan receives Syrian refugees within the framework of its Law of Residency and Foreigners' Affairs (Achilli 2015). For the LG, Syrians are referred to as "displaced people" (Nazihin), a status with a connotation of impermanence and without any legal meaning in the Lebanese body of laws (Naufal 2012; Janmyr 2016). Consequently, in both countries, Syrian refugees do not have any legal status. Even when they register with the UNHCR, they only become entitled to receive aid, but they continue to be subject to the risks of detention and expulsion (Centre Libanais des Droits Humains 2013).

5 Restrictive Measures and Their Consequences

In October 2014, Lebanon's council of ministers replaced the open door policy with new regulations that aim at dissuading Syrian refugees and almost forbidding Palestinian refugees from Syria (PRS) to seek protection in Lebanon, with the explicit goal of decreasing their number and protecting Lebanese employment (Onishi 2013; Turner 2015; World Bank 2013). Very specific and tightened conditions were also set for newcomers. However, with most people being unable to financially afford the renewal of their permits, statistics suggest that almost 70% of Syrian refugees and 90% of PRS do not benefit from a legal status in the country (Janmyr 2016). They lack access to services and healthcare and lose the moderate protection they enjoyed. They may also be detained and forced to return to Syria (Lenner and Scmetler 2016). Influx of Syrians has also strained Jordan's already overburdened infrastructure and scarce resources. Moreover, involved in the fight against IS, the Jordanian authorities were scared of reappraisals (Achilli 2015). Hence, in July 2014, the JG in turn imposed restrictions on the entry of Syrian refugees, and refugees without legal documents were forced into camps or, in some cases, compelled to return to Syria. Many refugees lost their mobility and access to aid and free services along with their legal papers (Lenner and Scmetler 2016; Achilli 2015). Thus, in both countries, the restrictive policies did not reduce the number of refugees, but only made more of them illegal settlers and subsequently increased their vulnerability (Turner 2015; Centre Libanais des Droits Humains 2013).

6 Camps in the Literature

Though the encampment has long been the cornerstone of the humanitarian response to displacement, only 25% of the world's refugees live in camps (Peteet 2005). An organized camp (OC) is usually created on public land according to preestablished plans by the UNHCR and with the government's authorization. It is created temporarily and for political reasons (McConnachie 2016). For humanitarian organizations, containment of refugees improves administrative efficiency and facilitates aid distribution (Verdirame and Harrell-Bond 2005). As for host states, rendering refugees visible helps attract funding (Harrell-Bond 1998). In some cases, encampment serves host states' economic goals by restricting poor population's access to labor market (Turner 2015). This point will serve as a basis for our discussion in analyzing the different responses in Lebanon and Jordan in relation to refugee settlement policies. In protracted situations, camps and cities, theoretically and practically, become linked and intertwined, which creates an urban ambiguity that Michel Agier named "camp city" (Agier 2001).

Refugee camps were not often considered as spaces for architectural and urban studies. Attention toward them as urban environments has increased with the pejorative politico-philosophical interpretations of camps as spaces of exclusion and exception (Agier 2011), spaces of bare life (as opposed to sociopolitical life) (Agamben 1998), non-places (Augé 1992), heterotopias (Foucault 1984), etc. Camps are also described as contradictory sites for distribution of humanitarian aid as well as containment, segregation, and control (Agier 2013). However, camps are also sites of agency, resilience, and appropriation (McConnachie 2016). In fact, while concepts related to camps describe a top-down approach, they neglect the many ways in which refugees negotiate, get self-organized, and adapt. These concepts also contribute to the elaboration of the paradigm that criticizes encampment. Through a comparative case study in OCs in Lebanon and Jordan, this paper aims to question this paradigm and demonstrate that living conditions in OCs vary enormously according to contextual conditions and that encampment, in some situations, could represent a better alternative than urban settlements.

7 Method: Comparative Case Study

This study explores the political and economic reasons that led to the adopted policies in settling refugees in Lebanon and Jordan and the consequences of these policies on urban refugees. But most importantly, it seeks to understand the living conditions of Syrian refugees in OCs. For that purpose, a comparative case study was conducted in the URDA camp in Lebanon and the Zaatari camp in Jordan. The case study is an empirical research approach that investigates a particular phenomenon in a real-life context taking place in uncontrolled environments (Gauthier 2009). It is known for the exploration of specific phenomena which are difficult to measure and interpretations that go beyond the limits of the case (Roy 2003). This strategy is adopted because of the importance of contextual conditions in relation to the phenomenon studied (Creswell 2007).

For the case study methodological approach, Yin (2003) recommends a research design, which is the logical sequence that connects empirical data to research questions and then to the ultimate conclusions in five steps: (1) what questions to study, (2) what data are relevant to answer these questions, (3) what data to collect, (4) what the link between the three previous steps is, (5) and finally, how to analyze the data. Thus, we established a plan for our research following Yin's recommendations in which we question the reasons that led to different policies in hosting refugees and we aim to understand refugees' living conditions in the studied contexts (Yin 2003). Yin also suggests validating the study by collecting data from different sources of evidence (Yin 2003) (see Table 1).

Table 1 Methods of data collection

Methods	Context and cases studies	Source	Year
Field observations	Urban refugees in Lebanon	Nabaa neighborhood in Beirut suburbs	2016
	URDA camp	Bar Elias, Bekaa	2015, 2016, 2017
	Zaatari camp	Mafraq, Jordan	2016
Group discussion	Urban refugees in Lebanon	12 refugees	2016
		1 employee from the municipality	
	URDA camp	12 refugees	2015, 2016, 2017
		2 managers	
	Zaatari camp	12 refugees	2016
		2 managers	
Study of reports and documents	Urban refugees in Lebanon	LG JG World Bank International Labor Organization UNHCR UN UNICEF REACH OCHA	2011 till date
	Urban refugees in Jordan		
	URDA camp		
	Zaatari camp		
Literature review on Syrian refugees, urban refugees, and OC	Urban refugees in Lebanon		2011 till date
	Urban refugees in Jordan		
	URDA camp		
	Zaatari camp		

As for the qualitative analysis of the data, we assessed the four studied contexts according to the dimensions that Yin (2003) refers to as the units of analysis: the political and economic context; housing; economy, services, and infrastructure; and social challenges. We inferred patterns and tendencies from our findings. We compared first the patterns from each camp with the patterns related to urban refugees in the same country. We then compared the patterns resulting from our studies in the OCs. This led us to conclusive results and a significant understanding regarding refugee settlement (see Fig 1).

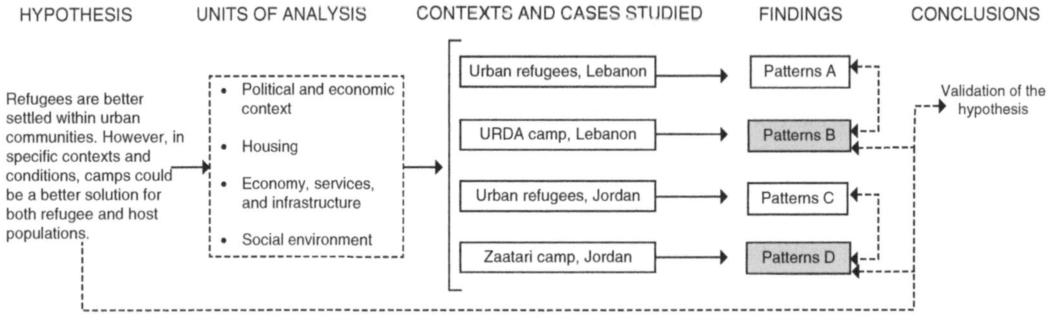

Fig. 1 Data analysis (According to Yin 2003)

8 Results

8.1 Urban Refugees in Lebanon: Consequences of the Non-encampment Policy

Lebanon's non-encampment policy has led refugees to spread informally in almost 2000 locations, without any prior planning or organization. They are specifically concentrated in the Bekaa, the North, and the capital Beirut (Hidalgo et al. 2015; Kikano et al. 2015). At first, many refugees settled with families or friends, but, with the prolonged crisis, they have resorted to rented accommodations where they pay high prices for small shelters or shared apartments with other families. When affordable housing became overfull and with resources drained, many of them moved to precarious shanty settlements, Palestinian camps, abandoned buildings, or informal tented settlements (ITS) (International Labour Organization 2014; Loveless 2013). Refugees face a range of challenges in their living accommodations, including lack of privacy, overcrowding, and exploitation. This, combined in many cases with the absence of formal rental contracts or agreements and difficulties in paying rent, has increased their vulnerability to possible evictions from their homes (UNHCR 2015). In urban neighborhoods with high density of refugees and in Palestinian camps hosting PRS, Lebanese owners are illicitly dividing apartments and adding illegal constructions on the roofs of existing buildings aiming to rent them at excessive prices for Syrians. These anarchic interventions are tragically altering the landscape in these areas (Kikano 2016).

As for the economy, prior the Syrian crisis, following the crisis, economic growth has slowed, private investments were reduced, the trade deficit has expanded, and real estate and tourism – the two most important sectors in the Lebanese economy– have declined (World Bank 2013). Furthermore, settling large numbers of refugees in poor regions has imposed enormous challenges on host communities (International Labour Organization 2014). In 2014, the offer on the job market was 50% higher, and the influx of refugees was projected to push 170,000 Lebanese into poverty and to double the unemployment rate (World Bank 2013; Chehade 2014). There is also an additional pressure on the already deficient healthcare and education services in terms of access and quality, and resources such as water supply and power are severely impacted (International Labour Organization 2014). Consequently, 40% of the Lebanese report negative feelings toward Syrian refugees. Syrians often undergo arbitrary arrests, face violence and racism, and see their fundamental rights violated (Centre Libanais des Droits Humains 2013). The longevity of the crisis has aggravated their vulnerability and exacerbated related protection risks (Zetter et al. 2014).

8.2 The URDA Camp, Bar Elias, Bekaa, Lebanon

In 2012, more than 100 humanitarian committees and organizations joined efforts under the umbrella of the Union of Relief and Development Associations (URDA) in order to support Syrian refugees. Despite the non-encampment policy, the URDA took the initiative of creating and managing a medium-size settlement for Syrian refugees. Authorized by the Ministry of Social Affairs, it was built on a private land of 21,000 m^2 owned by the Islamic group Al-Jamaa Al-Islamiyya, almost 10 km from the village of Bar Elias in the Bekaa Valley (Inter-Agency Coordination 2015). Almost 2000 refugees reside in the camp and every household pays a minor sum of USD 10 per month. There is a preference for hosting families, but the main condition for accepting refugees is that they do not belong to any armed militia.

The sheltering units (almost 600) are 6 × 12m modular caravans, aligned along predetermined axis and separated by internal roads or passages (see Fig. 2). They are produced in Lebanon and cost around USD 4000 each. Heated in the winter with a traditional stove, they are not insulated and are unfit for the high temperatures in the summer. It is a one-space shelter, and refugees lack intimacy and suffer in some cases of overcrowding. In order to overcome this problem, they often create additional spaces, which they connect to the main shelter.

The camp provides many services. A hospital, a pharmacy, a dental clinic, a physiotherapy clinic, a therapy clinic, an elementary school, a sewing workshop, a gym, and a mosque are built on the camp premises and offer almost free services to refugees. A communal kitchen also distributes free meals for households and offers cooking lessons for women. Larger and more sophisticated units that cost around USD 25,000 are used for these services. Only the schools are located in large tents. Unlike informal tented settlements (ITSs) that are often located on agriculture land and pollute the soil and the water table, and where electrical connections and water piping are illegal, the camp has very little environmental drawbacks; its sewage pipelines are connected to the public sewer, its electrical grid is independent from the government with a private generator that produces 24-h electricity, water is refined, and waste is recycled. Furthermore, an economic system has been created within the camp, targeting its sustainability. The camp managers designed and built a souk with a bakery, a butcher shop, clothes and shoe shops, a barbershop, a falafel shop, a vegetable and fruit shop, and so forth. The equipment is typically donated by NGOs.

Fig. 2 The URDA camp, Bar Elias, Lebanon (Source: Kikano 2017)

Differing from refugees in ITSs and in rental accommodations who often have former friendship or family ties among each other, refugees in the OC come from different localities in Syria, and consequently there is less conviviality between them. As for their relation with the host community, unlike other regions, the community in Bar Elias where the camp is located is relatively welcoming for refugees for two main reasons: First, refugees are all Sunnites like most local residents. Second, the creation of the camp's market is enhancing the local economy in the village and prevents refugees from being perceived as a burden. Despite a night curfew imposed by the municipality, refugees enjoy freedom of movement, and some of them work in the village. However, following the restrictive measures of the LG in 2014, many of the camp's residents became illegal settlers and remain in the camp out of fear of being arrested. This hinders their social integration with the local community and engenders feelings of isolation and segregation.

The managers named the URDA camp "the ideal village." Though the "idealization" could be a bit too pretentious, in general, refugees in the camp live in better conditions than urban refugees in Lebanon. They enjoy free services and an efficient and sustainable infrastructure. They are socially accepted by communities in surrounding villages and do not fear evictions or violence. The availability of work in the camp gives them a sense of dignity and self-respect (Kikano 2016).

8.3 Urban Refugees in Jordan

Prior to the creation of camps, early refugee arrivals settled in urban areas where they could engage their family and social networks (Lenner and Scmetler 2016). The northern governorates of Amman, Irbid, Mafraq, and Zarqa host more than the 80% of registered refugees outside camps (Lenner and Scmetler 2016). Consequently, in these areas, the high demand for housing seems to have had a negative impact on Jordan's housing sector: rent prices sometimes quadrupled through the capacity of refugees to afford higher rent by sharing housing (Achilli 2015). However, contrary to the situation in Lebanon where most refugees live in precarious housing solutions, only 20% of Syrian refugees in Jordan live in substandard accommodations, such as garages, chicken houses, and tents. A smaller number reside in ITSs (7000, which represent only 1% of Jordan's refugees, while in Lebanon they are 17% in ITSs), often lacking basic services such as health, education, water, and food (Achilli 2015).

Jordan's economy was impacted by the Syrian conflict less than in Lebanon. However, Jordan's international trade has been gravely affected by the loss of one of the principal points of access to regional trade through Syria. Moreover, the influx of refugees has intolerably increased the demand on school, sanitation, housing, food, energy, and water (Achilli 2015; Zetter et al. 2014). Until the inception of permissive measures regarding Syrians' participation in the job market in 2016, most Syrian refugees had no legal right to work in Jordan (Achilli 2015). Yet, Syrians often worked in the informal sector with very low wages and sometimes exploitative work conditions (Zetter and Ruaudel 2014; Achilli 2015). In 2016, the JG has committed to formalize access to the labor market for Syrians, based on the initiatives to replace other migrant labor with Syrians. The number of work permits distributed to Syrians doubled after the institution of a 3-month grace period. Thus, in Jordan, there is a glimpse of hope that there might be some prospects for a dignified and self-reliant life for urban refugees (Lenner and Scmetler 2016).

8.4 The Zaatari Camp, Mafraq, Jordan

Approximately 16% (100,000) of the entire refugee registered population resides in the five official refugee camps in Jordan (Fig. 3). The largest, the Zaatari, was opened in July 2012, over a year after the beginning of the conflict, in response to the arrival of large numbers of refugees (Lenner and Scmetler 2016). First planned for 10,000

Fig. 3 View inside the Zaatari camp (Source: Kikano 2015)

refugees, it hosted in less than a year almost 350,000 people (Hunter 2009). Harsh living conditions, insecurity, overcrowding, and the depletion of resources made the camp's conditions deplorable (Ward 2014). Many refugees have left. Some even went back to war zones in Syria, and the Zaatari is now home to over 80,000 people (Achilli 2015).

The Zaatari, a 5.3 km^2 settlement, the same size as the fourth largest city in Jordan, stands in the middle of the desert region of Mafraq, in complete isolation from inhabited and urbanized areas. It was built by the UNHCR and the Jordan Hashemite Charity Organization (JHCO) on Jordanian armed forces-owned land (Ledwith 2014). The flat white sand landscape is harsh and dry, with no water and no greenery in sight. Few spindly trees grow in front of the layers of concrete walls and barbed wires surrounding the camp. The site is controlled by the Jordanian army, and refugees need an exceptional authorization to be allowed out the camp even for a limited number of days. For visitors to access the camp, authorizations previously obtained from the Jordanian authorities are also necessary. The internal space is compartmented into sectors, each managed by a partner NGO. Sectors are also separated from each other with walls and metallic fences (see Fig. 3). All NGOs work under the umbrella of the UNHCR.

Infrastructure is inadequate and services are poor. The electrical network, primarily planned for temporary use, was implemented to ease the external space of the camp and few "public" buildings such as hospitals and schools. No provisions were made for the housing units. Yet, nearly three-quarters of households have electrical connections as refugees connect illegally to the common grid, which causes frequent and long power failures. Moreover, refugees receive a limited amount of water (60 L per day per household), which is hardly enough in the dry and hot desert climate. Communal toilets were primarily constructed outside the sheltering units, which is culturally inadequate, all adult women in the camp being veiled. Hence, refugees created toilets within their caravans and crafted makeshift septic tanks without connection to the public sewer. This make shift systems have proven to present many inconveniences and hygiene issues especially with frequent winter floods. Despite having three hospitals and many schools in the camp, refugees we interviewed complained that health and education services are of poor quality and insufficient (Kikano 2016).

Basic needs (that refugees often sell) which include staple foods, food vouchers, and nonfood items (NFIs) such as blankets and tents are distributed to refugees who possess the aid documentation obtained through formal registration

(Ledwith 2014). In July 2012, almost the entire refugee population lived in tents. Caravans, which consist of prefabricated modular units, inadequate for the hot climate, were later introduced when the displacement lasted longer than expected and more stable housing solutions were required.

Caravans serve multiple functions beyond housing. Camp residents buy, sell, and steal caravans to adapt them for commercial use or social space (Ledwith 2014). According to a shop owner, in the famous and mediated Champs-Elysées, the main shopping street in Zaatari that became similar to any souk in a Syrian city, the majority of shops in the camp are financed by Jordanian entrepreneurs who use the camp as a "tax-free" zone. They provide the necessary capital to buy goods and smuggle them inside the camp in cooperation with Syrian refugees who are willing to establish their own businesses in the camp or have already done so (Dalal 2015). The absence of regulations or any kind of supervision subjects shop owners to abuse as they have to pay extra fees for Jordanian security forces at the gate. In fact, despite the JG being officially on top of the hierarchical structure in the camp, informal power structures parallel the formal system. Hence, riots, smuggling, exploitation of labor, gang violence, rape, forced marriages, child trafficking, recruitment of child soldiers, and prostitution are frequent and exacerbate insecurity, fear, and inequalities among the camp's inhabitants (Ledwith 2014).

9 Discussion and Conclusions

The overall economic impact of the Syrian refugee crisis is negative, widespread, and severe in Lebanon and Jordan. These conditions, alongside capacity-stressed public services, are significantly diminishing the living standards and livelihood conditions for host and refugee populations (Zetter et al. 2014).

Lebanon's past experience with camps, the political internal divisions over the Syrian issue, and the weakness of the state led to the adoption of a "no camp" policy. Some theories link this policy to the country's economic need of a low-wage labor force (Turner 2015). Some others describe it as the "policy of no policy" (El Mufti 2014). In all cases, hosting large numbers of Syrian refugees in urban areas comes at a cost: sudden expansion of cheap labor has pushed down wages for Lebanese and Syrians alike, unemployment rate is increasing, and resources are being drained. Education and health services are insufficient. Affordable housing is overfull and many refugees are moving to precarious settlements (Loveless 2013). Humanitarian organizations face a great difficulty in distributing aid to the spread-out population. Tensions between communities are rising and refugees are often exposed to abuse and exploitation (Zetter et al., 2014). The extensive numbers put at risk the preservation of the cultural identity and the composition of the social and demographic fabric in areas with high refugee density.

The perception that settling refugees in camps is an incentive to make their stay permanent is false, just as false as the belief that restrictive policies may reduce their numbers. In Lebanon, the creation of more camps planned according to the URDA camp model is highly recommended. Despite it being a place of containment, refugees are not excluded due to the proximity to villages and to their freedom of movement. Settling refugees in camps would stop the chaotic urban expansion that the country is actually witnessing and would reduce the economic strain. Refugees in the URDA camp feel safe, accepted, and dignified. They benefit from free housing, free services, and job opportunities (in the camp market), and their living conditions are better on many levels than urban refugees. Another camp is still under construction, and it is an improved version of the existing one (e.g., shelters are flexible and adaptable by refugees to the number of household members and to other specific needs like the implementation of livelihood activities).

The Jordanian state is authoritative enough to monitor and control camps. Overwhelmed with large numbers of refugees, and in response to the protests concerning economic drawbacks, the JG created camps for Syrians. The biggest camp is the Zaatari. Despite its fast and remarkable urbanization, the creation of a sustainable market, and other initiatives undertaken by the refugee community, the Zaatari camp typifies the negative conceptual representation of a camp: it is an extraterritorial space of exclusion, exception, segregation, and control. The camp has indeed evolved into a city-like space. However, its occupants are bound by strict restrictions on their freedom of movement, and they live in very harsh conditions: housing is climatically and culturally unfit, there are significant shortages in electricity and water, and health and education services are very basic. Inequalities, different forms of violence, and corruption within the camp impact negatively refugees' well-being, their safety, and their quality of life.

The vast majority of Syrian refugees (80%) live in urban communities in Jordan (Zetter et al. 2014). Although opportunities for them are limited, following to the recent initiative of the government in 2016, which consists in formalizing and facilitating access to the labor market for Syrians, we presume that this is a step toward self-reliance and a decent life and conclude that refugees settled in urban areas hypothetically enjoy better living conditions than those in camps.

The protracted nature of most refugee crises has made "managing displacement in camp frameworks" a limited tool in its viability (Ward 2014). In most cases, camps are still created as temporary tools of containment, segregation, and exclusion for refugees who live in a harsh context of inequality and insecurity. In the literature, camps are depicted as spaces of biopolitics (tools for the control of specific populations) and nonsocial, nonhistorical, and noncultural extraterritorial spaces. Consequently, the humanitarian discourse tends to persecute encampment. Our study challenges this paradigm and demonstrates that, in very specific situations, when, for instance, the host country is unable to manage the refugee flow in urban settings, and when decent and dignifying living conditions are made available for refugees in camps, with accessibility to jobs and services and freedom of movement, camps could represent a better alternative than urban solutions. Aid would then be easily delivered, refugees would encounter fewer social tensions, and they would feel safe and protected. If camps are sustainable, they would be less costly and refugee populations would be less restraining on the host state's economy, services, and infrastructure. Encampment also avoids a chaotic and detrimental urban expansion.

References

Achilli, L. (2015). Syrian refugees in Jordan: A reality check. In M. P. Center (Ed.). Florence: European University Institute.

Agamben, G. (1998). *Homo sacer: Sovereign power and bare life*. Stanford: Stanford University Press.

Agier, M. (2001). De nouvelles villes: les camps de réfugiés. *Annales de la recherche urbaine*, 128–136.

Agier, M. (2008). *Gérer les indésirables – Des camps de réfugiés au gouvernement humanitaire*. Paris.

Agier, M. (2011). From refuge the Ghetto is born: Contemporary figures of heterotopias. *The Ghetto–Contemporary global issues and controversies*, p. 265.

Agier, M. (2013). Espaces et temps du gouvernement humanitaire. *Pouvoirs, 144*(1), 113–123.

Augé, M. (1992). *Non-lieux. Introduction à une anthropologie de la surmodernité*. Paris: Editions du Seuil.

Bernardot, M. (2008). *Camps d'étrangers*. Paris: Éditions du Croquant.

Centre Libanais des Droits Humains (2013). Travailleurs syriens au Liban. Une évaluation de leurs droits et de leur réalité Beyrouth.

Chalcraft, J. (2007). Labour in the levant. *New Left Review*, 45, 27–47.

Chalcraft, J. (2009). *The invisible cage: Syrian migrant workers in Lebanon*. Stanford: Stanford University Press.

Chehade, D. (2014). *L'impact de l'exode syrien sur le marché du travail libanais*. Beyrouth: Université Saint-Joseph.

Creswell, J. W. (2007). Five qualitative approaches to inquiry. *Qualitative inquiry and research design: Choosing among five approaches*, 53–84.

Dalal, A. (2015). A socio-economic perspective on the urbanisation of Zaatari camp in Jordan. *Migration Letters, 12*(3), 263.

Doraï, M. K. (2006). Les réfugiés palestiniens du Liban. Une géographie de l'exil. *Études internationales, 38*(3), 252.

El Mufti, K. (2014. Official response to the Syrian refugee crisis in Lebanon, the disastrous policy of no-policy. *Civil Society Knowledge Centre.*

Foucault, M. (1984). *Des espaces autres (Vol. IV, Dits et Écrits)*. Paris: Gallimard.

Gauthier, B. (2009). *Recherche sociale: de la problématique à la collecte de données*. Québec: Presses de l'Université du Québec.

Hanafi, S., & Long, T. (2010). Governance, governmentalities, and the state of exception in the Palestinian refugee camps of Lebanon. *Journal of Refugee Studies*, feq014.

Harrell-Bond, B. (1998). Camps: Literature review. *Forced Migration Review, 2*, 22–23.

Hidalgo, S., LaGuardia, D., Trudi, G., Sole, R., Ziad, M., Van, D., et al. (2015). *Beyond humanitarian assistance? UNHCR and the response to Syrian refugees in Jordan and Lebanon, January 2013–April 2014*. Transtec, UNHCR.

Humanitarian Aid and Civil Protection. (2017). *Lebanon: Syria Crisis*. Brussels: European Commission.

Hunter, M. (2009). The failure of self-reliance in refugee settlements. *Polis Journal, 2*, 1–46.

Inter-Agency Coordination (2015). Sheltor sector, monthly dashboard. Beirut: UNHCR, UNHABITAT.

International Labour Organization (2014). *Assessment of the impact of Syrian refugees in Lebanon and their employment profile*. Beirut: International Labour Organization Regional Office for the Arab States.

Jacobsen, K. (1996). Factors influencing the policy responses of host governments to mass refugee influxes. *International Migration Review*, 655–678.

Janmyr, M. (2016). *The legal status of Syrian refugees in Lebanon refugee research and policy*. Beirut: AUB Policy Institute.

Kikano, F. (2016). *Collecte de données (2), Unité de recherche AME 7106*. Montréal: Faculté de l'aménagement, Université de Montréal.

Kikano, F., Fayazi, M., & Lizarralde, G. (2015). Understanding forms of sheltering by (and for) Syrian Refugees in Lebanon. In *7th iRec conference 2015: Reconstruction and recovery in urban contexts, London, 2015*. London: iRec, Information and Research for Reconstruction.

Ledwith, A. (2014). Zaatari: The Instant City. *Affordable Housing Institute*.

Lenner, K., & Scmetler, S. (2016). Syrian refugees in Jordan and Lebanon: Between refuge and ongoing deprivation? In M. a. R. C. i. t. Mediterranean (Ed.), *IEMed. Mediterranean Yearbook 2016*. Florence.

Loveless, J. (2013). Crisis in Lebanon: Camps for Syrian refugees? *Forced Migration Review*, (43), 66–68.

McConnachie, K. (2016). Camps of containment: A genealogy of the refugee camp. *Humanity: An International Journal of Human Rights, Humanitarianism, and Development, 7*(3), 397–412.

Naufal, H. (2012). *Syrian refugees in lebanon, the humanitarian approach under political divisions*. Italy: MIGRATION POLICY CENTRE (MPC).

Onishi, N. (2013. Lebanon worries that housing will make syrian refugees stay. *New York times*.

Organisation des Nations Unies, O (1951). Convention et protocole relatifs au statut des réfugiés. Genève.

Peteet, J. M. (2005). *Landscape of hope and despair: Palestinian refugee camps*. Pennsylvania: University of Pennsylvania Press.

Roy, S. N. (2003). L'étude de cas. In *Recherche sociale: de la problématique à la collecte des données* (pp. 159–189). Sainte-Foy: Presses de l'Université du Québec.

Syrian Lebanese Higher Council, S (1993). Agreement on the regulation of transport of persons and goods between the Lebanese Republic and the Syrian Arab Republic.

Turner, L. (2015). Explaining the (non-) encampment of Syrian refugees: Security, class and the labour market in Lebanon and Jordan. *Mediterranean Politics, 20*(3), 386–404.

UNHCR. (2011). *UNHCR global appeal 2010–2011 – Lebanon* (pp. 34–37). Geneva: UNHCR.

UNHCR. (2014). UNHCR Global Appeal 2014–2015. Jordan. Geneva.

UNHCR. (2015). 3RP, Regional & Refugee Resilience Plan 2015–2016 in Response to the Syria Crisis. Geneva.

UNHCR. (2016). Regional Refugee & Resilience Plan 2016–2017 in Response of the Syria Crisis, 3RP. Geneva.

United Nations. (2016). *Syria conflict at 5 years: The biggest refugee and displacement crisis of our time demands a huge surge in solidarity*. New York: UNHCR.

Verdirame, G., & Harrell-Bond, B. E. (2005). *Rights in exile: Janus-faced humanitarianism* (Vol. 17). Oxford , New York: Berghahn Books.

Ward, P. (2014). Refugee cities: reflections on the development and impact of UNHCR urban refugee policy in the Middle East. *Refugee Survey Quarterly, 33*(1), 77–93. doi:https://doi.org/10.1093/rsq/hdt024

World Bank. (2013). *Lebanon - Economic and social impact assessment of the Syrian conflict*. New York: The World Bank Group.

Yin, R. K. (2003). *Case study research: Design and methods (Applied social research methods series)*. London, New Delhi: SAGE Publications.

Zetter, R., & Ruaudel, H. (2014). Development and protection challenges of the Syrian refugee crisis. *Forced Migration Review, 47*, 6–10.

Zetter, R., Ruaudel, H., Deardorff-Miller, S., Lyytinen, E., Thibos, C., & Pedersen, F. (2014). *The Syrian displacement crisis and a Regional Development and Protection Programme: Mapping and meta-analysis of existing studies of costs, impacts and protection*. Copenhagen: Ministry of Foreign Affairs of Denmark and Tana.

A Landscape Perspective on the Impact of Syrian Refugees in Lebanon

Maria Gabriella Trovato

Abstract

This reading of the Lebanese situation from the perspective of the refugee crisis is an occasion to describe the repercussions of the Syrian mass migration on the spatial assets of the country and to reveal the strong cultural and territorial component of this emergency. The paper argues for an understanding of the mechanisms that are at work in the shaping and changing of the country's landscape. In our reading, we used McClelland's definition, which considers 'landscape characteristics (as) the tangible evidence of the activities and habits of the people….[reflecting the] beliefs, attitudes, traditions, and values of these people' (McClelland et al., Nat Regist Bull 30:3, 1994). Thus, we used the landscape lens to interpret the new spatial organization of the land as an expression of imported cultures. The paper reports the findings of the study that was performed in two informal settlements (ISs) in Lebanon: Sarafand 009 and Al Tyliani.

Keywords

Syrian refugees · Landscape perspective · Displaced people · Informal settlements

M. G. Trovato (✉)
American University of Beirut, Beirut, Lebanon
e-mail: mt63@aub.edu.lb

1 Introduction

A new and unexpected network of displaced people's movement and flux is drawn everyday between the north and the south of the Mediterranean Sea. Along the Mediterranean Basin, a dynamic landscape is continuously being rebuilt, determining an overlapping of human timing and a range of identities. The daily transitional movements and population shift from one nation to another or within the same country transform and structure our world in economic, social, political, religious and especially spatial terms. Over time, these migrations reshape the morphological organization of the land and create a landscape as a self-organized territory. In their movements and relocations, communities adjust space to meet their housing needs in an accelerated process of settlement that considers individuals to be numbers and living simply to be an occupation of the land. According to Sassen, 'these flows may well be the merest beginnings of new histories and geographies made by men, women and children in desperate escape from unsustainable conditions' (Sassen 2016).

In most cases, the host countries are not prepared to accommodate the new arrivals in masses that infringe upon material and immaterial boundaries and establish new assets that require equal consideration of functional, productive and socio/cultural, aesthetic values because they are

A. Asgary (ed.), *Resettlement Challenges for Displaced Populations and Refugees*,
Sustainable Development Goals Series, https://doi.org/10.1007/978-3-319-92498-4_4

equally important in the organization of the land as they are as a product of political and cultural decision. The paper adopts the Cosgrove perspective of landscape as a social and cultural product, and it offers a general description of the Lebanese situation in light of the Syrian crisis and a classification of some of the most diffuse typologies of informal settlements in relation to the spatial and socio-economic context in which they are inserted (Cosgrove 1984). The second part of the article will focus on the two chosen case studies to illustrate the analysis that was conducted on site and the methodology that was used while assessing the informal settlements conditions.

2 The Syrian Refugee Situation in Lebanon

Every day, millions of people leave their country in search of better opportunities. According to the United Nations High Commissioner for Refugees (UNHCR), 5 years after the war started in Syria, 11 million refugees have fled across borders, 4.8 million escaped to Turkey, Lebanon, Jordan, Egypt and Iraq (Fig. 1), 6.6 million were displaced within Syria, and more than 212,000 people were trapped in besieged areas without access to humanitarian's assistance (www.syrianrefugees.eu 2016).

Lebanon is host to nearly 1.2 million Syrian refugees, and it represents approximately a quarter of the country's total population.

The massive influx, since the beginning of the Syrian crisis in March 2011, has seen refugees settling in every corner of the country, which has placed a huge strain on the country's already stretched services and infrastructure (Dunmore 2015). 'The countries in the region hosting Syrians do not grant proper refugee status to those fleeing violence in neighbouring countries, but instead they give refugees an ill-defined denomination as "guests". These "guests" are sometimes generously hosted and protected, but most of the time they are denied from all their

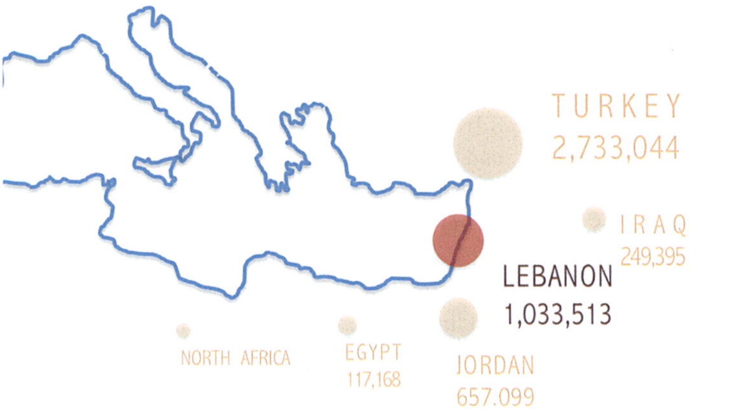

Fig. 1 Number of Syrian refugees in the Middle East and North Africa

basic rights that would make settlement an option (work, access to services, etc.), even though history shows that guests may wait a lifetime and never return home' (Fargues 2014).

Approximately 42,000 illegal tents (ITs) are scattered in the country in close to 1500 locations throughout Lebanon, mainly concentrated in the North and in the Bekaa Valley (712 ITs). The country represents a particular case in comparison with the other countries that have been affected by the Syrian migratory afflux. Indeed, Lebanon is the only country that, from the beginning, did not adopt a clear strategy of intervention and, until this day, refuses the establishment of formal camps as per UNHCR's suggestion and help, fearing a repetition of the country's experience with Palestinian refugee camps. Thus, the UNHCR shelter group was and is still unable to neither suggest nor change the location of the dispersed ISs throughout Lebanon. Their distribution in the Lebanese regions depends on family and religious affiliations, and it is also under the control of individuals who are known as 'Shawish', who initiate connections between Syrians refugees who are in search of a place to settle and Lebanese owners, who are ready to reorganize their land to accommodate informal shelters and to rent them to the newcomers. The majority of the Syrian informal settlements (ISs) are located in agricultural fields that, under the pressure of the mass migration, were illegally changed in terms of their use and converted by their owners into ISs without respecting the minimum standard requisites that are stated in the 'Handbook for Emergencies' (UNHCR 2007). Their dimensions vary from one place to another, and they range from small settlements, with less than 10 families, to the largest, which supports up to 100 families. UNHCR has tracked the geographical distribution of the displaced population in Lebanon, and it has found that the eviction movements and the relocation of displaced people in informal settlements have led to a remarkable 30% increase in the number of ISs (UNHCR 2016).

The new risk landscape, which is accentuated by exponential rates of Syrian immigration, represents one of the crises that is currently being faced by this part of the world.

The millions of displaced persons are transforming the landscape by introducing new patterns, new processes and new rhythms that require counterresponses to the new conditions.

While they are located on agricultural land (Fig. 2), the informal settlements are one of the major factors of change in the land use and land cover of Lebanon.

Fig. 2 Informal settlements in agricultural land

As per the Lebanon Environmental Assessment of the Syrian Conflict & Priority Interventions, 'Syrian refugees who live in ISs occupy more land than those who live outside…'. According to an ongoing assessment of 151 informal settlements in the Bekaa Valley (which harbours the highest number of ISs) housing 65,392 refugees, the average area of an IS is 10,000 m^2, (UNHCR 2014). This is equivalent to 23 m^2 per refugee, which confirms the observation that ISs are land greedy in comparison with other shelters' (MOE/EU/UNDP 2014, 101). As the number of Syrian refugees continues to rise, further ISs growth will inevitably encroach on agricultural lands and put those lands out of production unless they are designated by the Government of Lebanon (GOL) as exclusion zones.

At the environmental level, the dispersion of the clusters of tents throughout the productive land, and their location near water infrastructures, has furthered the degradation of the fertile soil and doubled the percentage of water, soil and air pollution. Sludge disposal on land and in streams contaminates ecosystems, especially in karstic terrain. The disposal of sludge in open lands, in dolines and in near streams will increase organic loads of these receptors and may alter the composition of aquatic life and riparian ecosystems (MOE/EU/UNDP 2014).

In addition, the spread out of informal settlements across the rural context has increased the percentage of built and concrete structures, especially second houses, on productive land, which determines new economic trends. In 2014, notwithstanding the Syrian refugee influx, the Ministry of Interior and Municipalities (MOIM) authorized municipalities and Qaem Maqam (the Governor of a Provincial District) to issue construction 'permits' to local residents in mostly rural areas. In theory, the permit allows an owner to build a 150 m^2 house, either at ground level or on top of an existing one-storey house, not to exceed 7 m in total. Nevertheless, the MOIM decision bypassed urban planning procedures. Although the policy was conceived to address continued socio-economic hardships, it has encouraged local residents to build new homes and/ or complete unfinished homes, thus imposing grave repercussions on public safety and on the urban environment if building standards (Law No. 464/2004) and urban planning requirements go unheeded. Haphazard and accelerated construction can be seen in all of the affected communities. Accelerated construction can also be detected from cement and clinker demand data. According to an industry professional, at least 10% of that increase is attributed to the Syrian refugee crisis (MOE/EU/UNDP 2014).

While not planned and not inserted into a strategic vision, all of these ISs, which are enclosed entities without connection to the urban structure, threaten the already compromised ecological, social and cultural stability of the country.

The transformation of land use brings with it consequences for social and/or economical assets, the cultures of these areas, and it has profoundly changed the identity, the formal composition and the relationship among the different elements of the land.

The human occupation of the Lebanese territory has drawn a new physical and intangible landscape, in which the formalization of living practices in agricultural or unused areas is expressed by the rich thematic variations of material composition of private, and settlement space versus private and productive land. The dynamic landscape system is evolving, and, under the pressure of the Syrian ISs, it is generating new patterns that are evident, specifically in the rural Lebanese landscape (Fig. 3).

With the intention to facilitate a better understanding of the phenomena in place, we assessed and mapped the different typologies of tent clusters in relation to the natural, infrastructural, productive and rural features. Organized in lines between agricultural fields, in groups of different dimensions, elongated along main roads, scattered in between rural settlements or green houses or located at the periphery of a marginalized urban and rural community, the tents are in nuce clusters of new informal cities (Figs. 4, 5 and 6), and they serve as 'motors of urbanization' (Herz 2013).

The diagrams in the pictures show the difference between organized ISs versus scattered and totally disorganized ones. The first typol-

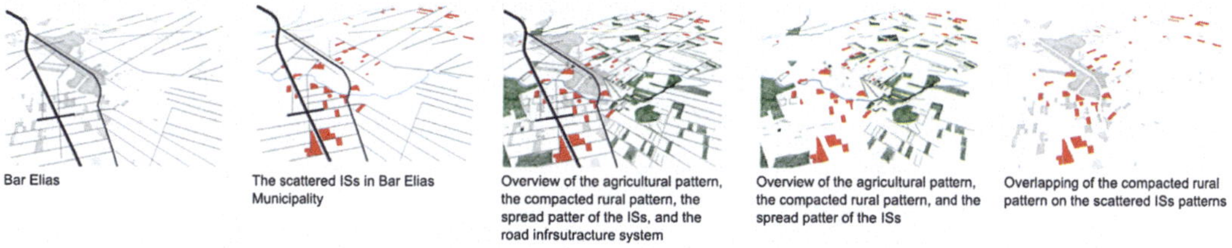

Fig. 3 New patterns generated by the Syrian ISs scattered in the rural Lebanese landscape

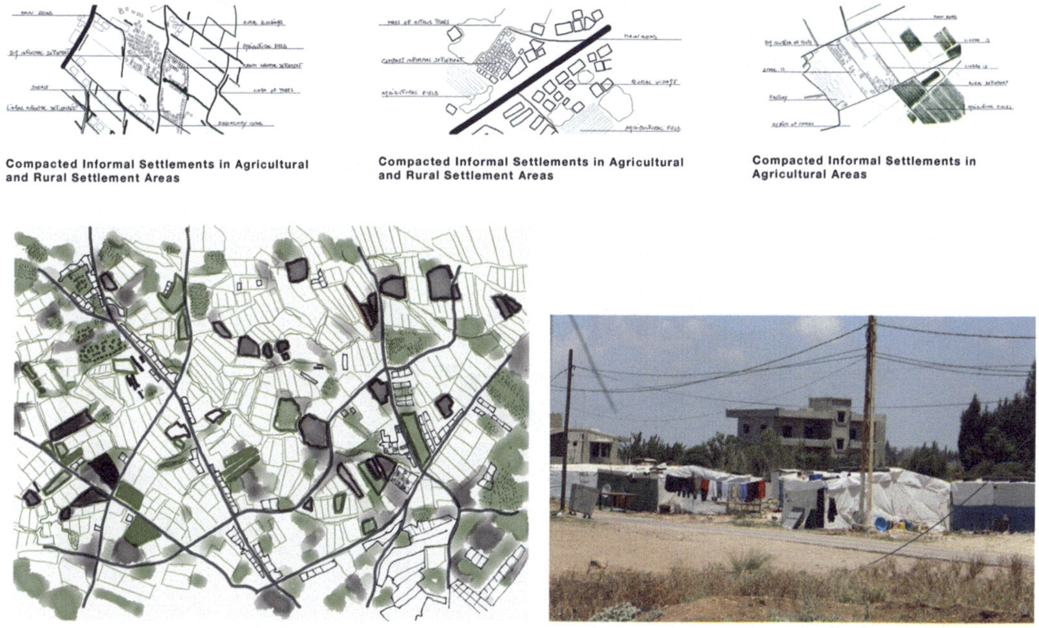

Fig. 4 Diagrams of some of the compacted ISs in Lebanon

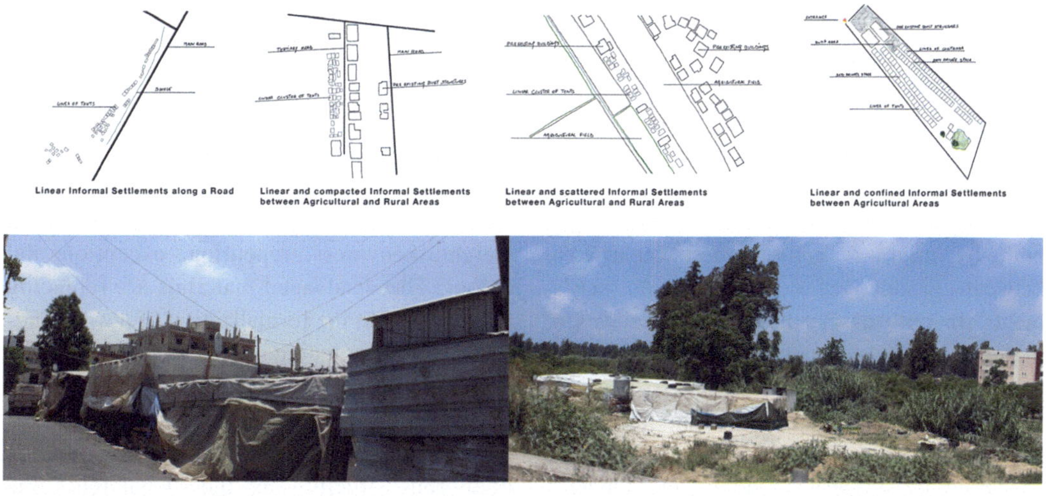

Fig. 5 Diagrams of some of the linear ISs in Lebanon

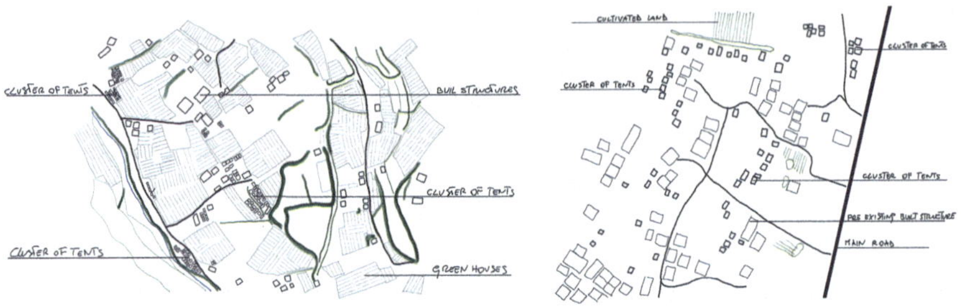

Scattered Informal Settlements in agricultural and green houses areas

Scattered Informal Settlements in agricultural and rural settlement areas

Fig. 6 Diagrams of some of the scattered ISs in the Lebanese landscape

Fig. 7 The pre-existing line of trees, all that remains of the previous agricultural use of the land, now replaced or doubled by the nonporous and built structures

ogy of informal settlements is mostly controlled and managed by land owners or religious associations, while the others are spontaneously built by migrants and/or guests nearby or between productive sources or rural settlements. In both cases, the ISs are enclaves of migrants who are in search of establishing communities and ground connections. As a result, the migrants are reshaping the land and building up layers of landscapes that exist and mutate over time; the process is readable in its materialization of limits. Plastic, metal, cartons and truck tires are forging new languages contexts and unexpected effects that are characterized by transparency, assembly, combination, height, weight, heaviness, repetition, exception and colour. These salvaged materials are replacing natural ones in the formalization of new boundaries and introducing urban artefacts into an agricultural context. The pre-existing lines of trees, which were used to delimit the agricultural fields, are now replaced or doubled by the nonporous structures of the illegal tents and clusters of ISs fences (Fig. 7).

2.1 The Two Study Areas: Sarafand and Al Tyliani

The paper reports the findings of the study that was performed in two Syrian Informal Settlements in Lebanon: Sarafand 009 in the south coast and Al Tyliani in the Bekaa Valley. The studies were conducted in 2015 as part of a third-year landscape design course at the Department of Landscape Design and Ecosystem Management, American University of Beirut (AUB) and as an exploration that was conducted by the 'Landscape in Emergency' research group that was formed at AUB during the international research workshop that was organized by the author in January 2015.

2.1.1 Methodology

The methodology that we used was based on cartographic interpretation, fieldwork, semi-structured interviews and analysis of the landscape uses that were determined through guided walk-throughs, readings and interpretations of the traces that were found on site and time evolution. The study was conducted by investigating different spatial scales, including regional, district and neighbourhood scales.

The regional analysis was based on reports and maps that were produced by local and international NGOs that work with Syrian refugees and especially on the materials that were made by UNHCR in the 5 years of work that was performed in response to the Syrian crisis (UNHCR, Shelter quarter 1 Dashboard).

At the district level, a desktop analysis was conducted to compare the land use map of two different years (2002–2010), overlapping the results on high-quality satellite imageries and on UNHCR maps (UNHCR, Shelter quarter 1 Dashboard). Later, a field survey was conducted to validate the data and to finalize our results (Figs. 8 and 9).

The neighbourhood investigation was based on the theory that was delineated by Meinig on *The Interpretation of Ordinary Landscapes* (Meinig 1979). Relying on Lewis' definition of landscape which provides strong evidence of the kind of people we are, and were, and are in the process of becoming (Lewis 1979), the landscape team spent time in situ, looking around, talking with women, in search of indicia and answers to the many problems that arise in a situation displacement.

The fieldwork focused on the assessment of the character and quality of spaces through the observation of all of the forms of appropriation and construction of lived space in the study areas (Fig. 10).

The mapping of all of the uses of the IS's open spaces was based on gender and age as a response to cultural and functional needs. The mapping exercise was a powerful tool to facilitate the understanding of the sociocultural reali-

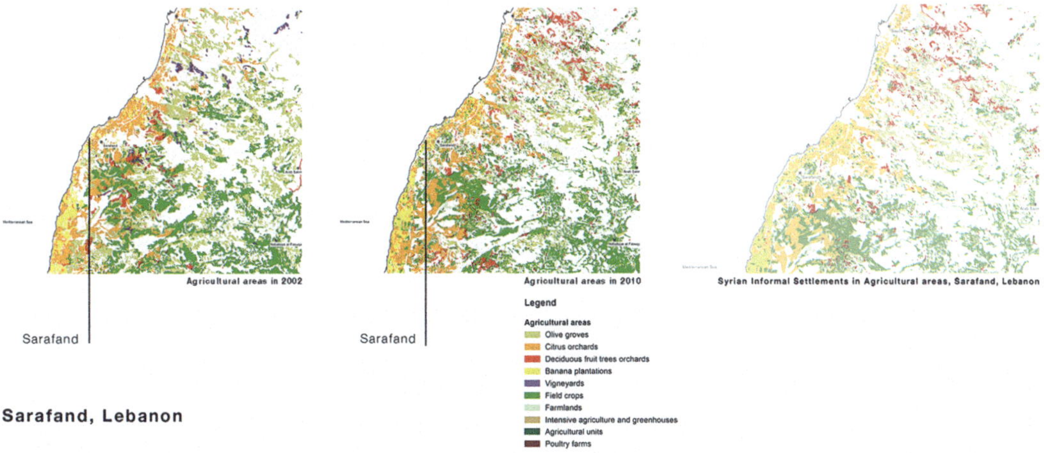

Fig. 8 Land use change in Sarafand Municipality and location of Saraband 009 IS in the agricultural area of the municipality

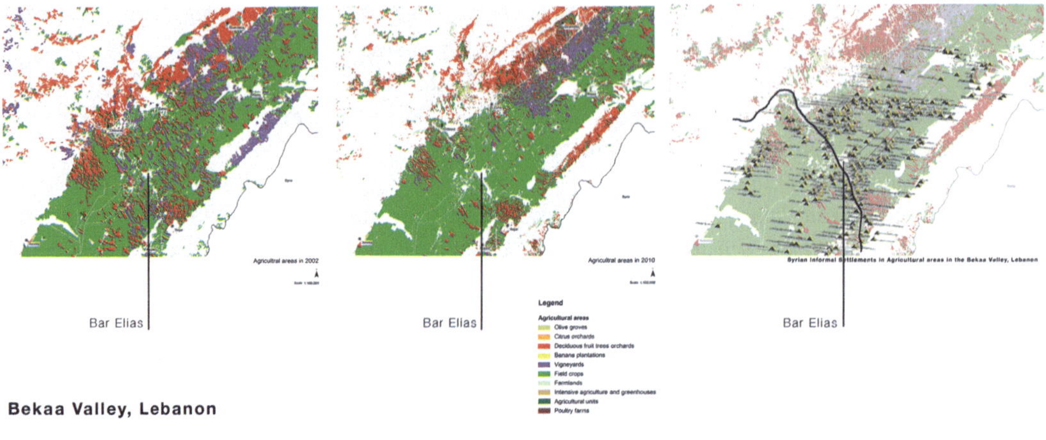

Fig. 9 Land use change in Zahle District and location of Al Tyliany IS in the agricultural land of this area

Fig. 10 The forms of appropriation and construction of the lived space in the ISs

ties of the involved communities, regions, landscapes and ecosystems (Ryan 2011). The situations of the two chosen Lebanese ISs were mapped first considering the spatial component, specifically the materials that were used, the ratio of open versus built, the vegetation in place and its role in the space. Then, the uses of the open areas were recorded and analysed through photographs, sections and diagrams with the aim of understanding the everyday rhythms, the activities that were taking place, the social constraints, the desires of the people, their memories and ways of living. No aesthetic and preformed judgements were expressed. Every single event, including trivial ones, was taken into consideration.

Cartographic representations of the multiple acts in place enabled the research to localize them and to identify strengths and weaknesses for possible interventions in the future.

Our reading of the maps and reports that were produced by the UNHCR at the regional scale revealed that the maximum concentration of refugees is located in the Saida and Akkar Districts on the coast and in the Zahle and Baalbek Districts in the Bekaa Valley (Figs. 11 and 12). Today, the Lebanese coast is an accumulation of multiple strata and layers. Some of them are still open to diffused and dispersed urbanity, and some are as yet part of the recent industrial heritage and the remains of past land use and economy. The Bekaa Valley rests between the Lebanon

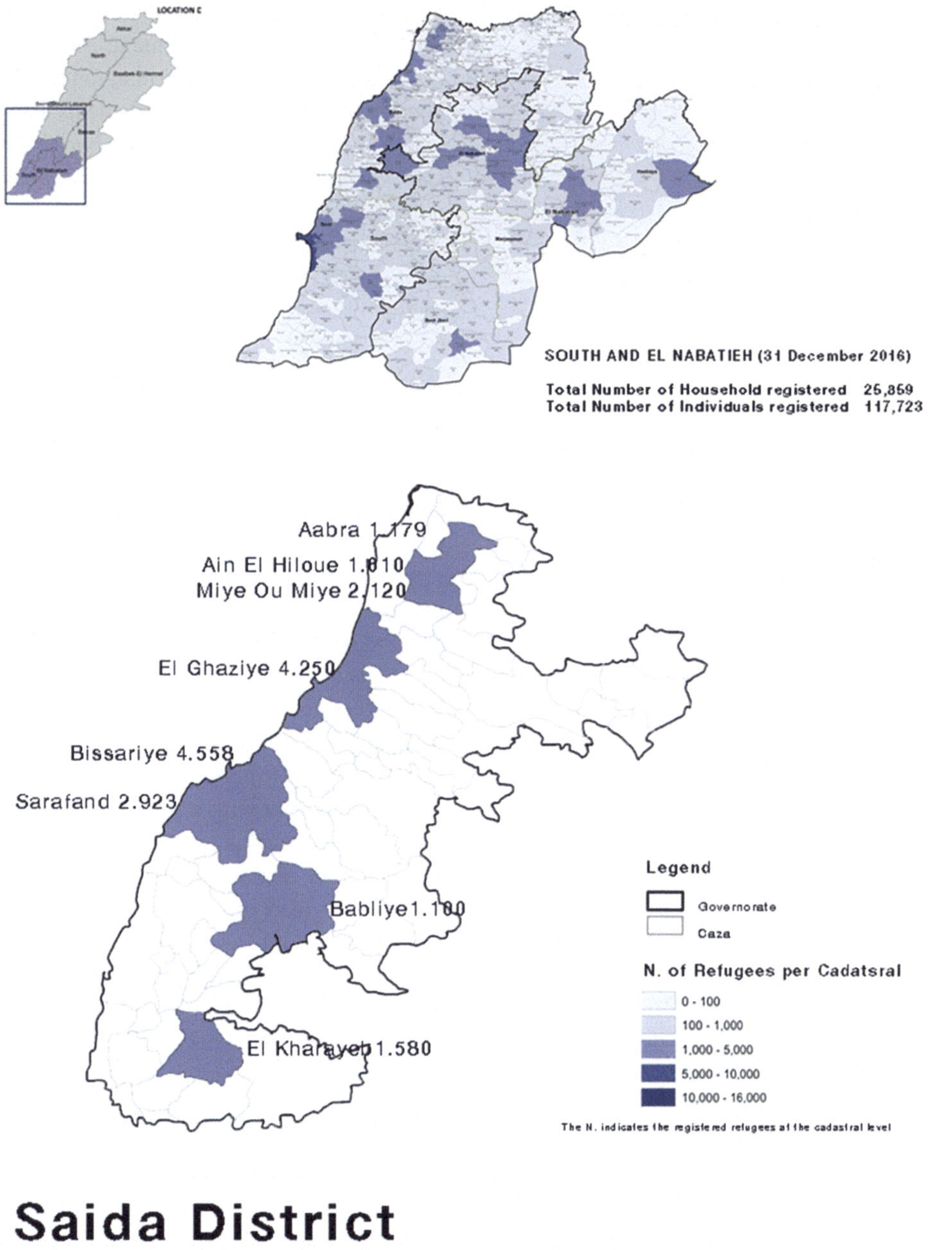

Fig. 11 Numbers of Syrian refugees in Sarafand Municipality in relation to the other municipalities in Saida and the South Districts

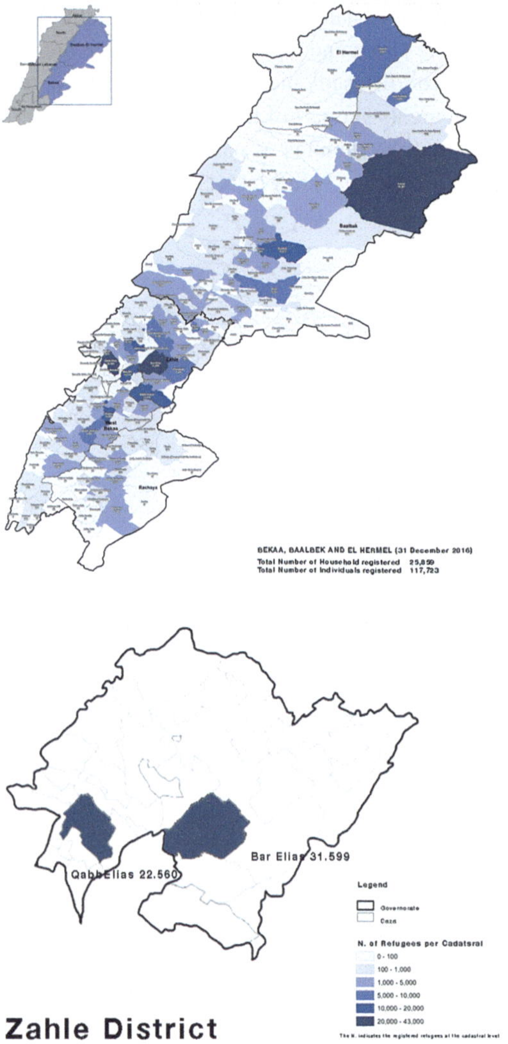

Fig. 12 Numbers of Syrian refugees in Bar Elias Municipality in relation to the other municipalities in Zahle and Bekaa Districts

Mountains on the Western side and the Anti-Lebanon range on the Eastern side near Syria. The valley is part of the Great Rift Valley system, and it is characterized by fertile soil of alluvial deposits and represents Lebanon's largest agricultural regions.

Thus, two municipalities with the highest numbers of Syrian refugees were selected in the two districts of Saida, on the Southern coast, and Zahle, in the Bekaa Valley. Our choice was also influenced by the easy access to the site and to the availability of data (map of the location of tents, gender, origin, social status) that the American University of Beirut, in partnership with other NGOs, had collected over the years.

In the Sarafand Municipality, Saida District, Sarafand 009 was selected because of the IS's relative small dimensions and its location on productive land. In the Bar Elias Municipality, Zahle District, Al Tyliani was chosen for its dimension with less than 100 families and its position at the periphery of the rural settlement. As the majority of ISs, they are both settled in agricultural neighbourhoods near poor rural settlements and private lands (mulk) that are managed by Syrian land brokers who are referred to as 'Shawish'. These individuals rent land from a Lebanese owner on an annual basis, and they collect monthly rent from each shelter unit. The average monthly rent per shelter is approximately LL60,000 ($40); however, it can be much higher (MOE/EU/UNDP 2014).

Sarafand 009, on the southern coast of Lebanon, is one of the small satellite encampments that are filling the void between the almost continuous linear coastal urbanization (Fig. 13).

Located in the middle of an area that is cultivated with bananas, the IS is compressed in a surface of 525 square metres with a total of 19 tents and 148 inhabitants (Fig. 14).

The settlement was built at the beginning of 2011, transforming the agricultural land into patches of concrete to support refugees' tents. The existing distribution of shelters with its 27.6 square metres per tents does not respect the minimum standards that were established by UN. 'Ideally the recommended minimum surface area is 45 square metres per person (including kitchen /vegetable gardening space). However, the actual surface area per person (excluding garden space) should not be less than 30 square metres per person. This minimum includes the area necessary for roads, foot paths, educational facilities, sanitation, security, firebreaks, administration, water storage and of course, plots for shelter' (UNHCR 2007).

Al Tyliani, which is situated in the Bekaa Valley, is one of the ISs that are located in an agricultural field at the edge of the village of Bar Elias (Fig. 15).

Fig. 13 Sarafand 009 IS location in the Southern Coast of Lebanon

It took its name from Mr. Mosaab Al Tyliani, who is the owner of the land. Before 2011, this property functioned as an agricultural field, producing vegetables such as lettuce and potatoes. The site has an average area of 9000 square metres with a total of 63 tents, hosting more than 60 families and 350 people (Fig. 16).

In 2014, the Kayany Foundation, in collaboration with the Centre for Civic Engagement and Community Service (CCECS/AUB) and private donors built 'Ghata', a portable school on a site that faces the settlement. 'Ghata' provides access to education for the children who reside in the informal tented settlements, and it plays a major role in the lives of the children who spend most of their day attending classes and playing in the playground. Several UN bodies such as UNICEF, alongside many NGOs, are involved in providing basic services to the settlement. Those concerned parties are helping the IS to become a healthier and cleaner space, donating outdoor bathrooms and water tanks and providing basic infrastructure and sewage systems to the site.

2.1.2 Mapping Sarafand 009 and Al Tyliani Informal Settlements

The investigation of Sarafand IS was first conducted by reading the available map of the settlement that the Civic Centre at AUB had provided to us.

The plan revealed how the IS is organized in three small and introverted clusters (Fig. 17): the first one is aligned with the road; the second placed is situated in the middle of the land; and the third one is located at the back, just at the foot of a sloped area. Each group of tents is structured as an autonomous neighbourhood with narrow 'alleys' between the shelters (Fig. 18) to recreate the rural feeling of its inhabitants' origins.

The IS is bounded by different typologies of fences that establish a sense of enclosure and disconnection from its surroundings (Fig. 19).

Along the road, the sequence of tents creates an opaque and coloured hedge and a pre-existing wall borders two sides of the settlement, while on the fourth side, a metallic and porous fence seems to extend the green patches of the neighbourhood inside the IS. During the fieldwork, we delved into the traces of the human occupation

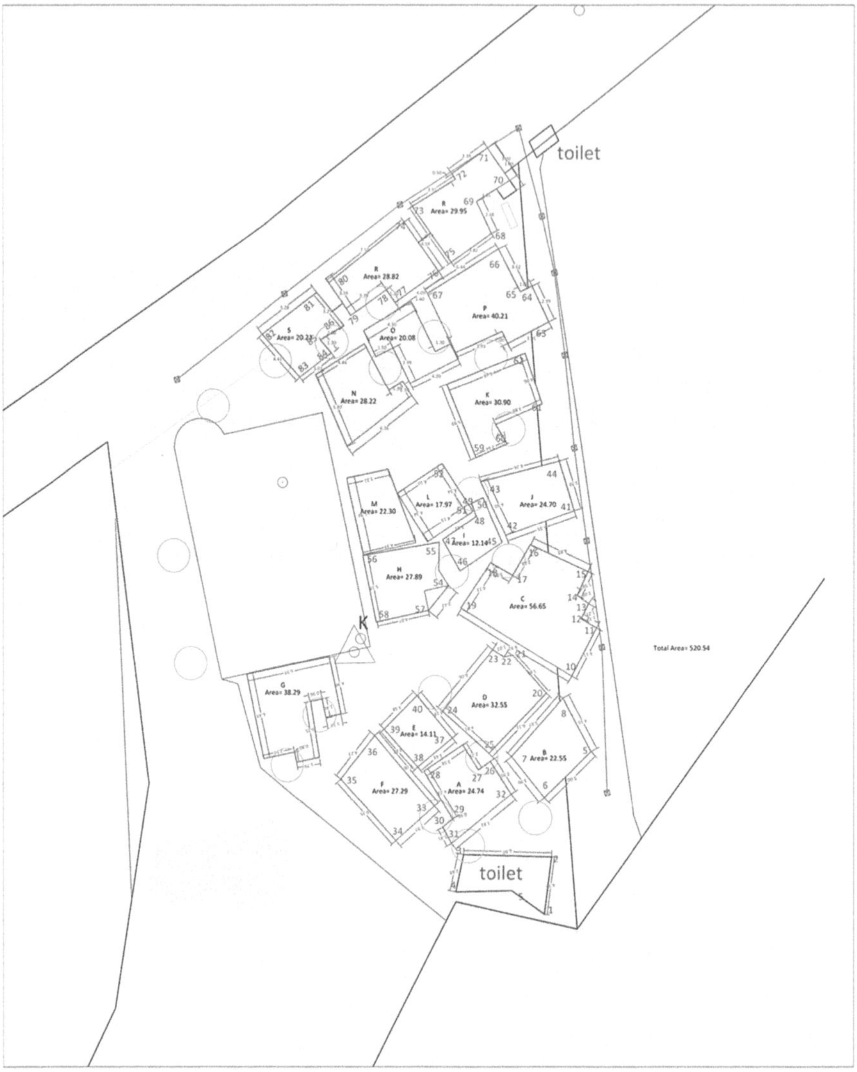

Fig. 14 Plan of Sarafand 009 IS

of the space by photographing and mapping this man-made landscape and sketching analytical diagrams to interlace the spatial component and/or organization of the open spaces with their uses.

We discovered that the very limited open areas between the tents (Fig. 20) are used to carry out daily activities, which include the drying of clothes, preparation and cooking of food and washing of linens, which forces the inhabitants to share the available space. During the day, these areas are full of life, colours, smells and the voices of women and children; however, they are quiet in the evening due to the lack of proper lighting (Fig. 21).

We also identified the trees in the main open spaces: at the entrance, in the middle and at the back of the IS. They are mostly unused and play the role of buffer zones between the three clusters of tents. In all of these spaces, there are scattered olive trees which shade them, transforming the trees into landmarks of the settlement.

The on-site work allowed us to discover how this man-made landscape is a set of organic form images whose interpretation is filtered through the knowledge of the culture of the people who

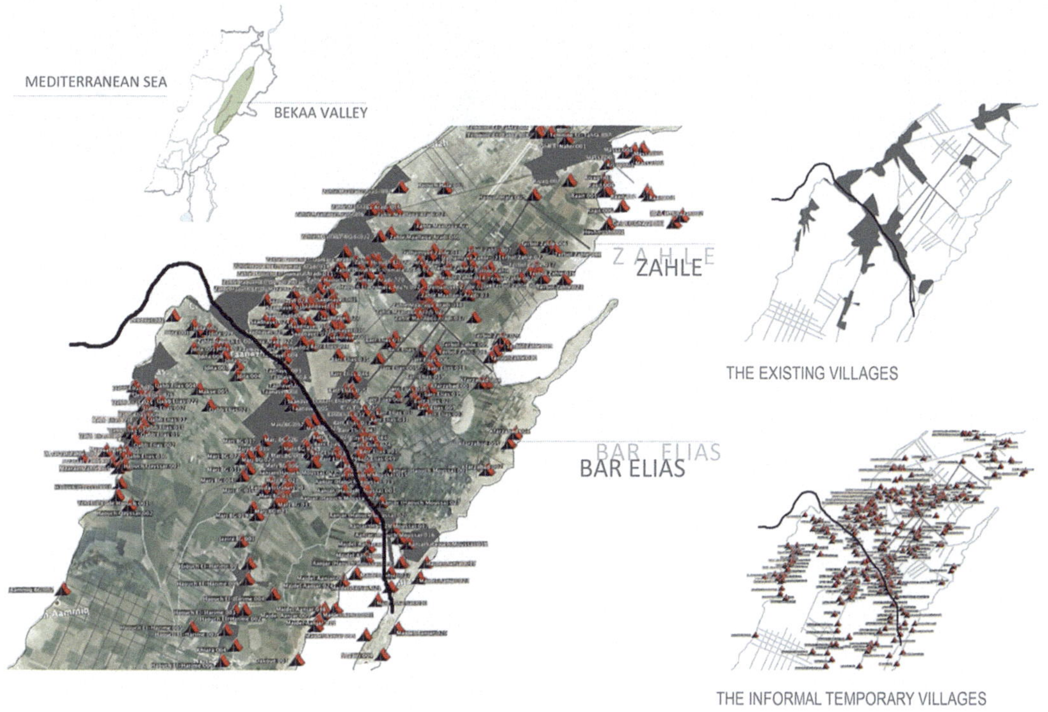

Fig. 15 Al Tyliani IS in Bar Elias, Bekaa Valley, Lebanon

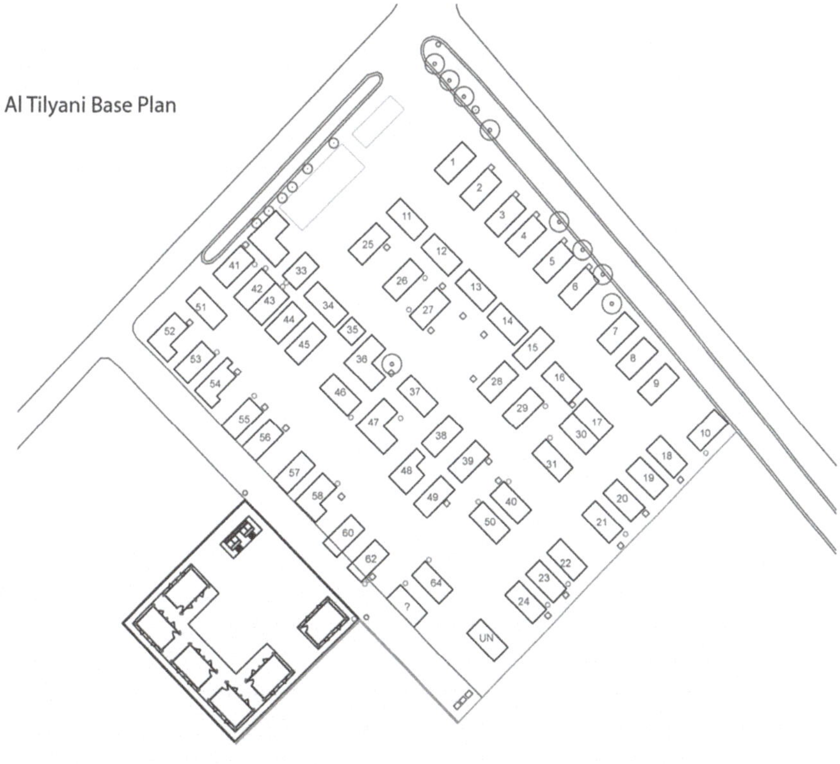

Fig. 16 Plan of Al Tyliani IS in Bar Elias

Fig. 17 Diagrams of Sarafand IS identifying the open spaces, the clusters of tents, the entrances and the pre-existing trees

Fig. 18 The narrow open spaces between the tents in Sarafand 009 IS

inhabit it. The findings of our analysis reported how the inhabitants engaged in a variety of landscape practices in the interstitial spaces of their proximity areas, growing medicinal and ornamental plants, collecting recyclable materials to reuse or to sell and organizing small spaces to allocate a few recreational activities. The in situ interviews with Syrian refugees, especially women, and the workshops that we ran with the children who live on the settlement helped us to better understand the social and psychological situation of the inhabitants of a transitional settlement. Hidden between the banana trees, this informal settlement is the sum of material and immaterial structures that do not relate to each other and are thus cannot create a sense of cohesion or 'community'. In Sarafand 009, the families are forced to live together and to share the same spaces; however, the spatial organization of the proximity area to tents, with the demarcation of the private versus the public and the additions of edges and semiprivate zones, indicates the desire to mark the inhabitants' individuality and the differences among them. The communal services that local NGOs collocated in the IS, such as sinks, toilets and showers, are not properly used or maintained, and thus they are mostly so dirty that they must be cleaned for them to become operative again. The overall quality of the settlement is poor, and it reflects the temporariness of the structures, although they are still standing after 5 years. The lack of any type of infrastructure is a determinant of technical problems that require intervention and innovative solutions.

The non-porous character of the ground does not allow the rainwater to be absorbed, thus creating rivulets through the few passages between the tents (Fig. 22).

In the Bekaa, the study was conducted in different steps: first, the area was investigated by the research group in Landscape in Emergency, during a theoretical workshop that was conducted in Lebanon in January 2015. At that time, we collected maps of the Zahle District, and we

Fig. 19 The metallic fence, bounding one of the sides of Sarafand IS, seems to extend the green patches of the neighbourhood inside the settlement

Fig. 20 The inhabitants of the IS engaged in a variety of landscape practices, such as growing medicinal and ornamental plants, in the spaces of their proximity areas

Fig. 21 Shaded areas in between the tents to create a sense of semiprivate space

Fig. 22 The rivulets, through the passages between the tents, created by the runoff on the non-porous ground of the IS

studied the situation of the ISs in the area by extrapolating typologies and comparing them to different locations and the distance from the rural villages (Fig. 23).

We met the local and international NGOs that work in the district and with the heads of some municipalities. Finally, the Kayani foundation and the CCECS at AUB shared with us the plans

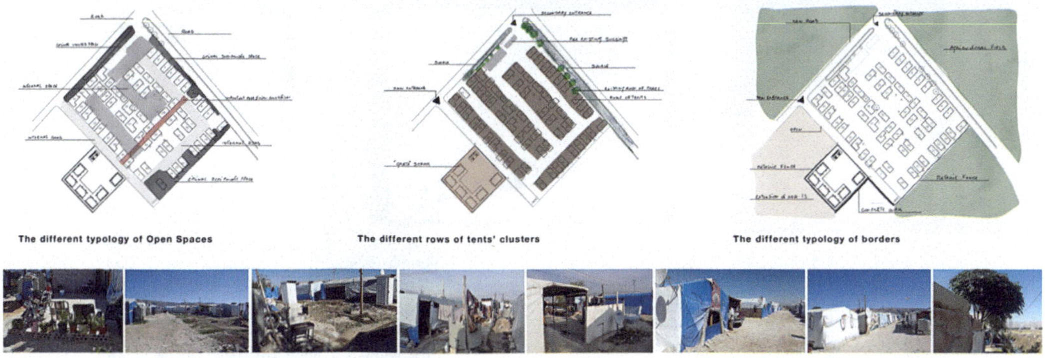

Fig. 23 Diagrams of Al Tyliani IS identifying the open spaces, the clusters of tents, the entrances and the pre-existing trees

of the settlement, and the socio-economic study was conducted on the site. The results indicated that the settlers come from different rural Syrian villages and cities such as Aleppo, Idlib, Daraa, Homs and Deir Azor and belonged to different families or clans. The inhabitants were farmers and workers, and the majority of them are illiterate, especially the women.

Then, we decided to concentrate our research in the assessment of the environmental and cultural state of the ISs and the repercussions of their conditions in the neighbourhood. For the cultural aspect, we decided to evaluate the culturally sensitive layout: the transitions from public to semiprivate to private places, the functional spaces (domestic work spaces and/or communal workspaces) and the recreational areas. For the environmental component, we focused our analysis on verifying the soil structure, fertility, erosion and compaction as well as the on-site water retention and/or detention option, the sewage system – dry sewage and/or composting toilet and/or other low-tech options (systems that are adapted to scale and/or population numbers), vegetation and food production.

From the analysis on the map, we noted that Al Tyliani is almost a square surface that is surrounded by agricultural fields. The two entrances are located along a secondary road that leads to the nearest village of Bar Elias. The IS is bounded on two sides by two swales, which, with the 5 years of the refugees' presence, has become a rubbish dump and created serious health problems for the residents. On the side in front of the entrance, a metallic fence defines the edge of a neighbourhood agricultural field. On the last side, the 'Ghata' school separates Al Tyliani from a newly established IS. The settlement was established in 2011 by the owner of the land who converted his agricultural field to accommodate the tents. In doing so, he compacted the soil, shaped 3 main roads and built 5 strips of concrete that are ready to house the 60 shelters. The newly produced landscape is the result of a new type of investment in the land, which was undertaken by the landlord. To control the expansion, the landowner overlaid a grid of vehicular roads to the ground, separating rows of a maximum of 10 tents, sized 4 m × 8 m. The grid was not fully followed. In fact, some tents are larger and/or smaller than the stated size, and some do not respect the alignment not to protrude or occupy communal space. The resulting configuration generated different typologies of the spaces between the tents that in some cases are used as alleys. In other cases, they serve as private or semiprivate space, as ornamental or kitchen gardens, or they are simply neglected. At one corner of the area, one Syrian refugee and a shawish opened a grocery store to provide people with all types of food and basic home supplies.

In May 2015, during the operative workshop, a group of 15 students and 10 professors worked for a week on the Al Tyliani IS. While we were

Fig. 24 The flexible and constantly changing proximity landscape of the tents in the IS

on site, we conducted an analysis based on sensitive and perceptive variables of the activities that were 'in place' in the settlement, considering the atmosphere of the private and open spaces with the aim of understanding and revealing the devices and skills that resulted from the organization of the landscape as a sensitive matter. This was a sensitive reading that was grounded on the daily practices of the space, on landscape perception and on the aesthetics of atmospheres (ambiances). Such an aesthetic analyses a different way of shaping the space that is related to the poetics of trash and junk art and to the aesthetic of open work (Eco 1989). By doing so, we were able to identify the adjustments to the rational and basic model of the settlement that had been made by the inhabitants to create places that could correspond to their needs.

We mapped all of the typologies of the space's appropriation that allowed the refugees to occupy (Fig. 24), control and organize the areas so they could consider them to be their own. They have put into practice new uses that have allowed them to make their stay in the IS as comfortable as possible. The forms through which the appropriations occur, and the associated traces, are sometimes permanent and visible, and other times they are ephemeral signs that are related to an event or occur on a seasonal basis (Fig. 25).

However, beyond the formal resolutions, the form of appropriation that is utilized in the settlement has allowed us to grasp the peculiarity of a space and its capacity to absorb changes and alterations. From the observation of the ordinary and everyday practices in place in various areas of the IS, we were able to abstract the categories that were used by the inhabitants to shape the new landscape.

The refugees used basic materials to build their shelters. An aerial view reveals a monochrome and monotone sequence of informal structures that are organized like a military campsite (Fig. 26).

In contrast, the life at the ground level is a complex sum of material and immaterial boundaries (Fig. 27), with an unclear definition and the hierarchy of spaces (public, collective, semiprivate or private). In 5 years, the initial shape of the tents was transformed by the addition of new rooms, internal courts and small gardens (Fig. 28).

Fig. 25 Permanent and/or ephemeral signs of appropriation occurring in the IS

A collection of plastic boxes, tires, satellite dishes and other materials that were collected and added to the roof form the profile of the settlement. A line of trees, all that remains of the previous agricultural use of the land, fences the IS on one side without creating any shape that benefits the residents. Due to a lack of proper disposal for liquid and solid waste, the latter is often burned on site, which generates an increase in the level of pollution and elevated environmental and health risks.

3 Results

The findings of the research show that at the regional scale, the dissemination of ISs in agricultural fields is creating changes in the equilibrium of the rural region that cannot be ignored. The increasing footprint of human settlements is eroding food-producing areas in many parts of the country, and food security is becoming an increasingly important issue. All of this is leading to the decline of the number of farmers and youth who may be interested in investing their future into agriculture in these areas.

Furthermore, the blooming of concrete structures on agricultural land, without control and planning, is degrading the landscape and changing the structure of the pre-existing Lebanese rural villages. Built at the margins of the rural settlements, the ISs are redesigning their borders and creating discontinuous fringes of scattered structures in an agricultural context. The rising number of legal and illegal construction will intensify the bacterial contamination of ground and surface water determined by the uncontrolled sewage disposal and by the appearance of more illegal open dumps that release leachate directly into the environment. Leachate will seep into the groundwater or runoff into the nearby watercourse, which increases the environmental crisis.

At the district level, the new risk which is caused by the massive impact of ISs on the landscape presents profound managerial and policy challenges for municipalities. Cities and rural governments are now obliged to manage the exponential rates of immigration, to protect and conserve the surrounding landscapes and ecosystem services that are sourced outside of their geopolitical boundaries, to ensure sufficient energy supplies for their industry and residents and to finance, construct and maintain hard infrastructure (Muller 2013).

According to the UNHCR data, with the escalating number of refugees in Lebanon, the tension between the host and guest communities, the poor quality of life in the informal settlements, the ecological and cultural impact on the territory and the lack of coordination of the local and international interventions are some of the pressing problems that the country is forced to address (UNHCR).

At the settlement scale, the investigation showed that life in the ISs is a struggle between the desire to return home, given the transitory nature of the refugee settlements, and the attempt to create a sense of stability to be able to engage in daily activities. The interiors of the tents are regularly cleaned and organized. Some movable furniture equips the rooms and subdivides them into small liveable islands (Fig. 29). Kitchen gardens and flowerpots that embellish the front

Fig. 26 A bird's eye view reveals a monochrome and monotone sequence of informal structures organized like a military campsite

Fig. 27 At the ground level, the life in the IS is a complex sum and sequence of material and immaterial boundaries, with an unclear definition and hierarchy of spaces

yard or the areas between the tents are the physical externalizations of the struggle that the refugees are going through (Fig. 30).

The communal open spaces are generally not paved and not equipped. They are no man's lands without the slightest form of communitarian sense that defines it as public space.

On the two private agricultural lands in Sarafand and Bar Elias, the individual tents were built without a real project, along informal roads. Nevertheless, the refugees have begun to ameliorate their tents, which have begun to look more like houses. Thus, the population finds itself in a conflict between the desire for individual expression of a certain level of comfort and beauty, on one hand, and the demonstration of the temporal nature of their presence in the IS, on the other (Herz 2013).

The progressive shaping and rebuilding of the tents is constantly changing and transforming the settlements' landscapes. The inhabitants try to reproduce the residential fabric of their original villages with their socio-spatial organization and division between public and/or private and open

Fig. 28 The initial shape of tents in the IS is continuously transformed by addition or subtraction of new rooms, internal courts and small gardens

Inside the tents

Fig. 29 The interior of the tents is regularly cleaned and organized; some movable furniture equips the room, subdividing it into small liveable islands

Fig. 30 Kitchen gardens and flowerpots blooming in the IS

and/or closed. The codes of use and appropriation of all of the spaces that are adjacent to the tent are tacit, and they regulate the extension of the domestic activity beyond the home. The resulting proximity landscape is flexible and constantly changing (Fig. 31).

Ephemeral landscapes are put in place at various times of the day to later disappear without leaving material traces of their existence. Permanent signs of appropriation are present where the ratio between the tents and the open spaces represent a danger to the respect and visibility of intimacy. Protection elements start from the narrow interiors and arrive at the threshold to prevent passers-by from getting too close to the intimate space. The limit defines the extension of familiar territory. Its mark designates a possession and the possibility of claiming a right. The threshold, which is a thin and first and foremost symbolic line of crossing (transition), marks more than ever the highest alienating nature between oneself and others, one's identity and otherness (Fig. 31). It shows that everyone could be on either side. The apparent flexibility of this line is governed by well-defined spatial and social codes that determine the nature of the allowed domestic floods.

4 Conclusion

Considering the Lebanese Syrian crisis through the lens of the 'landscape', which is eminently regarded as a cultural concept, the research was able to track down all of these spontaneous expressions, which are continuously shaping the Lebanese territory and present an overview of the situation that we face today. The results, while not exhaustive, present a situation in which scattered but numerous new cultural identities affect and transform the landscape by creating new ones from the addition and overlapping of native and imported models. The research is based on the assumption that cultures that are forced to give up a landscape and later to move and create a home elsewhere are likely to reinterpret old landscape values in different locations and to remould the new landscape to reflect those values.

Fig. 31 Example of extension of the tent interiors versus the communal area of the IS

Thus, landscape knowledge is transported, and new cultural landscapes are created. 'Place' is an important aspect of human existence, and it is an important source of security and identity. Places shape our memories and feelings, and in turn, people shape the landscapes around them through their experiences and actions. Unfortunately, in a globalized context, 'cultural landscape', as the assimilation of the ethnic, social, gender, economical and local characteristics of a built environment that is manufactured by human beings, is systematically neglected. The research presents some limitations due to the continuously changing situation and the difficulty of tracking and updating the data, the struggle to collect the correct information, especially from the socio and economical perspective, as well as the political and religious implications of the Syrian crisis and their repercussions in the spatial and social organization and location of the ISs. The massive presence of informal settlements in well-defined districts and governorates has a strong religious and/or political connotation that is evident in the available and published maps, but it is never clearly stated.

An important result of the research consists of the ascertainment that, while they were not planned, the threats to the environment which were caused by the ISs were not considered during the emergency phase of the refugees. In addition, they are still not accounted for after the environmental impact assessment that was conducted by the Ministry of Environment in 2014. However, the proliferation of unsustainable sprawl of illegal settlements could have unpredictable consequences for health and safety, causing huge environmental degradation. Natural resource management (NRM) must be considered at every level of planning, because damage to the environment and to relations with the local communities is difficult to repair (Corsellis and Vitale 2005).

To provide all people with an environment that is viable and worth living in, new planning models and approaches must be developed, both strategically and systemically, which consider landscape to be a tool to re-establish lost connections and to identify new typologies. We must begin the planning processes of integrated design with the effects on the landscape at different spatial and temporal scales. Additionally, the research was conducted with the aim that its results could serve as a foundation for the planning and territorial management of Lebanon. Knowing the existing conditions will help in defining a strategy of intervention that could help stakeholders at different levels, from municipalities to ministries, in decision-making and adopting an approach that better fits in their context.

References

Corsellis, T., & Vitale, A. (2005). *Transitional settlement, displaced populations*. Cambridge: University of Cambridge.

Cosgrove, D. (1984). *Social formation and symbolic landscape*. Madison: The University of Wisconsin Press.

Dunmore, C. (2015). UNHCR chief meets struggling Syrian refugees in Lebanon. http://www.unhcr.org/en-us/news/latest/2015/4/552e531d9/unhcr-chief-meets-struggling-syrian-refugees-lebanon.html. Accessed Mar 2016.

Eco, U. (1989). *The open work*. Cambridge, MA: Harvard University Press.

Fargues, P. (2014). Europe must take on its share of the Syrian refugee burden, but how? Migration Policy Center, Policy briefs. http://www.migrationpolicycentre.eu/. Accessed 10 Mar 2016.

Herz, M. (2013). The refugee camps of the Western Sahara. *Humanity, 4*, 365–391.

Lewis, P. (1979). Axioms for reading the landscape: Some guides to the American scene. In D. W. Meinig (Ed.), *The interpretation of ordinary landscapes: Geographical essays* (Vol. 15). New York: Oxford University Press.

Meinig, D. W. (1979). *The interpretation of ordinary landscapes: Geographical essays*. New York: Oxford University Press.

MOE/EU/UNDP. (2014). Lebanon environmental assessment of the Syrian conflict & priority interventions, report.

Muller, S. (2013). The next generation of infrastructure. *Online Journal Scenario 03: Rethinking Infrastructure*. http://scenariojournal.com/article/the-next-generation-of-infrastructure/. Accessed Mar 2016.

Ryan, R. L. (2011). The social landscape of planning: Integrating social and perceptual research with spatial planning information. *Landscape and Urban Planning, 100*, 361–363.

Sassen, S. (2016). *Massive loss of habitat triggers new array of migrations*. https://www.thrumanfactor.com. Accessed July 2016.

The Syrian refugee crisis and its repercussions for the EU, [website] http: www.syrianrefugees.eu. Accessed 17 Sept 2016.

UNHCR. (2007). *Handbook for emergencies* (3rd ed.). Geneva: United Nations High Commissioner for Refugees.

UNHCR. (2014). UNHCR monthly report. Community development. (Report). Retrieved from http://data.unhcr.org/syrianrefugees/country

UNHCR. (2016). Shelter quarter 1 dashboard, [report]. http://data.unhcr.org/syrianrefugees/country.php?id=122. Accessed 10 Oct 2016.

From Emergency to Integration? The Spatial Controversy of Reception Centres in Italy

Gisella Calcagno and Roberto Bologna

Abstract
The Mediterranean migrant/refugee crisis is stressing the capacity and capability of the EU reception system to guarantee adequate standards of living to asylum seekers, as for the provision of housing, threatening the protection provided by the right of asylum. The study analyses the spatial phenomenon of reception centres, formally instituted by host countries to accommodate needy applicants of international protection during the asylum procedure, which are spreading in the European borderscapes to immediately supply "n. places" for the "n. arrivals". Within the global challenge of housing and integrating migrants and refugees, the topic of reception centres is carried out characterizing the temporary status of the asylum seeker and the controversial nature of hospitality, affecting reception and its spatial practice. The mapping controversies method was used to read the spatiality of reception in the European and Italian contexts, where case studies were selected in the city of Florence. Results depict the heterogeneous spectrum of spatial solutions for reception, which reflects the lack of shared/qualitative standards and of strategic/contingency scenarios able to meet basic human rights and inclusive urban development. Finally, the study collected innovative spaces of reception, raising in the architectural discourse and looking at those directions.

Keywords
Asylum seekers · Reception centres · Mapping controversies · Urban reception · Integration

1 Introduction

Even larger populations are forced by human disasters to escape from their homes, and to start a certain displacement and/or an uncertain migration path, in search of safety: according to the international and human right of asylum, a person exposed to persecutions in the own State has the right to ask protection in another State, which has the duty to provide for it.

The global nature and dimension of causes and consequences of forced migration, and their respective exacerbation, is threatening the traditional durable solutions for the refugee displacement, as voluntary repatriation in countries of origin, local integration in host countries and resettlement in third countries (UNHCR 2015): displacement root causes are

G. Calcagno (✉) · R. Bologna
DIDA Department of Architecture,
Università degli Studi di Firenze, Florence, Italy
e-mail: gisella.calcagno@unifi.it; roberto.bologna@unifi.it

A. Asgary (ed.), *Resettlement Challenges for Displaced Populations and Refugees*,
Sustainable Development Goals Series, https://doi.org/10.1007/978-3-319-92498-4_5

becoming gangrenous in places of origin, transits across countries are even more difficult due to even more restrictive migration policies and the securitization, and resettlement, as selection and transfer, cannot cover the dimension of the phenomenon, remaining as the best solution.

In practice, only three no-durable solutions are available for refugees: encampment, destitution in urban areas or risking their lives in a perilous journey to another country (Betts 2016). The last is the forced choice of millions of displaced people in the Middle East and Central/ North Africa, who are fleeing emergencies crossing dangerously the Mediterranean Sea and reaching the Southern European borderscapes, where they are waiting for asylum.

The study focuses on reception centres, temporary settlement and housing solutions provided by arrival States to accommodate needy asylum seekers during the asylum procedure, spreading in the European Union context of the ongoing Mediterranean refugee/migrant crisis, and analysed in Italy.

Reception centres represent an intermediate settlement and housing solution in the transitional living of refugees, where ideally their emergency as forced migrants should end and their integration as new citizens could start. Nevertheless, they constitute a controversial space, young in the architectural discourse (Hauge et al. 2017), called to address both fundamental human rights in emergency and long-term integration perspectives (Fig. 1).

2 A Global and Urban Challenge

As huge human right deprivation, forced migration produces an immediate and protracted housing emergency: the transitional living of internally displaced people and, beyond borders, of asylum seekers and refugees.

Reaching the dramatic quote of more than 65 million of worldwide displaced people (UNHCR 2015), forced migration is even more recognized as a crucial global challenge for the sustainable development: considering the UN 2030 Sustainable Development Agenda, the theme is transversal to different Sustainable Development Goals, calling for interrelated actions covering broad social, economic and environmental aspects (Taran 2016). The increasing urban dimension of the refugee crisis (the 60% of worldwide refugees are settling in urban areas; UNHCR 2015) is particularly stressing the SDG 11 on "make cities and human settlements inclusive, safe, resilient and sustainable".

Addressing this goal, the New Urban Agenda adopted in United Nations Conference Habitat "Sustainable Cities and Human Settlements for All" in 2016 (UN HABITAT 2016) recognizes the need to handle migrants and refugees within the multiple forms of discrimination faced by vulnerable groups (Call for Action 20). Envisioning a sustainable urban development for social inclusion and ending poverty, the Agenda states the commitment:

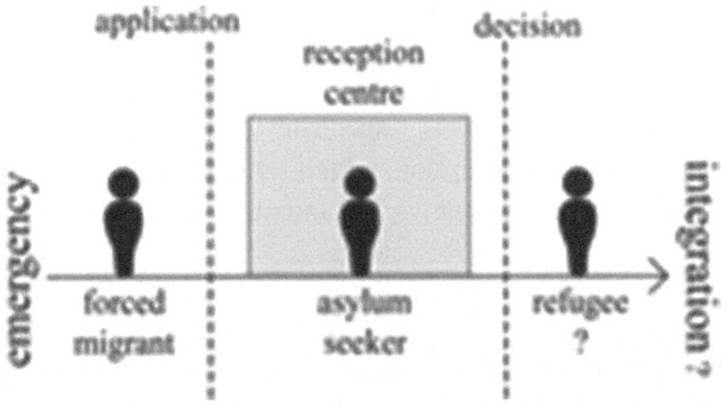

Fig. 1 Reception centres: asylum process, legal status and challenges

> to ensure the full respect for human rights and humane treatment of refugees, internally displaced persons, and migrants, regardless of migration status, and support their host cities in the spirit of international cooperation, taking into account national circumstances and recognizing that, although the movement of large populations into towns and cities poses a variety of challenges, it can also bring significant social, economic and cultural contributions to urban life (…) to strengthen synergies between international migration and development, at the global, regional, national, sub-national, and local levels by ensuring safe, orderly, and regular migration through planned and well-managed migration policies and to support local authorities in establishing frameworks that enable the positive contribution of migrants to cities and strengthened urban-rural linkages. (item 28, UN HABITAT 2016)

Focusing on the physical dimension of cities and settlements, the paragraph "Planning and managing urban spatial development" outlines the participatory and inclusive process of all inhabitants, not discriminating the migrant populations (item 104).

The issue of migrants and refugees in urban areas is specifically addressed prospecting the raising of a shared awareness of how inclusive planning can improve their management and protection in the access to services, opportunities and spaces: it is a call to looking forwards urban enabling environments that can contribute to the resilience of communities and drive to a sustainable urban development (HABITAT 2015).

In the Europe of the ongoing Mediterranean migrant/refugee crisis, the "Urban Agenda for the EU" – Pact of Amsterdam signed in 2016, and looking at the EU 2020 strategy for smart, sustainable and inclusive growth, considers as priority theme the inclusion of migrants and refugees. According to the Agenda:

> the objectives are to manage integration of incoming migrants and refugees (extra-EU) and to provide a framework for their inclusion based on willingness and integration capacity of local communities. This will cover: housing, cultural integration, provision of public services, social inclusion, education and labour market measures, chances of second-third generations, spatial segregation. (European Commission 2016)

Focusing on asylum seekers, Eurocities's "Statements on asylum in cities" (2015) recognizes the need to tackle asylum seeking not as an emergency but as a structural issue of human rights and the role of cities in arrival, transit and destination countries in Europe in addressing its social, humanitarian and financial challenges. As places where asylum seekers wait for a decision on their refugee status and where services are provided, cities must guarantee basic protection and the reception and integration of newcomers.

Looking at reception as an early integration phase, the Migration Policy Group (2016) identified the bottlenecks of urban reception in the implementation of EU reception standards, in the dispersal policies and in the knowledge/data gaps.

As a stress factor, forced migration is a challenge for urban resilience: cities, as agents of change, must embrace and incorporate human flows into urban planning, adapting and transforming for a successful reception and integration of newcomers, in long-term perspectives (100 RC 2016).

3 The Space of Reception

Following the influencing Lefebvre line of thought (1991), space is definitively a social product, with a necessary political nature: the social and political dependency appears particularly evident in the spatial production of reception. Even the Foucauldian biopolitical reading of space and architecture as power devices (Rabinow 1984), pervasive in the architectural discourse on total institutions (Brunon-Ernst 2016), suggests reception centres as "asylum devices" of nation States' policies. National asylum policies in fact can also produce detention centres for asylum seekers: in this case the objective of sheltering is covered by the incarceration one, resulting in more controversial spatial productions (Cairns 2004).

With the aim to characterize the distinguishing marks of reception centres, the political nature of asylum needs to be considered: the refugee status is a concession, such as reception conditions. According to the Italian philosopher Giorgio Agamben, considering human life in its double dimension of *zoe* (biological life) and *bios* (political life), the condition of the refugee, characterized by the lack of legal protection by any State (human rights are protected by law, and law is applied to citizens), creates a "space of exception", between inclusion and exclusion, where life is reduced to the biological one, a "bare life" (Darling 2009; John-Richards 2014). The deepness of such thought is entering in the broad architectural discourse (Boano 2017), but has already emerged in the space of exception par excellence: the refugee camp (Katz 2017; Tan 2016), immediate and protracted emergency settlement and housing solution, assuming today the dimension of cities.

Speaking on asylum, Derrida (2001) defines cities as refuges par excellence against insolvent and failed national States, where a more equal right of asylum can be guarantee, as in *sanctuary cities*. Nevertheless, cities are dooming refugees in informal settlements and substandard housing conditions, overlapping the existing ones of the urban crisis. Camps and cities are two opposite but converging spaces, which represent both the extremes and the core of the space of reception.

The focus on asylum seekers, more worrying than refugees (Baumann 2005), enriches the bio-political theme with the narratives of the "guest" and of the "illegal" (Rozakou 2012), origin of approaches between solidarity and hostility, resulting in a never unconditionally hospitality (Freise 2004). Referring to the provision of reception centres, Szczepanikova (2012) outlines their double nature of control and assistance, which produces prolonged spatial confinement and social segregation. Similarly, Darling (2016) denounced that asylum seekers in the global North are dispersed just to share the burden, to accommodate and not integrate, creating a policy-imposed liminality which results in forced networks with aliens in hostile new surroundings. The space of reception reflects the controversy (Fig. 2) of

> a shared uncertainty, a black boxed situation not stabilized and fixed but open to the negotiation of the actors. (Venturini 2010)

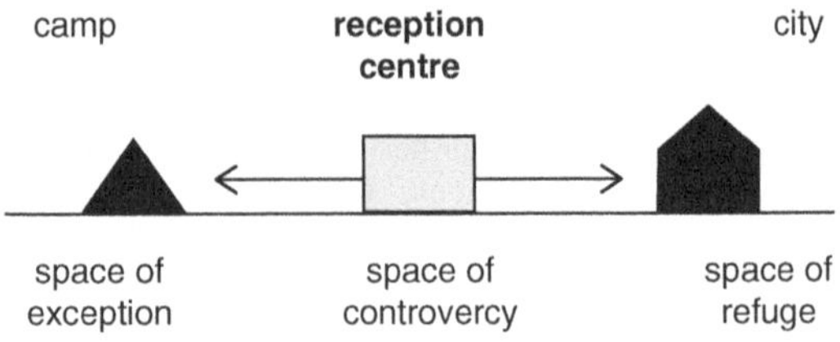

Fig. 2 Reception centre as space of controversy

4 Mapping the Spatial Controversy of Reception

Deriving from Science and Technology Studies, the "mapping controversies" method appeared as the most appropriate to uncover the controversy of reception and its spatial production.

The method derives from the actor-network theory, which has the advantage to consider inanimate artefacts as active actors influencing the social system: Latour himself looks at space as a mediator of the social system (Latour 2005). Its codification for the architectural discipline by Yaneva (2012), as a navigational tool to describe architectural objects, processes and practices, represented a unique occasion to look at architecture both as reflecting and generating society.

With the objective to reveal the spatial controversy of reception, the method was borrowed to follow the controversy of reception (socio-political aspects), from the European to the Italian level, and to document its spatial outcomes (reception centres): the ongoing mapping of reception centres in Florence has the objective to look at the controversy of reception at urban level.

4.1 European Union as Common Area of Asylum

In addressing the international right of asylum, European Union is a common area of protection, object of a 20-year-old process of harmonization of national asylum procedures and standards.

In the last years, European Union attested the peak of asylum applications, which reached the number of 1.2 million/year: most of the requests were directed to Germany (40%), followed by Italy (10%) (site: frontex).

The huge sea arrivals of summer 2015, whose dramatic images of disembarks and deaths highly impacted the public opinions, exacerbated the social and political debate on welcome refugees: Member States' national policies had the power to break the unity of the Union (Brexit), still united to augment controls (Hotspot approach) and externalize the problem (accord with Turkey), but definitively weak to sustain shared choices (internal EU borders, Schengen Area suspension) and each other's (relocations, as internal EU resettlement, are not working). The result is a multiform and diffuse humanitarian crisis, whose spatial production is becoming manifest across Europe in informal settlements of various temporariness and precariousness, appearing and disappearing along the forced migrants' routes.

The Directive 2013/33/EU (recast), laying down standards for the reception of applicants for international protection, states reception conditions to guarantee adequate standard of living for applicants during the period of examination of the procedure for international protection. Even if for the Directive states only basic material reception conditions as food, clothes and housing in kind or via vouchers and a daily pocket money (art. 2), it declares that they must be protective of the physical and mental (news of the recast) health of the applicants (art. 17).

The housing provision in kind can take form of premises at borders or in transit zones, accommodation centres (defined as any place used for the collective housing of applicants; art. 2) or private houses, flats or other premises adapted for housing applicants: in any case, housing solutions must consider the specific needs of vulnerable groups. Moreover, in duly justified cases and for a reasonable period, as short as possible, Member States may exceptionally set different modalities for reception, whatever covering the basic needs and vulnerabilities (art. 18).

Comparative reports at European level (ECRE 2016; FRAME 2015; EMN 2014) testified housing reception conditions differentiated (between and within States), inadequate (low standards of living) and insufficient (in term of capacity): moreover, the widespread turn on emergency solutions is worsened by the length of asylum procedures, which prolongs the permanence in reception (average of 1.5 years).

Looking at the AIDA database (site: asylumineurope) on Member States' housing conditions in reception, the interpretation of the Directive remains highly divergent (the same distinction between first-line and second-line reception is not provided by the Directive, while is reported in many States), producing a heterogeneous spectrum of housing solutions, scattering in the emergency facilities.

Reflecting the controversy, according to the last comparative report:

> emergency or makeshift accommodation solutions are not designed to guarantee an adequate standard of living for people engaging with the asylum process,

but at the same time

> the year 2015 has shown a commendable ability and readiness on the part of many receiving states to rapidly and creatively enlist spaces for the temporary accommodation of large numbers of newly arrived (ECRE 2016).

Moreover, even if the European Asylum Support Office (EASO 2016) on monitoring housing conditions recognizes as quality-factor location, distribution, safety, common spaces, sanitary conditions and maintenance, Member States do not have duties to report data: the information gap reveals a crucial shortcoming (ECRE 2016).

4.2 The Italian Reception System

For its geographical position of door of Europe from Africa, Italy is highly impacted by mass and mixed fluxes of forced migrants arriving mostly from the Central Mediterranean Route, the most dangerous and deadlier in the world, due to human traffick. As first arrival country, according to the EU Dublin Regulation III, Italy is responsible for the examination of the asylum procedures.

Fig. 3 Sea arrivals and instituted reception centres (1998–2015)

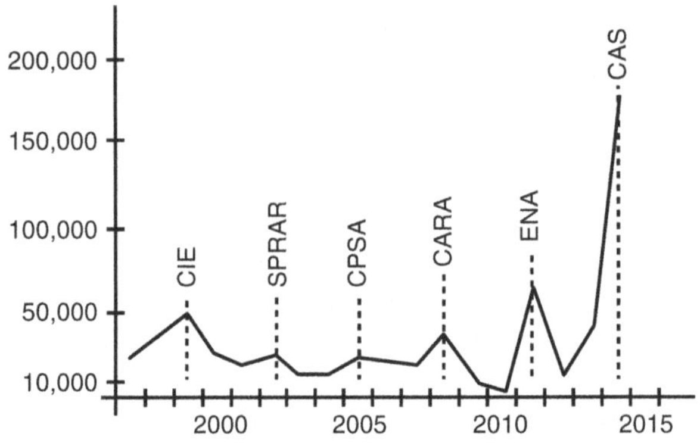

Fig. 4 Reception trends in Italy (2013–2016). (Source: interno.gov)

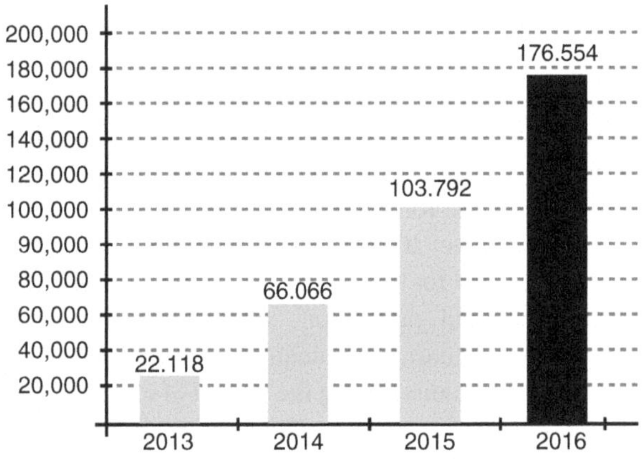

The Italian reception system is characterized by the juxtaposition of different reception models and centres, result of historically emergency-based asylum policies on reception, as mere answer to the sea arrivals (Fig. 3).

According to the last law on reception (D.lgs. 142/2015), addressing the 2013 EU Reception Directive, the Italian reception system is articulated in two ordinary levels: the first reception of arriving migrants takes place in governmental centres (art. 9), while a second level of reception (SPRAR), managed at local level, hosts applicants during the overall asylum procedure (art. 14). In case of lack of places in secondary reception, first-line centres can continue to supply for reception, and in case of temporary depletion of places also there, the government can provide for emergency facilities (art. 11).

Parallel with the huge sea arrivals of the last years, the capacity of the Italian reception system grew increasingly, reaching the quote of 176.554 places at the end of the 2016 (site: interno.gov) (Fig. 4).

4.2.1 First-Line Reception Centres

First-line reception takes place in governmental reception centres, activated by the Ministry of the Interior and managed, through public tenders, by the public or the private social sector: to this day, there are 15 of these centres hosting the 8% of the total presences (Asgi 2016).

Even if the Italian normative states different types of governmental centres (also the detention ones), public tenders follow a unique regulation (DM 21 November 2008) which defines generic reception standards, with no reference on minimal requirements for facilities (only a recent fire-prevention normative is specifically intended for reception centres).

Governmental centres are collective centres, former buildings of different types previously used for other purposes, and reconverted in collective accommodations (UNHCR 2010), for high capacities (hundreds/thousands) and usually located in border zones or peripheral urban areas.

During the last years, governmental centres are changing their nature and their denominations to follow the changing scenario of arrivals and asylum policies, also if more of them persists in the same facilities with the same problems.

The Hotspot approach of the 2015 EU Migration Agenda is threatening the aid nature of CPSAs (*Centri di Primo Soccorso e Accoglienza*), not changing the inadequacy of their spaces: the new Hotspot in Pozzallo, for example, was a CPSA when it was abandoned by NGOs for the unsustainable overcrowdings in the precarious physical conditions of the former dock in the harbour (MSF 2015).

CDAs (*Centri di Accoglienza*) and CARAs (*Centri di Accoglienza per Richiedenti Asilo*) persist as first-line centres, providing for the missing places in the secondary-line, even if recognized as negative reception models by a Parliament inquiry in 2014. One of the persisting worst practice, reported for the inquiry by Medu (2015), is the CARA of Mineo in Sicily, a former US military base isolated in the countryside (8 km from the nearest urban area). Beyond organizational problems, the facilities' inadequacy, together with the overcrowdings and the protracted staying, concurred to the human rights' violations faced by the guests: spatial segregation and assistance-based life are reducing people to numbers, nourishing violence, dangerous informal economies and link with the local black work (as the rural *Capolarato*).

The last national plan on reception (Ministero dell'Interno 2016) intended to overcome the model, proposing regional Hubs with a maximum capacity of 100 people.

4.2.2 Second-Line Reception Centres

The SPRAR reception system (*Sistema di Protezione per Richiedenti Asilo e Rifugiati*) was born as a spontaneous reception experience of voluntary organizations at local level to face the lack of appropriate governmental responses in the first nineties. The result of those initiatives is the ongoing solidification of a networked system, formally instituted in 2001 by the Ministry of the Interior, UNHCR and the Anci association of municipalities (*Associazione Nazionale Comuni Italiani*).

SPRAR system proposes an integrated reception based on the idea to go further the mere food-clothes-housing material supply, offering customized paths to autonomy and social integration, which comprise work and housing. The activation of a reception project is responsibility of an urban authority, which involves third sector associations to connect with the territorial and urban networks of social services and spaces.

The SPRAR operative manual details the reception standards, also referring to minimal requirements for the facilities, regarding the location in urban area, adequate hygienical services (1x6), maximum capacity for rooms (two to three in flats, four in collective centres), the presence of at least a common space for leisure activities and the possibility to personalize spaces (SPRAR 2017).

According to the annual report, the 83% of the SPRAR facilities are flats, mostly rented in the private market, followed by collective centres, with a total average of seven beneficiaries per facility (SPRAR 2016). Beyond housing, the other activities connected with the integrated reception (as language courses, cultural and legal mediation, work training and social integration) are distributed in the urban area, usually overlapping the existing social spaces (such as churches, schools and associations' centres).

The choice to activate a SPRAR at municipal level has two opposite outcomes: on the one hand, the local dimension is recognized as ideal to undertake integration objectives; on the other hand, it is clearly political: even if the trend is positive (+26% of places in the last year), less than half of the Italian municipalities are hosting the SPRAR, creating a huge imbalance (SPRAR 2016).

SPRAR system numbers many recognized best practices: according to the national plan on reception (Ministero dell'Interno 2016), it is the feather in the cap and the pivot of all the Italian reception system, to be implemented.

Nevertheless, at the end of 2016, SPRAR system hosted 24,000 people, covering only the 14% of the total presences (site: interno.gov).

4.2.3 Emergency-Line Reception Centres

The turn on emergency facilities is not new in the Italian tradition, persisting as the ordinary answer in case of mass arrivals, since the firsts in the nineties.

In 2011, the big influx from the MENA area (Middle East and North Africa) consequent to the Arab Springs was solved with the operation "North Africa Emergency" (ENA). According to Garelli and Tezzioli (2013), the planning was scattered, and the implementation decentralized, externalized and unregulated; moreover, the governmental choice to delegate the management of reception to the national civil protection represented an eloquent consideration of the migrants' arrival as a natural calamity. After two years of activities, ENA facilities were closed, and 19,000 guests dismissed with an exit bonus of 500 euros, remaining in the streets: Italy was in official breach of the right to housing (Swiss Refugee Council 2013).

In 2014, just one year later, the Minister of the Interior stated the "temporary" need of supplemental places for the reception of asylum seekers, and delegated its territorial offices (*Prefetture*) to find and activate centres with first-line reception standards (DM Novembre 2008), through public tenders or direct assignments. During the last years, CAS (*Centri di Accoglienza Straordinaria*) system grew increasingly in the whole Italian territory, sheltering the 80% of the national presences. CAS reception conditions were investigated by a voluntary monitoring activity, *InCAStrati* or "to get stuck", which revealed the diffuse lack of minimal sanitary requirements in the facilities and the negative answers of *Prefetture* to give data (LasciateCIEntrare 2016).

The distribution of places in the reception lines previously described (Table 1) shows the impressive incidence of CAS: the emergency line is ensuring the survival of the Italian reception system.

Furthermore, the insufficient capacity of the reception system (Asgi 2016) is emerging in the even more increasing number of asylum seekers and refugees living in informal settlements, from squatted buildings to camps, which are scattering in the entire national territory, in particular in big cities and near first-line reception centres (MSF 2016).

4.3 A Case Study: Reception Centres in Florence

The case studies analysis of reception centres in the city of Florence was intended to contextualize the urban challenges but also simplifies the variability of mesoscale contexts that can influence reception, as the regional one. Based on "the fair collaboration between the different levels of govern" (D.lgs. 142/2015, art. 8), regions are called to support the national reception system, managing the redistribution of migrants and the modalities of access to the crucial services in their territories (as the health system).

Since the North Africa Emergency in 2011, the Tuscan regional policy on reception was oriented to a wide distribution of migrants, avoiding the concentration in governmental reception centres and taking advantages of the

Table 1 Reception centres in Italy

Reception centres	First-line	Second-line	Emergency line
Denomination	Hotspot/Hubs/CARA	SPRAR	CAS
Main facility type	Collective centres	Flats	No data
N. of facilities	15	2.457	7.005
N. of guests	14.694	23.822	137.218
Distribution of guests	8%	14%	78%

Data retrieved from site: interno.gov and Asgi (2016)

well-rooted and diffused third sector network of associations. In the same year, the previously described CARA of Mineo in Sicily hosted alone more than the double of all the guests in the Tuscan region.

A monitoring activity promoted by the region produced important data to understand the spatiality of this previous emergency-line reception, as the georeferentiation of reception centres[1] and data on their capacity and locations. According to the report, in 2012 the Tuscan ENA emergency system hosted 1,500 people, distributed in 126 facilities: the 40.5% collective centres, the 36.5% flats and 23% touristic facilities (many available only seasonally). Operators declared a total capacity of 1,901 places, with an average of 15.3 people/facility, augmenting to 19.3 in collective centres. Regarding the locations, 1/4 of the facilities resulted characterized by a high distance from the urban areas: the report, recognizing the threat of isolation, also suggests how it can be overcome in the presence of well-organized reception model. Moreover, the high incidence of collective centres, showing the limits of decentralization, underlined the necessity of these facilities to cope with emergencies (Bracci 2012).

In the same period, the ordinary second-line reception of the Sprar system hosted less than 200 beneficiaries (SPRAR 2013): its inadequate capacity fuelled also in Tuscany the turn on CAS emergency-line reception when arrivals augmented in 2014. In line with the national data, the Tuscan CAS emergency line increased eightfold compared to the previous ENA, reaching 11,600 presences; in parallel, the SPRAR system augmented only fourfold its capacity to 860 places: Tuscan centres accommodate the 8% of the presences in the national reception system (site: interno.gov).

The Tuscan experience on diffuse reception was gathered in a "White Book on reception" (Regione Toscana 2017), which contains programmatic policies to address an integrated reception model, with SPRAR system as reception standard to adapt the CAS system.

Case studies were selected in Florence, the main city of the Tuscany region, famous touristic destination for its great Renaissance, circumstance that should enrich the controversy of reception with the refugee-tourist nexus (Röslmaierm 2016).

According to the SPRAR database[2], the ordinary second-line reception in the city of Florence is directed by two urban entities (the municipality itself and "Società della Salute"), responsible for the reception projects of 206 beneficiaries (1/4 of the regional presences). The SPRAR management involves several and specialized third sector associations, working in network to cover all the aspects of reception (as housing, legal assistance or integration). The 73% of the SPRAR beneficiaries is distributed in two reception centres, both identifiable as large-size collective centres and selected as case studies.

Looking at the emergency line in the city of Florence, CAS reception centres are directed by *Prefettura di Firenze* (local office of the Ministry of the Interior), which provided in the last three years with several and temporary public tenders to assign the service of reception. Beyond the weak first-line reception standards (DM Novembre 2008), the only reference to the spatial characteristics of reception centres regards the calculation of the number of guests per facility, retrieved from the regional standard for touristic reception and referring to the lower level (youth hostels). In line with the regional policy, a point mechanism considers as main influencer the "territorial sustainability", regarding the proportion between actual presences in reception and the resident population: no points are given for the spatial quality of the reception.

According to the data retrieved in May 2017 from *Prefettura di Firenze*, the CAS system is sheltering 622 asylum seekers in 23 facilities: the 72% of them are hosted in large-size collective centres, where the average capacity reaches the quote of 75 people/facility (Table 2). Another

[1] http://mappe.rete.toscana.it/webstat/index.html?area=emergenza_nordafrica

[2] http://www.sprar.it/progetti-territoriali

Table 2 CAS reception centres in Florence

Type of facility	N. of facilities	N. of guests	% of guests
Flat	12	86	14
Small collective centre	3	40	6
Medium collective centre	2	47	8
Large collective centre	6	449	72
Total	23	622	

Data gathering followed the SPRAR operative manual (2017), which defines the type of facilities according to the capacity and the organization model: flats (<10 people) and collective centres of different sizes (small <15 people, medium <30 people and large >30 people)

interesting data regards the public ownership of three facilities, large-size collective centres, ensuring the reception of 219 asylum seekers (1/3 of the total hosted by the city).

At the city level of Florence, reception revealed a high incidence of large-size collective centres both in the SPRAR and in the CAS system, representing the selection criterion for the case studies: Villa Pieragnoli, Ex Centro Paci, Centro Slataper and Villa Pepi.

Following the mapping controversies method, data were collected to look at actors and networks in the spatial controversy of reception at the urban scale, considering the facilities used for reception as focal no-human actors analysed.

A geo-referenced map (Fig. 5) localized the reception centres in the city of Florence, reporting data on reception line, typology, management, capacity, occupancy and property, and it is collecting information (photos, news) on the single facility (former use, conditions) and its urban context (neighbourhood, transport and public spaces).

For each case study, a field research was conducted to analyse the reception centre's documents (internal reports and regulations, technical drawings), visiting the centres (photos, notes) and interviewing the directors (history of the centre, management and organizational models). Data gathering (Fig. 6) is taking into account actors (broad users and facilities) and networks (activities and urban system) of reception centres, with the aim to identify who (users) and what (activities) both in the structure of reception (facility) and in its infrastructure (urban system).

4.4 Reading the Data

The Italian reception system has the double function to supply for the asylum seeker's housing emergency (first-line reception) and to mediate the refugee's housing integration (second-line reception): nevertheless, actual conditions reveal difficulties to meet both.

The insufficient expansion of the SPRAR system nourished and is still nourishing the proliferation of temporary facilities, revealing a reception system in a chronical housing emergency: the lack of and the undifferentiated use of reception centres represent a threat for the respect of fundamental human rights and for the achievement of the goals of inclusion of migrants and refugees. Low-quality standards and externalizations without monitoring are fomenting the idea of reception as a business and not as a public service, representing an obstacle to the transparency of public funds and an invite to the raising of populisms against welcome refugees.

The controversy of reception, with its strong basis on the biopolitical discourse origins a transitional space of exception, which has the possibility to become a space of refuge.

Reception centres as spaces of exception are total institutions where the private sphere cannot be protected, life is assistance-based and everyday life is void and repetitive. Empty spaces filled with beds and a food catering, or distance from urban areas, can produce alienation and segregation: the possibility to use these centres for the overall duration of the asylum procedure (years) represents a serious threat for the psychophysical well-being of the vulnerable person, as a perpetuation of the trauma he/she is escaping. Moreover, the fluctuation of arrivals provokes immediate overcrowding, worsening the conditions: centres are not ready to work in contingency also because facilities are not planned and designed to be flexible.

Fig. 5 Geo-referenced map of reception centres in the city of Florence

Actual first-line collective centres, and emergency facilities with the same standards, do not represent a sustainable and resilient solution: the inevitable spatial conflict has a negative impact both for the vulnerabilities and for the tricky social relations at work within and outside the centres.

The best practices of the diffuse reception in the SPRAR system are showing the positive side of reception as space of refuge: the integrated planning of local actors, experts of their territories, is connecting beneficiaries both with social services, opening the possibility to emancipate in work and housing, and with social spaces, where integration with local communities should physically start.

Recognized by the Tuscan region, the emerging need to drive a reconversion of the CAS emergency line to the SPRAR standards, for the creation of an integrated reception system, will require the necessity to tackle reception at the urban scale.

The analysis of reception centres in the urban context of Florence revealed the general predominance of large-size collective centres. The data was interpreted because of the limits of the rented flats housing solution in dense urban areas, where the access to house is more unaffordable: it should be verified analysing other urban contexts, as in shrinking cities, where the wide housing availability suggests different typologies of reception centres and models.

Fig. 6 Example of data sheets on a case study

Such consideration was suggested by the case study Ex Centro PACI (SPRAR), which was born within a national program to experiment reception centres in metropolitan cities. In these contexts (Ex Centro PACI is located in the inner centre of a new city expansion), the high capacity of the centre can be mediated by the presence of a well-structured urban system, whose services (as transport system) and spaces (as public spaces) can supply as reception infrastructure.

Still regarding the urban dimension, the SPRAR case study Villa Pieragnoli testifies how the problem of the distance from the urban area can be overcome with a well-rooted reception experience: the quiet community life of the isolated centre benefits by a territorial provision of reception services (such as language and work training), filling the everyday life of beneficiaries.

Considering the CAS case studies, the lack of standards for reception facilities immediately emerges in the overcrowding of the rooms, resulting in evident spatial conflicts. The limited activities provided in the emergency reception model increase the void time in the centre: this is the case of Villa Pepi, public-owned centre, whose potential spaces suffer the unplanning, inevitable due the short-term contracts in the emergency line.

The CAS case study Centro Slataper, resulting from a very controversial history, revealed an innovative typology of collective centre, characterized by the sharing of the same building for other housing emergencies (for homelessness) and social spaces (for associations): beyond standards, the cohabitation is enabling positive networks of encounter, also facilitated by the dense urban context in the proximity.

5 The Raising of New Spaces of Reception

Parallel with the mainstream production of reception centres, and in line with the described urban directions, experimental and innovative settlement models and housing solutions for asylum seekers and refugees are emerging, showing different ways to tackle the problem, even as an opportunity. The most representative is the platform Refugees Welcome (site: refugee-welcome), based on the idea to directly match the refugees housing demand with the citizens offer in their private houses.

In Europe, the theme of how to accommodate asylum seekers and integrate refugees was object of an increasing architectural attention, exposed in the Venice Biennale "Reporting from the Front" (Aravena 2016) and in the Oslo Architecture Triennale, "After belonging" (Casanovas et al. 2016). In the occasion of the Venice Biennale, the curators of three European states (Germany, Finland and Austria) proposed reflections that can be related to the specific theme of architecture of reception centres.

To face the influx of one million of refugees in 2015, German architects were called to design a huge number of temporary houses in emergency, also in special dispensation to regular housing standards: wooden prefabs, container settlements, and readaptation of pre-existing buildings were collected in a database organized by the DAM Architecture Museum (site: makingheimat). Another important contribution by the German Biennale team was the contextualization of the Doug Saunders' "Arrival City" in the German refugee/migrant crisis (Cachola Schmal et al. 2016), which suggests the importance to consider the spontaneous nature of migrant settlement processes.

The Finnish Museum of Architecture organized the architectural competition "From borders to Home" to find innovative temporary housing solutions for refugees, to contextualize in the short and long-term perspectives (site: mfa). The design outcomes, proposed at the Venice Biennale, speak of the advantages of temporary architecture, the readaptation of vacant offices for incremental housing solutions and the use of social network to connect refugees with the new urban context.

For Austria, the Biennale was the occasion to directly experiment design solutions in existing reception centres: the ephemeral solution of special parasols for interior showed that little spatial interventions can solve huge problems in collective centres, as the lack of privacy: the same objective was reached with the dislocation of wooden modules in former offices; another project showed the potential of furniture to address the social needs of cooking and working in reception centres (ortefuermenschen).

As first Italian innovative approach, Salus Space is a project in the municipality of Bologna, winner of a 5 million European fund in the Urban Innovative Action program (site: uia-initiative). UIA program had the objective to test innovative ideas for interconnected urban challenges: the first call matched the European Urban Agenda's priorities themes of urban poverty, integration of migrants and refugees, energetic transitions and local economy (European Commission 2016). Innovation, participation and transferability were considered as key factors for the selection. Salus Space is a project of urban regeneration based on the requalification and readaptation of the abandoned and squatted Villa Salus, near the city centre, to realize temporary housing and collective social spaces, in an experimental new space of welcome, welfare and well-being. The project is object of a process of participative design which involves a large range of heterogeneous stakeholders: the multiplicity of local points of view (municipal authorities and third sector agencies) is considered a unique occasion to experiment innovative social and spatial solutions (site: saluspace).

6 Conclusions

The structural phenomenon of forced migration will determine even more human flows in transit, just calling for the respect of their rights: it is a civic imperative offering safe spaces, welcoming and inclusive.

International, European, national and local organizations and institutions recognized the challenges posed and are envisioning urban strategies to tackle with the phenomenon: emergency-based logics should not be more justified.

Case studies testified both the emergency of reception and its potential for integration: echoing the converging narratives of camps and cities, reception is an irreducible space of exception but a possible space of refuge. The spatial controversy or reception, between emergency and integration, needs to be addressed solving the oppositions: contingency/necessity, informality/formality and conflict/mediation.

As the recent architectural incursions are showing, the general failure of the mainstream approaches can be overcome with the potential of the avant-garde ones: regenerative sustainability (Hes and Du Plessis 2015), temporary urbanism and architecture (Bishop and Williams 2012) and temporary uses (Oswalt et al. 2013), sustained by a diffuse design for social innovation (Manzini 2015), represent the perspectives to look at design as a game changer in the provision of urban spaces for reception.

As tactical urban pioneer, the need of reception has the potential to disclose the latent possibilities of the urban built environment to offer spaces both for hospitality, to address housing emergencies, and for social meeting, to sustain the formation of a new society. Such potential should be discovered integrating design experts in the definition of reception urban systems: a strategical evaluation of public underused facilities, as *reception structures*, and their relationship with urban services and public spaces, as *reception infrastructure*, should support a shared comprehension of urban reception scenarios between actors, as basis for planning, also in contingency, participative design processes and/or tactical design actions, involving the becoming community.

Reception should be a strategical space, where controversies can be mediated: after all, only the physical encounter reunites each other in the common humanity.

References

100 Resilient Cities. (2016). Global migration: Resilient cities at the forefront. Strategic actions to adapt and transform our cities in an age of migration. http://action.100resilientcities.org/page/-/100rc/pdfs/Global%20Migration_Resilient%20Cities%20At%20The%20Forefront_DIGITAL%20%28High%20Res%29.pdf. Accessed 15 Oct 2017.

Aravena, A. (Ed.). (2016). Reporting from the front. 15th Mostra Internazionale di Architettura. Marsilio, Venezia.

Associazione per gli Studi Giuridici sull'Immigrazione (ASGI). (2016). Aida country report: Italy. http://www.asylumineurope.org/sites/default/files/report-download/aida_it_2016update.pdf. Accessed 15 Oct 2017.

Baumann, Z. (2005). Vite di scarto. Edizioni Laterza, Bari.

Betts, A. (2016). TED Talks: Our refugee system is failing. Here's how we can fix it. https://www.ted.com/talks/alexander_betts_our_refugee_system_is_failing_here_s_how_we_can_fix_it. Accessed 15 Oct 2017.

Bishop, P., & Williams, L. (2012). *The temporary city*. London: Routledge.

Boano, C. (2017). *The ethics of a potential urbanism*. London: Routledge.

Bracci, F. (Ed.). (2012). *Emergenza Nord Africa. I percorsi di accoglienza diffusa: analisi e monitoraggio del sistema*. Pisa: Pisa University Press.

Brunon-Ernst, A. (2016). *Beyond Foucault. New perspectives on Bentham's Panopticon*. New York: Routledge.

Cachola Schmal, P., Elser, O., & Scheuermann, A. (2016). *Making Heimat. Germany, arrival country*. Berlin: Hatje Cantz.

Cairns, S. (Ed.). (2004). *Drifting: Architecture and migrancy*. London: Routledge.

Casanovas, L., Blanco, I., Galan, G., et al. (Eds.). (2016). *After belonging. The objects, spaces, and territories of the ways we stay in transit*. Zurich: Lars Müller.

Darling, J. (2009). Becoming bare life: Asylum, hospitality, and the politics of encampment. *Environment and Planning D: Society and Space, 27*, 649–665.

Darling, J. (2016). Forced migration and the city. *Progress in Human Geography, 41*(2), 178–198.

Derrida, J. (2001). *On cosmopolitanism and forgiveness*. London: Routledge.

ECRE European Council on Refugees and Exils (2016). Wrong counts and closing doors. The reception of refugees and asylum seekers in Europe. http://emn.ie/cat_publication_detail.jsp?clog=1&itemID=2653&t=6. Accessed 14 October 2017.

EMN European Migration Network (2014). The Organisation of Reception Facilities for Asylum Seekers in different Member States. http://emn.ie/cat_publication_detail.jsp?clog=1&itemID=2653&t=6. Accessed 13 October 2017.

Eurocities. (2015). EUROCITIES statement on asylum in cities. http://nws.eurocities.eu/MediaShell/media/EUROCITIES%20stmt_asylum_May%202015.pdf. Accessed 15 Oct 2017.

European Asylum Support Office (EASO). (2016). Guida dell'EASO alle condizioni di accoglienza: norme operative e indicatori. https://www.easo.europa.eu/sites/default/files/Guidance-on-ReceptionConditions-IT.pdf. Accessed 15 Oct 2017.

European Commission. (2016) Urban agenda for the EU. 'Pact of Amsterdam'. http://ec.europa.eu/regional_policy/sources/policy/themes/urban-development/agenda/pact-of-amsterdam.pdf. Accessed 15 Oct 2017.

FRAME Project (2015). Reception and Detention Conditions of applicants for international protection in light of the Charter of Fundamental Rights of the EU. http://www.refworld.org/pdfid/5506a3d44.pdf. Accessed 14 October 2017.

Freise, H. (2004). Spaces of hospitality. *Angelaki, 9*(2), 67–79.

Garelli, G., & Tazzioli, M. (2013). Arab Springs making space: Territoriality and moral geographies for asylum seekers in Italy. *Environment and Planning D: Society and Space, 31*(6), 1004–1021.

HABITAT III. (2015). HABITAT III issue papers 2 – migration and refugees in urban areas. http://habitat3.org/wp-content/uploads/Habitat-III-Issue-Paper-2_2_Migration-and-Refugees-in-Urban-Areas-2.0.pdf. Accessed 15 Oct 2017.

Hauge, Å. L., Støa, E., & Denizou, K. (2017). Framing outsidedness – Aspects of housing quality in decentralized reception centres for asylum seekers in Norway. *Housing, Theory and Society, 34*(1), 1–20.

Hes, D., & Du Plessis, C. (2015). *Designing for hope*. Abingdon: Routledge.

John-Richards, S. (2014). Asylum and the common: Mediations between Foucault, Agamben and Esposito. *Birkbeck Law Review, 2*(1), 13–36.

Katz, I. (2017). 'The common camp': Temporary settlements as a spatio-political instrument in Israel-Palestine. *The Journal of Architecture, 22*(1), 54–103.

LasciateCIEntrare. (2016). InCAStrati. http://www.cittadinanzattiva.it/files/primo_piano/giustizia/inCAStrati-report.pdf. Accessed 15 Oct 2017.

Latour, B. (2005). *Reassembling the social. An introduction to actor-network-theory*. New York: Oxford University Press.

Lefebvre, H. (1991). *The production of space*. Cambridge: Blackwell.

Manzini, E. (2015). *Design, when everybody designs*. Cambridge: MIT Press.

Medici per i Diritti Umani (MEDU). (2015). Report sulle condizioni di accoglienza. Cara Mineo. http://www.mediciperidirittiumani.org/pdf/REPORT_ACCOGLIENZA_MINEO_Giugno_2015.pdf. Accessed 15 Oct 2017.

Medici Senza Frontiere (MSF). (2015). Rapporto sulle condizioni di accoglienza nel CPSA di Pozzallo. http://archivio.medicisenzafrontiere.it/pdf/Rapporto_CPI_CPSA_Pozzallo_final.pdf. Accessed 15 Oct 2017.

Medici Senza Frontiere (MSF). (2016). Fuori Campo, Richiedenti asilo e rifugiati in Italia: insediamenti informali e marginalità sociale. http://fuoricampo.medicisenzafrontiere.it/Fuoricampo.pdf. Accessed 15 Oct 2017.

Migration Policy Group (MPG). (2016). Scoping paper on the urban reception of refugees. https://ec.europa.eu/futurium/sites/futurium/files/scoping-paper-on-the-urban-reception-of-refugees-.pdf. Accessed 15 Oct 2017.

Ministero dell'Interno. (2016). Piano di Accoglienza 2016. Tavolo di coordinamento Nazionale. http://www.vita.it/attachment/d601c9b0-b314-46ba-b708-d4341546c2d9/ Accessed 15 Oct 2017.

Oswalt, P., Overmeyer, K., & Misselwitz, P. (2013). *Urban catalyst*. Berlin: Dom Pub.

Rabinow, P. (1984). *Space, knowledge and power, in the foucault reader*. New York: Pantheon Books.

Regione Toscana. (2017). Libro Bianco sulle politiche di accoglienza dei richiedenti asilo politico e protezione internazionale. http://open.toscana.it/documents/438246/0/Libro+Bianco+dell%27accoglienza+in+Toscana/c57542bb-9550-4f6e-81dd-e626a57ca65e. Accessed 15 Oct 2017.

Röslmaierm, M. (2016) Exploring the tourism-refugee nexus. From global mobility to local immobility (MA Thesis), Mid Sweden University. http://miun.diva-portal.org/smash/get/diva2:938722/FULLTEXT01.pdf. Accessed 15 Oct 2017.

Rozakou, K. (2012). The biopolitics of hospitality in Greece: Humanitarianism and the management of refugees. *American Ethnologist, 39*(3), 562–577.

Sistema di Protezione per Richiedenti Asilo e Rifugiati (SPRAR). (2013). Rapporto annuale. http://www.cittalia.it/images/file/atlante_sprar_completo_2012_2013.pdf. Accessed 15 Oct 2017.

Sistema di Protezione per Richiedenti Asilo e Rifugiati (SPRAR). (2016) Rapporto annuale. http://www.sprar.it/wp-content/uploads/2017/06/Atlante-Sprar-2016-2017-RAPPORTO-leggero.pdf. Accessed 15 Oct 2017.

Sistema di Protezione per Richiedenti Asilo e Rifugiati (SPRAR). (2017) Manuale operativo. http://www.sprar.it/images/Documenti/Quaderni_servizio_centrale/manuale.pdf. Accessed 15 Oct 2017.

Swiss Refugee Council. (2013). Reception conditions in Italy. https://www.refugeecouncil.ch/assets/news/english-website/131213-osar-report-italy-english.pdf. Accessed 15 Oct 2017.

Szczepanikova, A. (2012). Between control and assistance: The problem of European accommodation centres for asylum seekers. *International Migration, 51*(4), 130–143.

Tan, P. (2016). Architecture in crisis: Exception as a form of decay. *ARQ, 92*, 118–125.

Taran, P. (Ed.). (2016). The sustainable development goals and migrants/migration. http://www.un.org/en/development/desa/population/migration/events/coordination/14/documents/backgrounddocs/GMPA_14CM.pdf. Accessed 15 Oct 2017.

UN HABITAT. (2016). New Urban Agenda. https://unhabitat.org/wp-content/uploads/2017/05/GA-Resolution-A-RES-71-256-New-Urban-Agenda.pdf. Accessed 15 Oct 2017.

UNHCR. (2010). Collective centre guidelines. https://www.globalcccmcluster.org/system/files/publications/doc18990-contenido.pdf. Accessed 15 Oct 2017.

UNHCR. (2015). Global trends, forced displacement in 2015. http://www.unhcr.org/576408cd7.pdf. Accessed 15 Oct 2017.

Venturini, T. (2010). Diving in magma: How to explore controversies with actor-network-theory. *Public Understanding of Science, 19*(3), 258–273.

Yaneva, A. (2012). *Mapping controversies in architecture*. London: Routledge.

Disaster Vulnerability of Displaced People in Rakhine State, Myanmar

Thomas Johnson, Jason von Meding, Thayaparan Gajendran, and Giuseppe Forino

Abstract

In Rakhine State (Myanmar), there has been intercommunal conflict and violence between the Rohingya ethnic minority and the Burmese population for decades. The conflict in 2012 resulted in the Rohingya population of the state confined to certain villages and internally displaced persons (IDP) camps. There are ongoing reports of violence in Rakhine State, including evidence of escalating state-sponsored violence in the 2016/2017 crack-down. Since the 1970s, over 1 million Rohingyas fled the country, and those who stayed live in constant fear. A further half a million people have left Rakhine after resurging violence between August and October 2017. Affected Rohingyas in Rakhine live in makeshift housing. They have no access to basic needs; their trade opportunities are limited and have movement restrictions. These oppressive realities have left Rohingya communities in an exceptionally vulnerable and critically unstable situation. Although these communities are frequently at risk from flooding, they face extremely destructive cyclones from time to time (e.g. Cyclone Nargis in 2008 killed over 100,000 people in Myanmar). The research aims to identify the key contributing factors to cyclone vulnerability in Rakhine State. We studied three different communities that are vulnerable to cyclones: an IDP camp, a village with a predominantly Rohingya population and a village with a predominantly Burmese population. These communities were chosen to ensure that the findings are representative of the various living conditions in Rakhine and represent the unique characteristics of each community. The research project adopted a mixed methods approach, involving case studies with multiple data sources. This paper presents preliminary results from a part of the study: a quantitative analysis of the questionnaire data ($N = 156$) gathered from the three communities. Our data identifies clear differences between the three communities and provides evidence of the key drivers of vulnerability in Rakhine State. The research will make a valuable contribution to the effectiveness of disaster risk reduction efforts in Rakhine State and will contribute to ongoing advocacy effort for the Rohingya people of Myanmar.

Keywords

Vulnerability · IDP · Rohingya · Myanmar · Refugees

T. Johnson · J. von Meding (✉) · T. Gajendran
G. Forino
University of Newcastle, Newcastle, Australia
e-mail: thomas.johnson@uon.edu.au;
jason.vonmeding@newcastle.edu.au;
thayaparan.gajendran@newcastle.edu.au;
giuseppe.forino@newcastle.edu.au

A. Asgary (ed.), *Resettlement Challenges for Displaced Populations and Refugees*,
Sustainable Development Goals Series, https://doi.org/10.1007/978-3-319-92498-4_6

1 Introduction

In the far western corner of Myanmar, bordering with Bangladesh on the west and running along the Bay of Bengal is the state of Rakhine. It is a unique area and differs greatly from the rest of the nation in terms of ethnicities, economy, environment and religion. These factors, along with the implications from the communal violence, have left the population disproportionately vulnerable to floods and cyclones. Natural triggers such as floods and cyclones are common in the region; however, it is the human interaction with these triggers that creates the high frequency of disasters.

Disasters occur due to a combination of vulnerability and hazard. A disaster triggered by an environmental hazard is often mistakenly referred to as a natural disaster. These hazards include events such as volcanic activity, seismic activity, precipitation, microbial pathogen or climatic trends (Kelman et al. 2016). These hazards can become disasters when combined with vulnerability, which refers to the propensity of a system to be harmed. Some of the leading sources of vulnerability are population growth, rapid urbanisation and unplanned land use (Comfort et al. 1999). All of these causes are present in Rakhine, creating an increasingly vulnerable population, and one of the leading causes being the widespread unplanned land use. Whilst vulnerability in the area has root causes to be found in colonial times and religious and political tensions (Ibrahim 2016), such vulnerability has dramatically increased in the last 5 years as a result of the 2012 communal conflict which left an estimated 300,000 people displaced (APHR 2015). Many of these people have since been living in makeshift housing in IDP camps. The people within these camps are lacking basic infrastructure, job opportunities, food and water security, freedom of movement and resource access (including health and medicines). This makes the population extremely vulnerable to the effects of hazards, including frequent floods and cyclones.

Floods are the most common and widespread hazard in the world. They affect millions of people and damage housing, infrastructure and the natural environment (Vojinovic 2015). On a global scale, floods account for more than 55% of all disaster-related fatalities. In the last 20 years floods have caused 153,722 deaths (EM-DAT 2015). Additionally, they account for more than 30% of economic loss from disasters triggered by natural hazards (Hallegatte et al. 2013). Asia has the highest frequency and magnitude of disasters compared to any other continent (Fatemi et al. 2016). The region of Southeast Asia is particularly prone to flood and cyclone casualties. From 1996 to 2016, the region suffered 178,414 casualties caused by floods or storms (EM-DAT 2015). Myanmar alone accounts for 139,205 of these deaths from just 26 major events (EM-DAT 2015). It is possible that Myanmar could face hazards of increased frequency and intensity with the potentially growing effects of climate change. The country ranks 2nd out of 187 countries in the Global Climate Risk Index (UNOCHA 2016) and 9th out of 191 in the Index of Risk Management (INFORM).

The high level of displacement in combination with frequent hazards in Rakhine State creates a unique opportunity to study the effects of displacement on vulnerability. This paper aims to determine the key contributing factors to flood and cyclone vulnerability. To do this, the paper explores the vulnerability of three different communities in Rakhine State with drastically different living conditions. The study uses a mixed method approach, utilising data obtained through household questionnaires and semi-structured interviews. The data from three different communities will be compared to form the discussion. The findings show clear differences between the three communities and provide evidence of the key drivers of vulnerability. The conclusion will highlight the key differences in vulnerability and present a scope for further research in this area.

2 Vulnerability Theory

In order to effectively evaluate the vulnerability of a region, it is important to have a clear understanding of the various definitions and theories. The term vulnerability has many definitions, varying greatly across disciplines and in different regions. For example, within climate research, vulnerability is described as the degree to which a system is unable to cope with the adverse effects of climate change (IPCC 2001), whilst in the environmental hazard context, it is described as the conditions determined by physical, social, economic and environmental factors or processes which increase the susceptibility of an individual, a community, assets or systems to the impacts of hazards (UNISDR 2017). Furthermore, understanding vulnerability is complicated by temporal and spatial changes. Cutter and Finch (2008) highlight the dynamic nature of social vulnerability by demonstrating considerable variability across different regions of the United States over five decades. Hence, applying a generic vulnerability theory on a global scale could be problematic. Rufat et al. (2015) demonstrate in a meta-analysis of 67 flood case studies that the leading challenge in measuring vulnerability is aligning the output metrics with the context in which the vulnerability exists.

The drivers of vulnerability can be divided into three main categories: exposure, susceptibility and resilience. The exposure category includes physical and environmental drivers such as hazard frequency, proximity to urban centres, age of buildings and housing type. The susceptibility category includes socio-economic drivers such as levels of sanitation, waste management, health facilities, insurances, income, education, occupation and age. The resilience category includes institutional drivers such as warning systems, development control, risk awareness, culture and disaster perception (Salami et al. 2017). These drivers can be helpful in assessing vulnerability; however, they do not provide a holistic view. It is crucial to use a combination of quantitative and qualitative methods for assessing vulnerability (Birkmann 2007). Kelman (2009) points out that much research into disaster vulnerability neglects the comprehensive view by assuming that vulnerability is quantitative, objective, absolute and non-contextual. In order to gain a holistic view of long-term vulnerability, it is important to understand that vulnerability is qualitative, subjective, proportional and contextual and is a process rather than a snapshot (Oliver-Smith et al. 2016). This is particularly relevant to this paper, given the unique set of circumstances in Rakhine.

Within the literature, it is commonly agreed that among the most important factors of vulnerability of communities exposed to disaster risk are public health conditions, population density, socio-economic status and gender discrimination (Burton 2010; Li et al. 2010; Zhang and Huang 2013; Wolkin et al. 2015). Recent literature reports that in Myanmar flood risk reduction measures have been implemented in the past decade (Kawasaki et al. 2017); however, almost all the aforementioned factors of vulnerability can be easily identified and are still underexplored. Myanmar has relatively poor public health conditions particularly in rural areas. For example, gender discrimination is ingrained into society; the patriarchal structure, referred to as *phon*, teaches that men are born with power, glory and holiness whilst women are not. This has resulted in a gender hierarchy which often leaves women marginalised, oppressed and discriminated in political, social and religious realms (Gravers 1993).

Each of these sources can be broken down to several different indicators. For example, socio-economic indicators could include household income (Burton 2010), unemployment rates (Roshti 2010), means of transportation (Adger 1999), tenancy status (Zhang and Huang 2013), reliance on agriculture (Muyambo et al. 2017) and number of long-term residents (Cutter et al. 2010). In Rakhine State, unemployment rates (10.4%) are the highest of the whole country (DPMIP 2016). Following the 2012 conflict, many of these indicators have been further negatively affected. This is demonstrated in the primary data of this study and is discussed in the subsequent sections.

Current conditions of individuals and communities strongly characterise their vulnerability; however, Lewis and Kelman (2012) argue that vulnerability is also created by a wide variety of dynamic and long-term social, economic and political processes which are often neglected in top-down international documents. In fact, there are two main clusters of processes that cause vulnerability: endangerment and impoverishment. The endangerment cluster involves processes such as discrimination, displacement and environmental degradation. On the other hand, impoverishment involves processes such as corruption, denial of access to resources and siphoning of public money (Lewis and Kelman 2012).

In Rakhine State, all of these long-term processes are prevalent. The Rohingya in particular fall victim to increased vulnerability fuelled by these factors. For example, their access to resources has been partially denied by the Government through restrictions on trade and movement. Many of the Rohingya within Rakhine have been displaced from their homes for over 5 years with no foreseeable reintegration possibilities. They live in temporary shelter and often rely on relief projects such as food and clothing distributions for survival. There is a clear need for a transition from relief to development projects. The long-term lack of schools, hospitals, infrastructure and employment opportunities is contributing to increased vulnerability in the IDP camps.

Discrimination against this ethnic group can be traced back to before the Burmese independence in 1948 (Ibrahim 2016). After independence, the Rohingya were accepted as one of the 135 official Burmese minorities. However, in 1982 they were stripped of this title and hence their citizenship. In its place, they were issued white cards to confirm their identity and rights within the state of Rakhine. In more recent years, the Rohingya are being stripped of their identity cards and are faced with even stricter restrictions on their movement and trade (Green et al. 2015). This makes travelling to any other state in Myanmar illegal for Rohingya people. Additionally, they face great difficulty in seeking refuge internationally, including in Bangladesh, despite the Rohingya being often incorrectly described as 'Bengali immigrants'.

These conditions have raised attention within the humanitarian sector. UN agencies, the IFRC societies and numerous NGOs are operating in Rakhine to provide aid to people affected by the conflict. Relief projects have always been subject to a number of restrictions imposed by the government and the military. However, these restrictions were exacerbated following violent protests by Nationalist groups against foreign aid in March 2014. During these protests, all foreigners were encouraged to leave the state and many NGO offices were looted and burned. Following the protest, all aid agencies were evicted from central Sittwe (the capital of Rakhine State) and were only permitted to lease land in the designated 'grey area' outside of the city centre. Additionally, the organisation Doctors Without Borders was ordered to suspend all of their operations within Rakhine as they were accused of bias in favour of Muslims (Reuters 2015). It is difficult for organisations to maintain an unbiased appearance given that a majority of humanitarian assistance is needed within the Rohingya population.

3 Methodology

The study adopts a cases study design, investigating three different communities in Rakhine State. The selected communities include two villages and one IDP camp which all differ greatly in living conditions. One of the villages and the IDP camp are populated predominantly by Rohingya people. The other village is populated predominantly by Buddhist Arakanese. The three communities vary in shelter, infrastructure, access to materials and socio-economic factors. This is to ensure that the findings are representative of the diverse living conditions in Rakhine. The three communities are located within 30km of Sittwe Township with populations between 785 and 1500 people. The Arakanese village, Pyin Chaung, has a population of 1500 people making up approximately 180 households. The riverfront village is prone to flood damage during monsoon season. The Rohingya village, Lat Tha Mar, is home to approximately 1090 people making up 120 households. Located on the coast of the Bay of Bengal, the village experiences flood damage every year and cyclone damage approximately every 3 years. The IDP camp, Boe Du Pha, has a population of 785 people making up approximately 90 households. These three communities

Fig. 1 Locations of target communities: Boe Du Pha, Lat Tha Mar and Pyin Chaung (Google Earth 2016)

were chosen for the study as they are faced with a similar scale and frequency of natural hazards. Yet they are hypothesised to differ greatly in vulnerability given the extremely different living conditions.

The fieldwork in the three selected communities was done by the leading author between June and August 2016. To obtain primary data, a mixed method approach was selected. A voluntary questionnaire was submitted to 156 residents (52 from each community), and then semi-structured interviews with 9 residents (3 from each community) were realised (Fig. 1).

The research analyses household questionnaires in relation to themes identified in the literature. The questionnaire was divided into four sections to explore vulnerability in the social, physical, economic and institutional areas.

The adult literacy rate in Rakhine is only 75% and is even lower in women (72%) and in rural areas of the state (70%) (UNICEF 2011); this rate was reflected in the selected villages. Most of the residents in the villages speak either Burmese (the national language) or solely Rohingya. The study was therefore designed to ensure that the data was representative of all languages and literacy levels. A local NGO provided the translation services for both languages. Employees from this organisation provided one-on-one assistance to those who could not complete the questionnaire individually. A translator from the organisation was present throughout the interviews. The NGO provided the translation service free of charge as the findings can assist in their humanitarian project design. The semi-structured interviews had a duration of approximately 20 min. The interview was guided by a set of 11 questions to help identify emerging themes within each community. The first five questions aimed to explore the respondents' previous experiences with disasters. Further questions consider social support structures, ability to self-recover and disaster education. Interviews were recorded and then verbatim transcribed by the leading author.

The data obtained from questionnaires are analysed using the software SPSS. The analysis utilises cross-tabulation and expected counts to identify key patterns and relationships to form the discussion. The qualitative data were analysed in order to retrieve data which supported quantitative analysis.

The results from the quantitative and qualitative analysis are contextualised with the current literature on vulnerability in Myanmar to form the discussions. This forms a pilot study which will inform the design of a larger quantitative study.

4 Findings and Discussion

Table 1 reports the main demographic details of the investigated communities.

Women represented the majority of the respondents, making up 87 of the 156 total participants. Women are often more vulnerable due to fewer employment and education opportunities (Ahsan and Warner 2014; Cutter et al. 2003; Fordham 1998). This was reflected in the data with significant disparities between genders in most categories. A majority of the respondents were married; however, there was a significant number of respondents that reported being widowed. This group is particularly vulnerable and may be more susceptible to flood and cyclone impacts (Cutter et al. 2009).

Most of the respondents reported relying on agriculture or fishing for their livelihood. This can be a source of vulnerability as these jobs are often more susceptible to severe impacts by hazards (Muyambo et al. 2017). In an interview with man who makes a living from fish farming, he says "it can take up to 6 months for the business to return to regular productivity… NGOs will only help repair the housing but do not help to restore businesses". In Pyin Chaung and Lat Tha Mar, a majority (60% and 52%, respectively) of respondents were reliant on fishing and agriculture. Boe Du Pha was less reliant on this industry (8%); however, this is likely because of a high unemployment rate and a dependence on humanitarian aid. Unemployment rates also have a significant impact on overall vulnerability. Aside from the obvious financial benefit, employment can also reduce vulnerability through an increase of social capital and resource access (Sanyal and Routray 2016). In Boe Du Pha 69% of the respondents were unemployed compared to only 17% in Lat Tha Mar and 2% in Pyin Chaung.

Table 1 Demographic snapshot of data

	Pyin Chaung	Lat Tha Mar	Boe Du Pha
What is your gender?			
Male	33%	35%	65%
Female	67%	65%	35%
How old are you?			
18–30	31%	0%	37%
31–40	29%	19%	37%
41–50	25%	25%	25%
51–65	12%	33%	17%
Older than 65	4%	26%	6%
How many children are in your household?			
0	6%	0%	0%
1–3	52%	8%	19%
4–6	31%	46%	73%
7–9	10%	25%	8%
10 or more	2%	21%	0%
Which best describes your employment status?			
Employed full time	2%	6%	2%
Employed part time	40%	21%	29%
Self-employed	54%	54%	0%
Retired	0%	2%	0%
Student	1%	0%	0%
Unemployed	2%	17%	69%
What is your monthly household income? (1350 kyats ≈ $1 USD)			
Less than 10,000	13%	17%	71%
10,001–50,000	15%	2%	0%
50,001–100,000	31%	67%	29%
100,001–150,000	25%	6%	0%
150,001–200,000	2%	6%	0%
More than 200,000	13%	2%	0%

A key difference between the villages is the access to transport. Lat Tha Mar and Boe Du Pha have no access to motorised transport and only 8 respondents have access to a bicycle. This makes it difficult for them to sustain their livelihood and increases immediate risk to disasters as it is much more difficult to seek safety in higher ground or in cyclone shelters (Mazumdar and Paul 2016). Their movement is further inhibited by the poor access roads to the villages; they are congested, rough and easily damaged by floodwaters. Figure 2 demonstrates how easily the roads are damaged from floodwaters. The photo was taken following average rainfall for the month of June

Fig. 2 Main access road to an IDP camp after a small storm (Author Provided)

2016. Meanwhile, the situation is much different in Pyin Chaung with 40% of respondents having access to a motorcycle and main access roads are sealed.

Education levels are extremely low in Lat Tha Mar and Boe Du Pha with a majority of respondents claiming they have no formal education (83% and 73% respectively), whereas in Pyin Chaung 79% of respondents have completed at least primary education. The education levels are particularly interesting when cross-tabulated with gender. Within Lat Tha Mar and Boe Du Pha, almost all women have no formal education (100% and 89%, respectively). A lower education level limits job opportunities and can be a constraint on one's ability to understand warnings and information related to evacuation. Additionally, it is important for women to be educated as they can better teach and train their family and neighbours about the safety measures (Yadav and Barve 2017). In an interview with a Rohingya man living in an IDP camp near Sittwe, he states that "what the community needs from NGOs operating in post-disaster reconstruction is not money but materials and technicians … the money is useless because they lack access to materials and the technical capacity". He also mentions that a key bottleneck to recovery is the government restrictions on local residents and NGOs.

Most participants in the survey claim to have not received any early warnings prior to major floods or cyclones. This was particularly evident in Lat Tha Mar (71%) and Boe Du Pha (65%). Early warning systems can reduce fatalities by allowing people enough time to prepare for evacuation to the nearest shelter (Hossain 2015). However, in the case of Boe Du Pha, it is debatable whether the early warning systems would have a significant effect. Residents of this community report no training to prepare for floods or cyclones and lack any means of transport to evacuate. Additionally, in an interview with a resident of Boe Du Pha, he states that "we do not have access to a cyclone shelter because it is too far away, the closer shelter is still under construction".

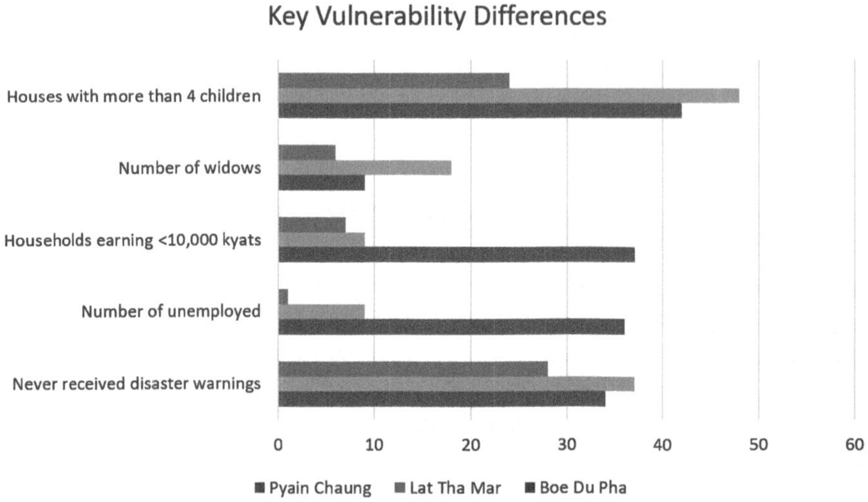

Fig. 3 Notable differences in vulnerability between communities

There are noticeable differences in the size of families between the selected communities. In Boe Du Pha 79% of households have more than 4 children and in Lat Tha Mar 92% have more than 4 children with a notable 21% of households having more than 10 children. This is significant compared to Pyin Chaung with 58% of households having less than 4 children. Additionally, a larger percentage of households in Lat Tha Mar have elderly over the age of 65 (23%). Households with a larger number of children and elderly are more vulnerable as these groups require a greater level of care in disaster situations (Ahsan and Warner 2014). These data are particularly alarming when cross-tabulated with the number of households with a monthly income of less than 10,000 kyats (approximately $8.30 USD). In Lat Tha Mar, 15% of households have more than four children and a monthly income of less than 10,000 kyats. In Boe Du Pha, 55% of households have more than four children and have a monthly income of less than 10,000 kyats. The figures in the IDP camp are higher but are possibly less significant given the dependence on humanitarian aid.

It is commonly acknowledged in the literature that communities with a large number of long-term residents are less vulnerable (Cutter et al. 2010). This is due to an increase of social capital such as societal relationships and memberships of organisations. A large majority (90%) of respondents in Pyin Chaung have lived as part of the community for more than 10 years compared to only 44% in Lat Tha Mar. In Boe Du Pha, no respondents have lived as part of the community for more than 6 years which is to be expected as many of the IDP camps (including Boe Du Pha) were only created following the 2012 conflict (Fig. 3).

5 Conclusions and Scope for Further Research

The study was undertaken to identify and analyse the flood and cyclone vulnerability of three different communities in Rakhine State, Myanmar, using data obtained from a household questionnaire and interviews. Results found that the flood and cyclone vulnerability in the region is very high. Additionally, results reveal that the vulnerability greatly varies between the different common living conditions. Residents of the IDP camp were unsurprisingly more vulnerable in a number of categories compared to the other communities included in the study. Furthermore, the Rohingya village of Lat Tha Mar appeared to be more vulnerable than the Arakanese village of Pyin Chaung. The most notable findings on vulnerability which highlight the disparity between the villages include:

- Transport – Motorised transport is very limited in the region with the data from one community revealing that no respondents had access to private transport. Additionally, the roads leading to two of the communities are easily damaged during storms. This undermines their capacity for sustaining their livelihood.
- Time in community – Long-term residents have a greater opportunity to build social capital to reduce vulnerability. The study revealed large differences between the communities in the number of long-term residents.
- Restrictions on trade/movement – Government restrictions on trade and movement have a substantial effect on disaster vulnerability. It limits livelihood, employment and education opportunities and restricts populations from moving to locations with greater environmental protection and access to resources.
- Education – A lack of education can create vulnerability through limiting employment options and potentially limiting one's ability to understand early warning systems or disaster preparedness training. The study found that education levels were particularly low in the female participants.
- Employment – Employment rates can improve vulnerability through financial stability and an increase in social capital. Additionally, the employment type is important as industries reliant on the natural environment can be more heavily affected by disasters. The study found large differences in employment rates and types of employment between the three communities.

The study has some limitations. The small sample size used in the study has limited the potential findings. Also, although many drivers of vulnerability were identified in the region, the study was unable to link these to their root causes. It is hypothesised that many of these factors are directly affected by the 2012 conflict, the 2016/2017 military crackdown and the ongoing marginalisation and discrimination of the Rohingya population. Notwithstanding the limitations, the study represents a pilot study for further research. It provides a background for further studies exploring disaster vulnerability factors in Myanmar and the differences across individual and communities existing in the variegated cultural and social landscape of the country. Particularly, further research in linking these vulnerabilities to sources derived from marginalisation, conflict and displacement would be valuable for exploring root causes of vulnerability.

Acknowledgements The authors thank Myittar Resource Foundation (MRF) for supporting this project. Their ongoing support helped to gain access to the key villages in the region. Additionally, MRF staff provided the translation services (for both Burmese and Rohingya languages) necessary to complete the surveys and interviews. The authors also thank the local residents who participated voluntarily in the survey and interviews.

References

Ahsan, M., & Warner, J. (2014). The socioeconomic vulnerability index: A pragmatic approach for assessing climate change led risks – A case study in the southwestern coastal Bangladesh. *International Journal of Disaster Risk Reduction, 8*, 32–49.

Adger W Neil, (1999) Social Vulnerability to Climate Change and Extremes in Coastal Vietnam. World Development 27 (2):249–269.

APHR. (2015). The Rohingya crisis and the risk of atrocities in Myanmar: An ASEAN challenge and call to action, ASEAN Parliamentarians for Human Rights, viewed 1 May 2017, http://aseanmp.org/wp-content/uploads/2015/08/The-Rohingya-Crisis-and-the-Risk-of-Atrocities-in-Myanmar-An-ASEAN-Challenge-and-Call-to-Action.pdf

Birkmann, J. (2007). Risk and vulnerability indicators at different scales: Applicability, usefulness and policy implications. *Environmental Hazards, 7*, 20–31 https://doi.org/10.1016/j.envhaz.2007.04.002.

Burton, C. (2010). Social vulnerability and hurricane impact modelling. *Natural Hazards Review, 11*, 58–68.

Comfort, L., Wisner, B., Cutter, S., Pulwarty, R., Hewitt, K., Oliver-Smith, A., Wiener, J., Fordham, M., Peacock, W., & Krimgold, F. (1999). Reframing disaster policy: The global evolution of vulnerable communities. *Environmental Hazards, 1*, 39–44.

Cutter, S., & Finch, C. (2008). Temporal and spatial changes in social vulnerability to natural hazards. *Proceedings of the National Academy of Sciences, 105*(7), 2301–2306.

Cutter, S., Boruff, B., & Shirley, W. (2003). Social vulnerability to environmental hazards. *Social Science Quarterly, 84*(2), 242–261.

Cutter, S., Emrich, T., Webb, J. & Morath, D. (2009). *Social vulnerability to climate vulnerability hazards: A review of the literature, Hazards and vulnerability research Institute*. Final report to Oxfam America, Department of Geography, University of South Carolina, Columbia.

Cutter, S., Burton, G., & Emrich, T. (2010). Disaster resilience indicators for benchmarking baseline conditions. *Journal of Homeland Security and Emergency Management, 7*(1), 1–22.

DPMIP. (2016). The 2014 Myanmar population and housing census: The union report: Occupation and Industry, viewed 5 April 2017, http://www.themimu.info/sites/themimu.info/files/documents/Report_Occupation_Industry_-_Census_Report_V2-B_DOP_Mar2016_ENG.pdf

EM-DAT. (2015). Disaster database, online database, viewed 28 April 2017, http://emdat.be/emdat_db/

Fatemi, F., Ardalan, A., Aguirre, B., Mansouri, N., & Mohammadfam, I. (2016). Social vulnerability indicators in disasters: Findings from a systematic review. *International Journal of Disaster Risk Reduction, 22*, 219–227. https://doi.org/10.1016/j.ijdrr.2016.09.006.

Fordham, M. H. (1998). Making women visible in disasters: Problematising the private domain. *Disasters, 22*(2), 126–143.

Google Earth. (2016). Satellite map, viewed 28 April 2017, https://earth.google.com/web/@20.21056717,92.86386803,2.23527176a,63671.96419361d,35y,0h,0t,0r

Gravers, M. (1993). *Nationalism as political paranoia in Burma: An essay on the historical practice of power*. Surrey: Curzon Press.

Green, P., MacManus, T. & Venning, A. (2015). Countdown to Annihilation: Genocide in Myanmar, International State Crime Initiative, viewed 28 April 2017, http://statecrime.org/data/2015/10/ISCI-Rohingya-Report-PUBLISHED-VERSION.pdf

Hallegatte, S., Green, C., Nicholls, R., & Corfee-Morlot, J. (2013). Future flood losses in major coastal cities. *Nature Climate Change, 3*, 802–806.

Hossain, M. (2015). Analysis of human vulnerability to cyclones and storm surges based on influencing physical and socioeconomic factors: Evidences from coastal Bangladesh. *International Journal of Disaster Risk Reduction, 13*, 66–75.

Ibrahim, A. (2016). *The Rohingyas: Inside Myanmar's hidden genocide*. London: Hurst Publishers.

Intergovernmental Panel on Climate Change. (2001). *Climate change 2001: The scientific basis*. Cambridge, UK: Cambridge University Press.

Kawasaki, A., Ichihara, N., Ochii, Y., Acierto, R. A., Kodaka, A., & Zin, W. W. (2017). Disaster response and river infrastructure management during the 2015 Myanmar floods: A case in the Bago River Basin. *International Journal of Disaster Risk Reduction, 24*, 151–159.

Kelman, I. (2009). Understanding Vulnerability to Understand Disasters, viewed 28 April 2017, http://www.island vulnerability.org/docs/vulneres.pdf

Kelman, I., Gaillard, J., Lewis, J., & Mercer, J. (2016). Learning from the history of disaster vulnerability and resilience research and practice for climate change. *Natural Hazards, 82*, 129–143.

Lewis, J. & Kelman, I. (2012). The good, the bad and the ugly: Disaster risk reduction versus disaster risk creation. *PLOS Currents Disasters*. https://doi.org/10.1371/4f8d4eaec6af8.

Li, F., Bi, J., Huang, L., Qu, C., Yang, J., & Bu, Q. (2010). Mapping human vulnerability to chemical accidents in the vicinity of chemical industry parks. *Journal of Hazardous Materials, 179*, 500–506.

Mazumdar, J., & Paul, S. (2016). Socioeconomic and infrastructural vulnerability indices for cyclones in the eastern coastal states of India. *Natural Hazards, 82*(3), 1621–1643.

Muyambo, F., Jordaan, A., & Bahta, Y. (2017). Assessing social vulnerability to drought in South Africa: Policy implementation for drought risk reduction. *Jamba: Journal of Disaster Risk Studies, 9*(1), 326. https://doi.org/10.4102/jamba.v9i1.326.

Oliver-Smith, A., Alcantara-Ayala, I., Burton, I. & Lavell, A. (2016). Forensic Investigations of Disasters (FORIN): a conceptual framework and guide to research, viewed 7 October 2017, http://www.irdrinternational.org/wp-content/uploads/2016/01/FORIN-2-29022016.pdf

Reuters. (2015). 'MSF resumes work in Myanmar state after government ban' viewed 28 April 2017, http://www.reuters.com/article/us-myanmar-msf-idUSKBN0KU22F20150121

Roshti, A. (2010). Social Vulnerability Assessment to Earthquakes in Cities. Case study: Zanjan city. Municipal and regional researches and Studies, 7, 71–90.

Rufat, S., Tate, E., Burton, C., & Maroof, A. (2015). Social vulnerability to floods: Review of case studies and implications for measurement. *International Journal of Disaster Risk Reduction, 14*, 470–486.

Salami, R., von Meding, J., & Giggins, H. (2017). 'Urban settlements' vulnerability to flood risks in African cities: A conceptual framework. *Jamba: Journal of Disaster Risk Studies, 9*(1), 370. https://doi.org/10.4102/jamba.v9i1.370.

Sanyal, S., & Routray, J. (2016). Social capital for disaster risk reduction and management with empirical evidences from Sundarbans of India. *International Journal of Disaster Risk Reduction, 19*, 101–111.

UNICEF. (2011). MMR_MNPED/SIDA/UNDP/UNICEF,Integrated Household Living Conditions Survey in Myanmar, Poverty Profile, 2009-2010_2011.

UNISDR. (2017). Vulnerability, viewed 26 October 2017, http://www.preventionweb.net/english/professional/terminology/v.php?id=508

UNOCHA. (2016). Myanmar: Natural disaster risks and past events, viewed 28 April 2017, http://reliefweb.int/report/myanmar/asia-and-pacific-weekly-regional-humanitarian-snapshot-16-22-may-2017

Vojinovic, Z. (2015). *Flood risk: The holistic perspective: From integrated to interactive planning for flood resilience*. London: IWA Publishing.

Wolkin, A., Patterson, R., Harris, S., Soler, E., Burrer, S., & McGeehin, M. (2015). Reducing public health risk during disasters: Identifying social vulnerabilities. *Journal of Homeland Security and Emergency Management, 12*, 809–822.

Yardav, D., & Barve, A. (2017). Analysis of socioeconomic vulnerability for cyclone-affected communities in coastal Odisha, India. International Journal of Disaster Risk Reduction, 22, 387–396.

Zhang, N., & Haung, H. (2013). Social vulnerability for public safety: A case study of Beijing, China. *Chinese Science Bulletin, 58*, 2387–2394.

A Conceptual Framework for Understanding Transformation: Transformative Adaption of Refugees in Nakivale Refugee Settlement

Oluwadunsin Ajulo, Jason von Meding, and Patrick Tang

Abstract

Transformation has been identified as a better way to adapt to disasters and build resilience. This concept advocates for a paradigm shift in the way things are normally done in communities and governance and for equitable distribution of wealth and justice in communities. There are some examples of communities within which transformation is evident, and these changes have been observed to occur in three dimensions: political, social and technological. The aim of this paper is to consider a framework for transformative adaptation and to understand the practicality of transformation among displaced populations. This is important because study has shown that transformation, although recommended by researchers for adaptation, has so far been ambiguous and difficult in its implementation. Among the factors that determine this, there is no clear-cut standard upon which transformation is judged. The paper is based on secondary data sources from book, journals and Internet sources. We consider the case study of a refugee settlement in Uganda that reflects characteristics of transformative adaptation. Through this lens we develop a conceptual framework for understanding the transformation process in this context. We argue that the transformation starts through the emergence of knowledge, which is a form of social adaptation. This serves as bedrock for social, political and technological transformations. Further studies are needed to develop a basis upon which the success of transformation can be measured.

Keywords

Transformation · Adaptation · Resilience · Refugees

1 Introduction

Human lives, social ties and the natural and built environment are constantly subjected to the disaster impacts in its various forms. Communities suffer losses due to these impacts, some of which are quantifiable while the others are not. There has been a lot of attention on the issue of climate change, but according to Ziervogel et al. (2016), the current situation is in fact far worse than what it used to be. Exposure to hazards is not evenly spread across communities and likewise is the

O. Ajulo (✉) · J. von Meding · P. Tang
University of Newcastle, Newcastle, Australia
e-mail: Oluwadunsin.ajulo@uon.edu.au; Jason.vonmeding@newcastle.edu.au; Patrick.tang@newcastle.edu.au

A. Asgary (ed.), *Resettlement Challenges for Displaced Populations and Refugees*, Sustainable Development Goals Series, https://doi.org/10.1007/978-3-319-92498-4_7

ability to cope with the negative impacts of hazards on communities. Research indicates that developing countries are at the most risk of disaster risk (MRG 2002; UNISDR 2012) despite their very negligible contribution to greenhouse gas emission (Mearns and Norton 2010).

The impacts of climate change are imminent and happening already, and according to Dow et al. (2013), communities all around the world are beginning to reach limits in their capacity to adapt to climate change and variability which are associated with increasing losses, behavioural changes and migration. The persistent increase in the occurrence of disasters is posing a big threat on the implementation of sustainable development and has resulted in massive displacement of population (UNISDR 2001). Disasters are responsible for a large percentage of global migration, and sometimes places where displaced people arrive at have associated vulnerabilities and risks (Harrison and Williams 2016). Although environmental factors are a main influence on human migration, they are however shaped by complex social, economic and cultural processes within communities (Raleigh and Jordan 2010). In 2015, countries around the world made an agreement to adopt a new set of goals for the earth's sustainability, the Sustainable Development Goals (SDGs). Among these are the goal to end poverty, to reduce inequalities and to allow for peace, justice and strong institutions (United Nations 2017). Poverty, inequality and injustice are major characteristics associated with displaced population and refugees.

Adaptation has been identified as the appropriate response to climate change. Coping, incremental and transformative approaches are three distinct approaches to adaptation (European Environment Agency 2016). Beyond adaptation, the transformative adaptation approach considers long-term impacts of climate change and *transforms* the environment on political, economic and technological scales (Lonsdale et al. 2015). O'Brien and Sygna (2013) explained that transformation occurs in the practical, political and personal spheres. Transformation is a process that leads to fundamental positive changes in the social and political order, addressing systemic issues that might otherwise lead to vulnerability. It is characterised by the redefinition of societal objectives by the various stake holders and is a necessary ingredient for effective adaptation (Dow et al. 2013; Dinshaw 2014).

The desired change and resilient management of the environment can be achieved by setting transformational adaptation strategies in motion (Dow et al. 2013). Transformational change is vital for effective adaptation because other adaptation strategies have been very limited in their ability to protect the most vulnerable and tend to tackle disasters for only short periods (Dinshaw 2014).

Transformation in literature has also been identified as a product of accumulated incremental adaptive responses over a period of time (Kates et al. 2012). In this paper, transformation will sometimes be referred to as adaptive responses. It is important to note that although they are referred to as adaptive responses, these responses are considered in this paper because they resulted in transformation. Changes associated with transformative adaptation have greater depth and scale in comparison to incremental or coping adaptation. In response to disasters and climate-induced risks, an increasing wealth of literature acknowledges migration as a way of adapting (Ferris 2014; Martin et al. 2013). Vulnerability and exposure to climate change impacts fundamentally relate to inequalities, discrimination and uneven development processes (Demetriades and Esplen 2010).

Three dimensions to transformative adaptation are identified in this research, technological, social and political transformations, and the broad concept of transformative adaptation will be examined using as a case study the Nakivale (Uganda) refugee settlement.

2 Theoretical Position

Many theories have been developed in the study of climate change, resilience, vulnerability, adaptive capacity and adaptation. Adaptation is a broad concept that encompasses the totality of human living. Adaptation has been identified to have social and economic implications and result in behaviour and lifestyle change, political change and technological change; hence, the variables to be considered in adaptation research are dependent on the aim of the research. Brown et al. (2011) emphasised the lack of a unified and singular approach to understanding and framing adaptation; instead, divers approach have their focus and areas of strengths and weaknesses.

Resilience is a concept used in the understanding of adaptation. The concept is centred around the notion that the resilience of a community (the unflinching characteristic of a community in the face of disturbance(s) giving the community the ability to persist and continue with her usual activities) determines its adaptive capacity, hence adaptation. In simpler terms, the more resilient a community is, the better chance the community has to adapt. In literature, both quantitative and qualitative approaches are used in understanding resilience and adaptation.

The initial reference to resilience identified in literature is in the field of ecology and has subsequently been used in fields like geography, engineering, medicine, psychiatry and psychology and in disaster risk reduction studies (Alexander 2013; Weichselgartner and Kelman 2015). Findings from the research carried out by Lewis (2013) in Wittenberge suggested that resilience is a function of preconditioned innate abilities which is further affected by available resources. Additionally, the author stated that favourable political and social conditions, ability and capacity of population, clear and common goals and good planning are preconditions for building resilience.

Some researchers in environmental science and disaster management have defined resilience as the capacity to recover from shocks. Resilience is sometimes understood as a function of absorptive, adaptive and transformative capacities (Bene et al. 2012). Dovers and Handmer (1992) argue that resilience covers resistance to change, incremental adjustments and flexibility to change. Resilience is the personal, community and economic characteristic the exposure unit has to adapt and recover from shocks (Regional Australia Institute 2013).

Bene et al. (2012) argued that resilience can be good or bad and hence should not be used a normative concept for government interventions which may result in abuse and negligence of its potential disadvantages. Lewis (2013) however argued that resilience does not grow of its own accord, neither can it be demanded but a product of good governance and authoritative direction. This led to a more comprehensive definition for *good resilience* given by Bene et al. (2012) as "the ability of a system to accommodate positively adverse changes and shocks, simultaneously at different scales and with consideration of all the different components and the agents of the system, through the complementaries of its absorptive, adaptive and transformative capacities" (p. 48).

The transition theory is an important theory used in the study of climate change adaptation. In literature, transition and transformation are sometimes used as synonyms, which may create an assumption that transition theory is well-suited for the study of transformative adaptation. This idea was discussed in detail by Rotmans et al. (2001). They defined transition as transformation processes that bring about fundamental changes to a society (Rotmans et al. 2001). However, the article highlighted a distinction between transition and transformative adaptation by seeing transition as a planned and flexible regime shift that is not necessarily triggered by disaster risk. This theory is limited in its application to the understanding of transformative adaptation for the following reasons:

1. It encapsulates societal shift of regimes due to new inventions and ideas or for the purpose of climate change mitigation.
2. Due to the fact that it does not address any existing environmental risk, the theory does address the time factor because transition does not usually occur with a sense of urgency. Those changes are addressed as long-term goals (Loorbach 2010).
3. The main and only actor identified is the government which indicates that transitions are mainly government initiative to improve livelihood although the need for public support

was emphasised since they are at the receiving end.

The theoretical positioning of the study is derived from the anatomy of adaptation by Smit et al. (2000), the action theory of adaptation developed by Einsenack and Stecker (2010) and various perspectives on transformation/transition/transformative adaptation. Transformative adaptation is a concept that stems out of adaptation seeking to take advantage of disaster risks as opportunities for change and long-term adaptation. These frameworks were adopted and modified, bearing in mind the meaning of transformative adaptation, to address all the variables involved in transformative adaptation.

Smit et al. (2000) stated that a good understanding and implementation of adaptive responses can be achieved by addressing three core questions: adapt to what, who or what adapts and how does adaptation occur? The three main variables in evaluating adaptation according to Smit et al. (2000) are stimuli, system and cost.

1. *Stimuli*: Wamsler and Brink (2015) defined stimuli as the risk drivers within natural and societal systems. Smit et al. (2000) explained that stimuli in adaptation can be climate, an extreme weather event, recurring abnormal climatic conditions, perceived changes in climate and impacts and effects of climate. In addition to identifying the stimuli, the authors proposed that speed of rate of change and the frequency of extreme events, predictability of future changes, prediction period and spatial characteristics of stimuli (whether it affects a few local settlements or a wider area) are stimuli-related factors that influence adaptation and adaptation response.
2. *System*: A system in adaptation can be referred to as the human, social, ecological, economic or political entities that undergo some changes in the adaptation process (Smit et al. 2000). The authors acknowledged the fact that a party in the adaptation process may only be influential in making changes by ensuring the *system* affected by the *stimuli* is able to adapt but were not however clear if such parties are part of the *system* in the adaptation process. Additionally, Smit et al. (2000) stated that factors such as sensitivity, vulnerability and adaptability are characteristics of the system that influences the occurrence and nature of adaptation.

   Einsenack and Stecker (2010) further subdivided this group into:
   (i) *Operator*: This a general term carefully used by the authors a collective term to describe the social entity(ies) involved in responding to the stimulus. Under this category are the actors; however, not only the actors fall into this category. The authors define actors as the social entities that make conscious efforts (actions) in responding to stimuli, while other operators may be involved in other social phenomena not carried out purposefully in response to stimuli (processes).
   (ii) *Receptor and Exposure Unit*: This is the social system or biophysical entity to whom the actions of the actor is addressed. The exposure unit includes the social system or biophysical entity that is susceptible to stimuli. The exposure unit is usually a subcategory of the receptors

   The two categories of systems are an intrinsic part of the system and variables needed to be studied in the analysis of adaptation.
3. *Cost*: Adaptation costs relate to the assessment of the impact of the adaptation stimuli—benefits and damages (Smit et al. 2000).

   Einsenack and Stecker referred to cost as means. Means encompasses all the resources needed by an operator in responding to stimuli. This includes financial, knowledge, information, intellectual, legal and other resources needed (Einsenack and Stecker 2010). Means used in adaptation analysis are subdivided into the following:
   (i) Available means: the resources accessible and at the disposal of the operator in response to stimuli
   (ii) Employed means: the resources used by the operator for adaptation
   (iii) Necessary means: the resources required for optimum adaptation

Another term used by Frankenberger et al. (2013) in describing means is community asset. They defined it as tangible and intangible resources that are necessary for meeting the need of a community which affects its capacity to adapt. In measuring the means at a community's disposal, the extent and quality of asset, the controlling institutions, population ratio of asset ownership and the effect the assets have on the community's livelihood are factors to be considered (Frankenberger et al. 2013).

Adaptation actions are dependent on the control the operator has over the available resources. Actions and resources that cannot be controlled by the operator are called conditions (Einsenack and Stecker 2010).

3 Conceptual Framework of Transformative Adaptation

Transformation is a common word used in our day-to-day lives which delineates a significant change. Transformative adaptation with respect to climate change can easily be termed as making conscious changes to our usual way of living and interaction with the environment to foster our ability to live with the constantly changing climate. By this understanding of transformation, it can be said that for a community to be said to have transformed:

(i) There has to be an evident change.
(ii) Change should have been triggered by something (e.g. climate change).
(iii) The change was an informed decision (will and information).
(iv) It involved a process (time).
(v) The process was influenced and monitored by someone (systems and stakeholders).

This research focuses on understanding transformative adaptation by carrying out case studies on a transforming communities. For this to be done, it is important to have variables to be measured to know what it takes for transformation to occur and transfer the knowledge to other communities that need to undergo transformation. A framework for measuring transformative adaptation was developed for the study (Fig. 1), and this framework is a product of the influence of existing literature and theories on adaptation, transformation and resilience (Einsenack and Stecker 2010; Smit et al. 2000; Frankenberger et al. 2013; Colloff et al. 2016) and authors' own idea of transformation. Researchers like O'Brien and Sygna (2013) and Colloff et al. (2016) have emphasised the relationship between the societal values and behaviour, political will and backing and technology in transformative adaptation. For the purpose of this study, transformative adaptation was examined under the variables below.

1. *Trigger*: Every form of change is triggered by something. In the case of adaptation to climate change, the trigger may be climate, the impacts of the change and extreme weather events. Other authors have referred to triggers as stimuli (Smit et al. 2000; Einsenack and Stecker 2010; Wamsler and Brink 2015). Resilience Alliance (2010) described triggers as disturbances, anything that disrupts the proper functioning of a system for which resilience, adaptation or transformation is required.
2. *Operators*: These are the stakeholders involved in influencing the transformation process. They may be individuals, professionals, government agencies and policymakers.
3. *Receptors*: The receptor is the human component exposed to the effect of the trigger. The characteristics of the receptors will be measured based on these three concepts: adaptive capacity, transformative capacity and resilience.
4. *Exposure unit*: Exposure unit are the biophysical and social components affected by the trigger. This includes built environment, natural components and social relations that may be affected by the trigger. Exposure unit will be measured based on sensitivity and vulnerability.
5. *Change*: This is the product of the transformation process. Lonsdale et al. (2015) opined that transformation occurs on political, economic and technological scales, but for this

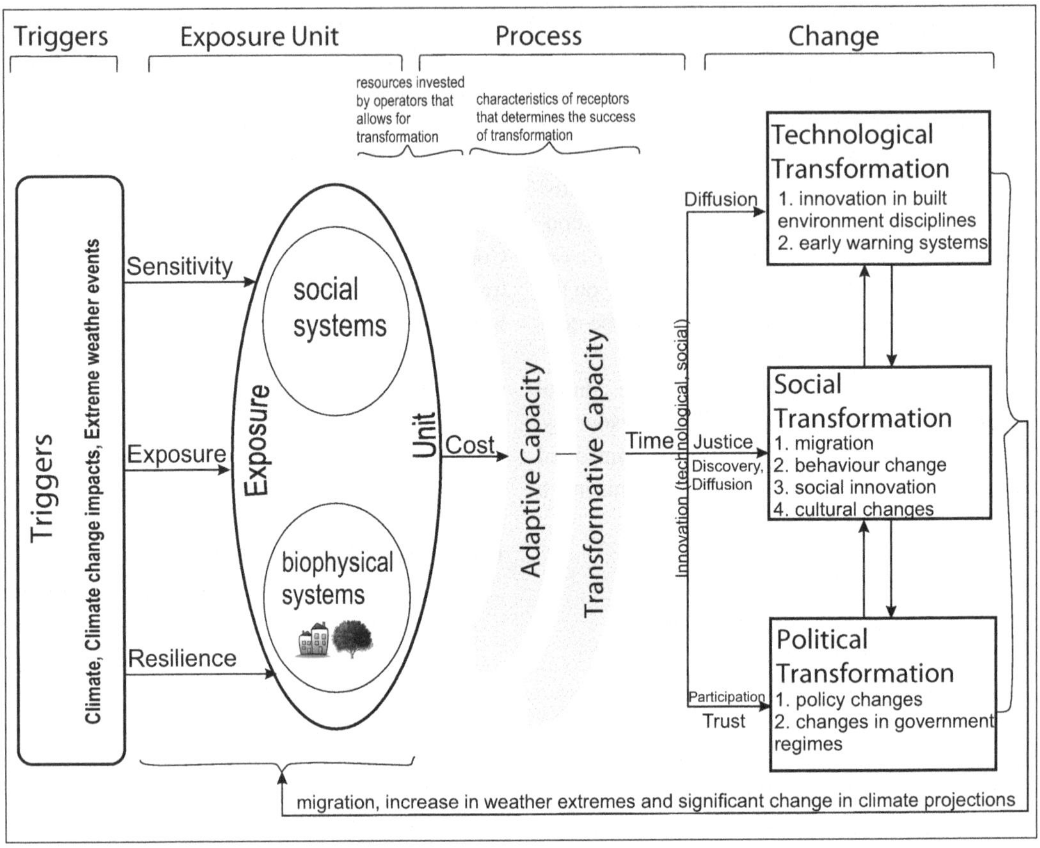

Fig. 1 A conceptual framework for transformative adaptation

research, social transformation will be measured rather than economic transformation. In this research, the economic scale of a community is seen as a part of governance and the social scales.

6. *Process*: The process of transformation is one of the most important factors in transformative adaptation because it determines the outcome. This research will investigate the transformation process and determine if transformation is phased or simply continuous, the stages involved in transformation and when and what happens in the process of transformation.
7. *Time*: This research will investigate how much time transformative adaptation takes.
8. *Cost*: The cost of adaptation will be measured in terms of opportunity cost of the social and political scales in achieving transformation. The financial, intellectual and human investments in the process will also considered under cost.

3.1 Technological Transformation

More often than before, climate change adaptation has been integrated in daily thinking, and this has resulted in an outburst of developments over the past few years (Rutherford et al. 2016). Technological transformation was described as the practical sphere of transformation, usually regarded as the core, because the results are tangible and visible (O'Brien and Sygna 2013). Technological transformative adaptations are new discoveries, incremental improvements in technology, government support and unprecedented low-probability climate change risks (Wilbanks 2011).

3.1.1 Technology

In simple terms, technology can be referred to as inventions, products and application of knowledge. The popular view of technology is in its tangible form. Skolnikoff (1993) posited that technology means more than hardware but a broad knowledge base that is peculiar with the creation or invention of a product (hardware).

3.2 Political Transformation

Policies are outcomes of governance actions or inactions that influence the interaction between humans and the ecosystem (Orach and Schluter 2016). Policies are an important part of governance. Orach and Schluter (2016) defined social transformations as basic changes in structures and processes within a system brought about by policy changes. This is achieved by understanding political processes to identify processes that promote or impede change and work with them. Savini and Dembski (2016) suggested that politics is an integral part of planning because policies are made in response to social demands.

Transformation is a political process, and according to Manuel-Navarrete and Pelling, political influences on adaptation happen in political spaces. These political spaces are defined as social spaces for reflecting, contesting and acting against inequality and vulnerability and opening up choices for better policy formulation (Manuel-Navarrete and Pelling 2015).

3.3 Social Transformation

Social transformations are gradual or sudden changes in societal attributes, values, norms and culture (Zbarauskaite et al. 2015). Social transformation is a fundamental shift in the organisation of a community that transcends continuous incremental changes (Castles 2010). Khondker and Schuerkens (2014) emphasised the contrast between social transformation and social change when they described social transformation as fundamental societal changes and social change as gradual and incremental changes in a society which occurs over a period of time. An important concept to be considered in the study of social transformation is environmental justice which conceptualises the relationship between the environment, the power relations and the social aspects of a community (Breukers et al. 2017). For every technological breakthrough, there is a resultant effect on culture and cognition. Messner (2015) stated that there is an established relationship between technological change and cultural changes.

Literature indicates a linkage between environmental justice and social transformations with regard to climate change and disasters. The components of environmental justice indicated by Breukers et al. (2017) and the indicators of social change suggest that more justice in the society will lead to a positive social change. Castles (2010) identified migration as one of the indicators of social change. He categorised migration into two: internal migration (relocation within a country) and international migration (relocation to another country). Reser et al. (2012) indicates that psychological adaptation variables are indicators of social adaptation. Some of the variables identified are trust, connection to the environment, behaviour, risk perception, belief and concern for climate change, responsibility and willingness to act and objective knowledge.

4 The Concept of Global Justice

For decades, environmental justice has been a topic for environmental researchers and activists (Davoudi and Brooks 2014). Cameron (2012) opined that climate change has a strong connection to past and present colonial practices, more so, colonial activities such as evicting indigenous dwellers for activities that emits greenhouse gases. There is an existing inequality in the access people have to processes that produce greenhouse gases, and this in turn is responsible for poverty and inequality among population (Ribot 2013). Rich countries consume about three-quarters of the world's resource and have a per capita consumption that is about 15–20 times that of poor countries (Trainer 2010). Those with minimum and almost negligible contribution to climate change are made to suffer and adapt to the effects of climate change. The poor in developing countries with very little contribution to climate change suffer from the adverse effects of climate change (Mearns and Norton 2010). They described climate change as a multiplier of existing vulnerabilities and a true definition and reflection of the global justice issues of our time.

Beyond the global context of justice, there are evidences of injustice even within countries and communities. Political systems in most communities have a way of marginalising a certain group of people especially the poor. As stated by Ribot (2010), those in political position tend give little priority to the weak within society. He posited that in order to get rid of the bias against the poor in the society, approaches to vulnerability analysis should be pro-poor.

Breukers et al. (2017) stated that most literature on justice focuses purely on equity in the distribution of resources rather than examining other constitutive elements of justice. Five dimensions or constituents of justice were identified in literature (Breukers et al. 2017; Davoudi and Brooks 2014):

(i) Distribution: it relates to fairness in sharing the available resources.
(ii) Recognition: it relates to social structures with regard to who matters in the society.
(iii) Participation: it relates to the rights individuals have to engage in decision-making and governance.
(iv) Capabilities: it refers to abilities, freedom and opportunities members of a community are exposed to.
(v) Responsibility: it refers to the duties of members of a community to one another and to the environment.

Communities around the globe are making bottom-up efforts to put these elements of justice in place. These transformations are triggered by environmental, political and social factors leading to change in beliefs, behaviour and culture.

5 Case Study: Nakivale Refugee Settlement, Uganda

Sub-Saharan Africa houses about two million refugees displaced by conflict (O'Laughlin et al. 2013). Nakivale refugee settlement is one of the oldest and largest in Uganda, accommodating a large influx of refugees. It was established in 1960 to house refugees from Rwanda and is situated in the southwestern part of the country. The settlement which spans about 28 square miles now accommodates about 56,000 displaced populations from Democratic Republic of Congo, Somalia, Rwanda and Burundi (Adam-Bradford et al. 2016; O'Laughlin et al. 2013).

5.1 Transformative Adaptation in Nakivale

This case study was selected based on peculiar traits in the Nakivale refugee settlement that is representative of the transformation process. The case study will be analysed using the lens of the conceptual framework.

Transformative adaptation in social settings are characterised with systemic changes aimed at reducing inequalities and fostering social inclusion. Nakivale refugee settlement is well-regarded for the inclusive treatment of refugees. Living as a refugee is often associated with difficulties, and in the case of Nakivale, access to food is a major challenge for residents (O'Laughlin et al. 2013).

The political transformation (changes in policies) and social transformation (education, social inclusion and participation) have resulted in significant changes in behaviour, culture and values of the residents. We will discuss some of the key observations about Nakivale in relation to transformative adaptation.

5.1.1 Greening Innovation

Greening innovation is a change in the living processes that creates more ecologically resilient landscape and living spaces (Adam-Bradford et al. 2016). Associated with climate change risks is the problem of food shortage. Eradication of hunger is one of the sustainable development goals (SDGs) of the United Nation, and this places responsibility on not just government but individuals likewise. Responsibility is one of the dimensions of justice standing Breukers et al. (2017) which is what has been demonstrated by the Nakivale refugee settlement in Uganda. A common problem in refugee camps in Uganda is food insecurity which is partly minimised through subsistence farming (Adam-Bradford et al. 2016), and this practice is promoted and adopted as a policy for reducing food insecurity (Adam-Bradford 2016). According to Adam-Bradford (2016), agroforestry is encouraged and used in Nakivale which provides energy and reduces soil erosion. Beyond the obvious advantages this practice has on the environment, growing food enables refugees to earn a living and promotes sustainability (Adam-Bradford et al. 2016).

5.1.2 Social Independence and Innovation

Social innovation is the new way of creating social change (van der Have and Rubalcaba 2016). It is a rebirth of the social aspects of innovations created from preconceived actions aimed at encouraging social change and creating new pathways for social development. The concept covers changes in social systems and solutions to social problems. Activities such as adult education and English literacy classes are conducted to overcome the existing language barrier within the settlement and create oneness among residents thereby facilitating community engagement, productivity and education. Social innovation within the settlement includes education, initiatives for leadership and support, community-based organisations for human rights and empowerment of residents and music groups, among others (Betts et al. 2015).

Despite the challenges of being away from home and separation from family ties which some of the refugees experience, community organisations and religious groups and meetings have been harnessed by members to create social ties and cohesion among themselves (Tollebrandt and Wrede 2013). Also, the use of social incentives is used by the government and NGOs to encourage positive attitudes towards social well-being among residents. According to O'Laughlin et al. (2013), incentives such as food, cultivating land and shelter are used to encourage health awareness of residents.

The power structure within the settlement makes it possible for refugees to be leaders of the settlement, hence giving them the power to raise issues and solve problems (Tollebrandt and Wrede 2013). They added that the high proportion of refugees within the settlement gives the refugees more right and power to influence decisions.

5.1.3 Policy Changes

A very influential aspect of governance is policies. Policies are representations of political, ethical and programmatic viewpoints that are used for decision-making and resource allocation (Tableman 2005). Changes in policies are not simply based on needs and scientific evidence but also on politics and power relations (Hallsworth et al. 2011). According to Tollebrandt and Wrede (2013), up until 2006, the intake and administration of refugees in Uganda were guided by the Control of Alien Refugees Act which was enacted in 1960 by the British. The Refugee Act was however enacted in 2006 by the Ugandan government. The act gives refugees the right of movement, the right to education and the right to earn a living (The Refugees Act 2006).

More than is possible through social cohesion, policies play a more important role in wielding social power to residents (Tollebrandt and Wrede 2013) especially refugees who have probably lost a sense of identity. The significant change in policy and systemic structures allowing for access to housing and land has resulted in an increase in the social power and social independence of refugees (Patton 2016).

5.2 What Do We Learn from Nakivale?

The case study indicate how policies and education can be used a means to shape human behaviour. According to Akello and Dobbs (2009), the Ugandan Refugee Act is a model for Africa and has resulted in major changes in the rights of refugees. The refugee policy has played an important role in providing a favourable condition for refugees in Nakivale settlement. Agroforestry in the Nakivale refugee settlement was further facilitated through the introduction of policies to promote that practice. Refugees in Nakivale are not just surviving but living and making things happen for both themselves and the environment. Community groups and organisations have also helped in improving the social wellbeing of refugees. This has created a sense of community and participation leading to empowerment, enlightenment, cultural changes and change in value of residents.

The complex and interconnectedness of the various parts of the society is clearly spelt out from the case study. Politics, power, knowledge, social networks and communication have played significant roles in the transformative adaptation of refugees in Nakivale. An important lesson here is that individuals, organisations and governments equally have the responsibility of creating environments for transformation to thrive.

6 Conclusion

The transformation process is undoubtedly a complex process. Having studied theoretical perspectives and a real-life case of transformation in Nakivale refugee camp, we begin to consider a new conceptual framework for understanding the transformation process. The framework can be used to examine the complexity of the transformation process. From the case study, it is clear that transformation does not occur in a single dimension but basically in the social, political and technological dimensions.

The case study indicates that transformation occurs based on the social (behaviour, culture, knowledge), political (political will and organisation) and economic (available resources) peculiarities of a country or community. However, from this study, it can be inferred that transformation starts through social adaptation. For change to occur, knowledge is important. Transformation starts by a few people knowing what the problem is and finding means within their capacity to solve it. Knowledge, will and resources can lead to one or all of political (knowledge and will of those in power), social (knowledge, educating others about the risks) and technological transformations (applying the knowledge for innovation with the use of resources).

References

Adam-Bradford, A. (2016). Agroforestry for refugee camps. *Agriculture for Development, 28*(5), 29–31.

Adam-Bradford, A., et al. (2016). *Transforming land, transforming lives: Greening innovation and urban agriculture in the context of forced displacement*. Dallas, USA: The Lemon Trust.

Akello, V., & Dobbs, L. (2009). Uganda's Progressive Refugee Act becomes operational. [Online]. Available at: http://www.unhcr.org/en-au/news/latest/2009/6/4a3f9e076/ugandas-progressive-refugee-act-becomes-operational.html. Accessed 26 Oct 2017.

Alexander, D. (2013). Resilience and disaster risk reduction: An etymological journey. *Natural Hazards and Earth Sysytem Sciences, 13*, 2707–2716.

Bene, C., et al. (2012). Resilience: New Utopia or New Tyranny? Reflection about the potentials and limits of the concept of resilience in relation to vulnerability reduction programmes. s.l.: IDS Working Papers.

Betts, A., Bloom, L., & Weaver, N. (2015). 3 Ways refugee innovation is being supported in Uganda. [Online]. Available at: http://www.unhcr.org/innovation/3-ways-refugee-innovation-is-being-supported-in-uganda. Accessed 20 Oct 2017.

Breukers, S., et al. (2017). Institutional 'Lock-out' towards local self-governance? Environmental justice and sustainable transformations in Dutch social housing neighbourhoods. *Energy Reserach and Social Science, 23*, 148–158.

Brown, A., et al. (2011). *Managing adaptation: Linking theory and practice*. Oxford, UK: UK Climate Impacts Programme.

Cameron, E. (2012). Securing Indigenous politics: A critique of the vulnerability and adaptation approach to the human dimensions of climate change in the Canadian Arctic. *Global Environmental Change, 22*, 103–114.

Castles, S. (2010). Understanding global migration: A social transformation perspective. *Journal of Ethnic and Migration Studies, 36*(10), 1565–1586.

Colloff, M., et al. (2016). An integrative research framework for enabling transformative. *Environmental Science and Policy, 68*, 87–96.

Davoudi, S., & Brooks, E. (2014). When does unequal become unfair? Judging claims of environmental injustice. *Environment and Planning, 46*, 2686–2702.

Demetriades, J., & Esplen, E. (2010). *The gender dimensions of poverty and climate change adaptation* (pp. 133–143). Washington, DC: The World Bank.

Dinshaw, A. (2014). What is the role for transformation in adaptation?. [Online]. Available at: http://www.wri.org/blog/2014/04/what-role-transformation-adaptation. Accessed 4 Nov 2016.

Dovers, S., & Handmer, J. (1992). Uncertainty, sustainability and change. *Global Environmental Change, 2*(4), 262–276.

Dow, K., et al. (2013). Limits to adaptation to climate change: A risk approach. *Current Opinion in Environmental Sustainability, 5*, 384–391.

Einsenack, K., & Stecker, R. (2010). An action theory of adaptation to climate change. Berlin: s.n.

European Environment Agency. (2016). *Urban adaptation to climate change in Europe 2016 – Transforming cities in a changing climate*. Luxembourg: Publications Office of the European Union.

Ferris, E. (2014). Planned Relocations, Disasters and Climate Change: Consolidating Good Practices and Preparing for the Future (Background Document). Geneva: UNHCR. Retrieved October 26, 2017, from http://www.unhcr.org/53c4d6f99.pdf

Frankenberger, T., et al. (2013). *Community resilience: Conceptual framework and measurement feed the future learning Agenda*. Rockville, MD: Westat.

Hallsworth, M., Parker, S., & Rutter, J. (2011). *Policy making in the real world*. London, England: Institute for Government.

Harrison, C., & Williams, P. (2016). A systems approach to natural disaster resilience. *Simulation Modelling Practice and Theory, 65*, 11–31.

Kates, R., Travis, W., & Wilbanks, T. (2012). Transformational adaptation when incremental adaptations to climate change are insufficient. *Proceedings of the National Academy of Sciences, 109*(19), 7156–7161.

Khondker, H., & Schuerkens, U. (2014). Social transformation, development and globalisation. Sociopedia.isa.

Lewis, J. (2013). Some realities of resilience: A case-study of Wittenberge. *Disaster Prevention and Management, 22*(1), 48–62.

Lonsdale, K., Pringle, P., & Turner, B. (2015). *Transformative adaptation: What it is, why it matters and what is needed*. Oxford, UK: UK Climate Impacts Programme, University of Oxford.

Loorbach, D. (2010). Transition management for sustainable development: A perspective, complexity-based government framework. *Governance: An International Journal of Policy, Administration, and Institutions, 23*(1), 161–183.

Manuel-Navarrete, D., & Pelling, M. (2015). Subjectivity and the politics of transformation in response to development and environmental change. *Global Environmental Change, 35*, 558–569.

Martin, M. et al (2013). Policy analysis: Climate change and migration Bangladesh, Dhaka, Bangladesh: Refugee and Migratory Movements Research Unit (RMMRU).

Mearns, R., & Norton, A. (2010). *Equity and vulnerability in a warming world: Introduction and overview* (pp. 1–44). Washington, DC: The World Bank.

Messner, D. (2015). A social contract for low carbon and sustainable development: Reflections on non-linear dynamics of social realignments and technological innovations in transformation processes. Technological Forecasting & Social Change, 98, 260–270. doi:10.1016/j.techfore.2015.05.013

MRG (2002). Annual review: Natural catastrophes 2002. S.L.: Munich Re Group.

O'Brien, K., & Sygna, L. (2013). *Responding to climate change: The three spheres of transformation* (pp. 16–23). Oslo, Norway: University of Oslo.

O'Laughlin, K. N., et al. (2013). Testing experiences of HIV positive refugees in Nakivale refugee settlement in Uganda: Informing interventions to encourage priority shifting. *Conflict and Health, 7*, 2.

Orach, K., & Schluter, M. (2016). Uncovering the political dimension of socio-ecological systems: Contributions from policy process frameworks. *Global Environmental Change, 40*, 13–25.

Patton, A. (2016). In Uganda, a new approach for refugees – And for Ugandans. [Online]. Available at: https://www.devex.com/news/in-uganda-a-new-approach-for-refugees-and-for-ugandans-88774. Accessed 26 Oct 2017.

Raleigh, C., & Jordan, L. (2010). *Climate change and migration: Emerging patterns in the developing world* (pp. 103–131). Washington, DC: World Bank.

Regional Australia Institute. (2013). *From disaster to renewal: The centrality of business recovery to community resilience*. Canberra: Regional Australia Institute.

Reser, J., et al. (2012). *Public risk perceptions, understandings and responses to climate change in Australia and Great Britain, Gold Coast*. Australia: National Climate Change Adaptation Research Facility (NCCARF).

Resilience Alliance (2010). Assessing resilience in socio-ecological systems: Workbook for practitioners. s.l.: s.n.

Ribot, J. (2010). *Vulnerability does not fall from the sky: Toward multiscale, pro-poor climate policy* (pp. 47–74). Washington, DC: The World Bank.

Ribot, J. (2013). Risk and blame in the anthropocene: Multi-scale climate change analysis. USA: s.n.

Rotmans, J., Kemp, R., & van Asselt, M. (2001). More evolution than revolution: Transition management in public policy. *The Journal of Future Studies, Strategic Thinking and Policy, 3*(1), 15–31.

Rutherford, V., Hills, J., & Le Tissier, M. (2016). Comparative analysis of adaptation strategies for coastal climate change in North West Europe. Marine Policy (Article in Press).

Savini, F., & Dembski, S. (2016). Manufacturing the Crearive City: symbols and politics of Amsterdam North. *Cities, 55*, 139–147.

Skolnikoff, E. (1993). The Elusive Transformation: Science, Technology, and the Evolution of International Politics. New Jersey: Princeton University Press.

Smit, B., et al. (2000). An anatomy of adaptation to climate change and variability. *Climate Change, 45*, 223–251.

Tableman, B. (2005). How governmental policy is made. Best Practice Briefs, November, Volume 34.

The Refugees Act. (2006). *Uganda: The Refugee Act 2006*. Uganda: National Legislative Bodies.

Tollebrandt, S., & Wrede, S. (2013). *Power distribution between refugees and host population – A case study of the Nakivale refugee settlement*. Sweden: Linnaeus University.

Trainer, T. (2010). The global predicament: Radical implications for design. *Architectural Science Review, 53*(1), 29–36.

UNISDR (2001). Natural disasters and sustainable development: Understanding the links between development, environment and natural disasters. Johannesburg: United Nations Department of Economic and Social Affairs.

UNISDR (2012). Towards a Post-2015 framework for disaster risk reduction. s.l.: UNISDR.

United Nations (2017). Sustainable development goals. [Online]. Available at: http://www.un.org/sustainabledevelopment/sustainable-development-goals. Accessed 20 Oct 2017.

van der Have, R., & Rubalcaba, L. (2016). Social innovation research: An emerging area of innovation studies? *Research Policy, 45*, 1923–1935.

Wamsler, C., & Brink, E. (2015). From adaptive capacity to adaptation: Rethinking the Interface between Institutions and Citizens. s.l.: UNISDR.

Weichselgartner, J., & Kelman, I. (2015). Geographies of resilience: Challenges and opportunities of a descriptive concept. *Progress in Human Geography, 39*(3), 249–267.

Wilbanks, T. (2011). Inducing transformational energy technological change. *Energy Economics, 33*, 699–708.

Zbarauskaite, A., Grigutyte, N., & Gailiene, D. (2015). Minority ethnic identity and discrimination experience in a context of social transformations. *Procedia- Social and Behavioural Sciences, 165*, 121–130.

Ziervogel, G., Cowen, A., & Ziniades, J. (2016). Moving from adaptive to transformative capacity: Building foundations for inclusive, thriving, and regenerative urban settlements. *Sustainability, 8*, 955.

Multilevel Informality: Migration, Spatial Organization and Control in the Toi Market in Kibera (Nairobi)

Georgia Cardosi, Gonzalo Lizarralde, Susan N. Kibue, and Lisa Bornstein

Abstract

For decades, urban growth in developing countries has been examined through a formal-informal dichotomy. Conceptions of the "informal" urban development have evolved: from a temporary phenomenon to be eradicated, to a solution to unmet housing and services demands, and to a recognition of interdependent practices that overlap with the "formal" ones. Based on empirical evidence from Nairobi, Kenya, this paper examines how informality works and analyzes the governance of shared contested spaces. A long-term longitudinal case study is presented, the informal Toi Market, where data was collected between 2004 and 2016. Semi-structured interviews, focus groups, mapping, and observation from different periods provide insight into the market spatial organization, shedding light on informal-formal links, their underlying dynamics, and governance. Results indicate that urban informality consists of multilevel alliances among diverse stakeholders and is an integral part of, and not opposed to, formal urban development. Those alliances shape both horizontal and vertical structures in which community groups, external groups and institutions operate. Results further suggest that migration and the resulting dynamics of governance networks rooted in complex, and sometimes conflictual identity politics of urban dwellers, including ethnic affiliations, needs to be factored into the matrix of informal urbanism. The paper thus proposes adding the term *multilevel informality* to the literature on informality, suggesting the use of a concept that better reflects such complexity.

Keywords

Informal urbanism · Informal markets · Spatial organization · Migration · Governance · Longitudinal case study · Nairobi

G. Cardosi (✉)
The Canadian Disaster Resilience and Sustainable Reconstruction Research Alliance (Oeuvre durable), The Fayolle-Magil Construction Research Chair in Architecture, the Built Environment and Sustainability, Université de Montréal, Montréal, Canada
e-mail: georgia.cardosi@umontreal.ca

G. Lizarralde
The Royal Society of Canada, Professeur École d'architecture, Université de Montréal, Titulaire de la Chaire Fayolle-Magil Construction en architecture, bâtiment et durabilité, Directeur du Groupe de recherche IF (grif) et OEuvre durable, Montreal, Canada
e-mail: gonzalo.lizarralde@umontreal.ca

S. N. Kibue
Jomo Kenyatta University of Agriculture and Technology, Nairobi, Kenya
e-mail: skibue@jkuat.ac.ke

L. Bornstein
The Canadian Disaster Resilience and Sustainable Reconstruction Research Alliance (Oeuvre durable), McGill University, Montréal, Canada
e-mail: lisa.bornstein@mcgill.ca

A. Asgary (ed.), *Resettlement Challenges for Displaced Populations and Refugees*, Sustainable Development Goals Series, https://doi.org/10.1007/978-3-319-92498-4_8

1 The Need for a New Framework for Understanding Urban Informality

This paper analyzes common approaches to informal urbanism based on the *formal-informal dichotomy* and proposes an analytical framework that focuses on the governance of shared contested spaces, where tight informal-formal links and underlying dynamics, particularly migration and ethnic affiliations, play a role.

Over the last decades, Nairobi has experienced rapid urbanization mostly fed by rural-urban migration (Rahbaran and Herz 2014). Kibera, Nairobi's and East Africa's largest slum has historically been home to thousands of migrants. We explore the informal Toi Market (Fig. 1), located at the north border of Kibera (Fig. 2), through a longitudinal study that investigated the relationships between processes of spatial organization and adaptive capacities in contested urban spaces, at the community scale.

Adaptive capacities are networked dynamic resources such as community competencies and aspects of the built environment (Fayazi and Lizarralde 2013; Norris et al. 2008) that enhance processes of community formation and consolidation. In particular, we examined processes of decision-making and governance (a community competency) related to spatial organization. Our study suggests that *multilevel informality*, the main topic of this paper, conceptually best explains the Toi Market's governance structures and corresponding processes of spatial control and transformation. Migration and related ethnic dynamics emerged as context-specific factors affecting socio-spatial transformations.

Studies on urbanism in developing countries have long sought to demystify the nature of informal settlements. For decades, these studies have typically misrepresented informality as existing in opposition to "formal" development, presuming a *formal-informal* dichotomy. Understanding how informal settlements form, grow, and transform remains a priority for the Sustainable Development agenda, which promotes sustain-

Fig. 1 Toi Market, October 2016 (Photo credits: Patrizia Piras 2016)

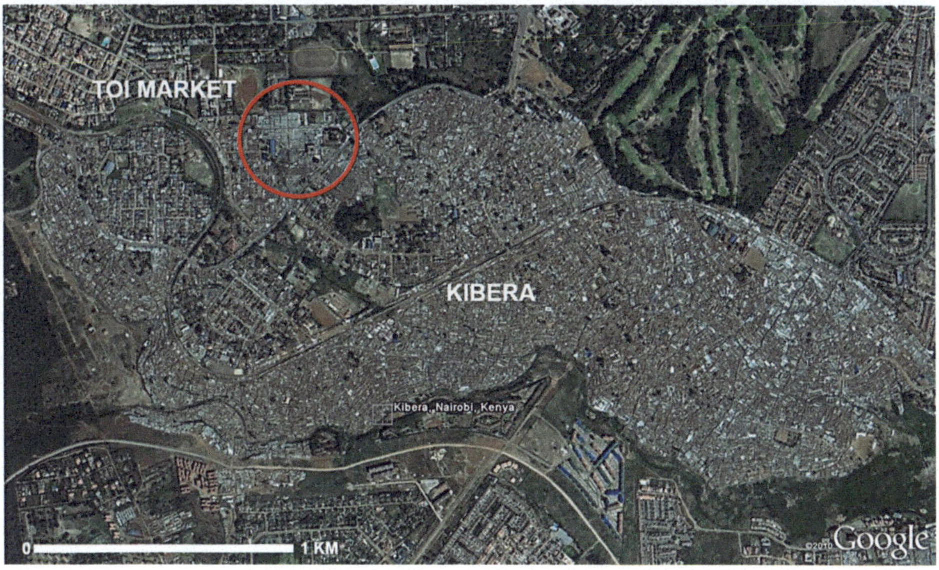

Fig. 2 Toi Market location (Google Earth 2010)

able urbanization through slum upgrading, inclusive urbanization, and "the capacity for participatory, integrated and sustainable human settlement planning and management" (Sustainable Development Goal (SDG) N.11). However, in order to develop appropriate urban policies, especially in Sub-Saharan Africa, new theoretical approaches are needed to understand informal urbanism practices and their relationships with the "formal" at the community scale.

The first section of the paper reviews theories applied to informality and introduces concepts of order, growth, and integration, which later are used to analyze socio-spatial and governance structures in the Toi Market. The second section describes the longitudinal case study and methods used for data collection and analysis. The third section presents the research results on the impact of migration on Nairobi's informal settlements formation, living conditions and ethnicity issues in Kibera, and the Toi Market. Results outline the structure of informality, how it can be interpreted, and how it shapes and transforms the urban space. The concept of *multilevel informality* is introduced to explain multiple alliances among a variety of actors that have affected the market form. Links between these alliances and political-ethnic affiliations are shown, which ultimately shed light on the impact of migration on "informal" urban development.

2 Approaches to Informality

Initially considered a temporary phenomenon that should be eradicated, informal settlement proliferation was later conceptualized as the natural response to the gap left by the state in both services and housing provision to an ever-growing urban population (Turner 1972). Despite continuing eradication of informal settlements by local administrations in many countries, informality was increasingly seen by experts as a solution rather than as a problem, and as the manifestation of a process by which marginalized residents claim their right to the city (Elsheshtawy 2011; Lefebvre 2000). It is now clear that almost all informal settlements become permanent and extend to accommodate a growing number of migrants and low-income citizens. On the one hand, informal settlements are considered so integrated into their urban contexts that cities could not function without them (Dovey and King 2011); on the other hand, living conditions in informal settlements, especially in Sub-Saharan Africa, have been deteriorating due to rapid migration, densification, and environmental decay.

Demographic projections foresee the doubling of urban population by 2040 in Sub-Saharan Africa (UN-Habitat 2016). Decades of urban planning interventions have not slowed the emergence of informal settlements (Dovey and

King 2011). Consequently, there is growing interest by decision-makers to devise new urban policies and planning tools that recognize slum dwellers' rights and improve their living conditions. New approaches can best be devised through a deep understanding of the interrelations between formal and informal actors, processes, and structures. Although their interdependence has been conceptually acknowledged, it is not yet clear how informality works and how, and to what degree, it interacts with formality.

2.1 On Formal/Informal Dichotomies

Formal and informal urbanism in cities in developing countries have typically been differentiated based on informal settlements' instability, impermanence, and lack of land ownership, tenure security, and infrastructure. In parallel, other authors have documented high levels of resilience and self-organization capacities among slum dwellers, notably in the face of urban vulnerability, social hostility, and risk. Moreover, slums have been progressively studied through notions of order, growth patterns, and integration that are examined here below.

2.2 Order in Informal Settlements

Recent literature has stressed the importance of acknowledging "order" in informal settlements. Chaos and Complex Systems theories provide a basis to reflect on patterns and cycles, and ultimately, types of order within slums, beyond their apparently random and unpredictable growth behavior. In this vein, Arefi (2011) analyzes Mandelbrot's (1983) differentiation between "characteristic" (implicit) and "systematic" (explicit) order. Characteristic order is conceptualized as ad hoc, flexible, dynamic, fluid, short-term, unfixed, and achieved by means of "local knowledge." Ad hoc and formal planning differ here in the opposition between "local" vs "expert" knowledge. As we will see later, the concept of ad hoc planning allows looking at spatial organization from outside a professional framework, as practices carried out by the Toi Market members. Other dichotomies have also tried to explain patterns in the use of urban spaces such as "everyday urban practices" versus functional urbanism's typical rigid order (Crawford 2005; Lefebvre 2000). Similarly, the notion of the "kinetic" city, typically associated with dynamic adaptive order, is opposed to the notion of the "static" city, associated with functional and planned order (Mehrotra 2005). This paper stresses the need to look beyond dichotomies and deeper into processes of urban transformations.

2.3 Rhizome Versus Hierarchy

Discussions about order facilitate our understanding of human settlements and stress the presence of consistent inner organization, thus challenging stereotypes of slums – as places of chaos and despair – that too often drive public authority's actions. For example, Dovey and King (2011) explain Asian informal settlement formation and growth patterns through a morphological analysis. Three main processes of formation – *settling*, *inserting*, and *attaching* – are recognized, as well as eight settlement morpho-typologies: *districts*, *waterfronts*, *escarpments*, *easements*, *sidewalks*, *adherences*, *backstages*, and *enclosures*. These clarify how settlements physically leak into the emptiness left by the formal city. This process of infiltration is conceptualized as "rhizomic growth," referred to Deleuze and Guattari's (1980) concept of *rhizome*, used to support an epistemological model that challenges rational hierarchical reasoning. A rhizome is in fact characterized by *connectivity* and *heterogeneity* (any point can be connected at any other), *multiplicity* (having neither subject nor object, only magnitude and dimensions), *asignificant rupture* (it can be ruptured at any point but it can create relationships again), and *cartography* and *decalcomania* (always open and accessible at any point, like a map) (Deleuze and Guattari 1980). Dovey and King (2011) portray informal urbanism as a rhizomic, rather than hierarchical, as an order of accretion that originates from years of

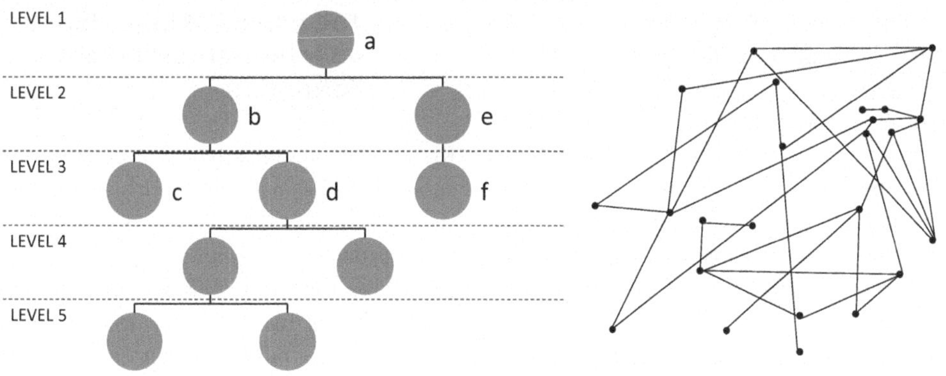

Fig. 3 Comparison between a tree-like and a rhizomic structure. Source: G.Cardosi, 2017

trial and error. So, informal urbanism would be made up of dynamic, interconnected multiplicities that are assembled without rigid, pre-established vertical tree-like hierarchies (Fig. 3). In this way, the authors strengthen the informal/formal distinction suggested by the rhizome/tree opposition (Fig. 3).

2.4 Integration

Integration is first considered in terms of interdependence between informal settlements' physical features and their social organization. Pinches (1994) explains, for instance, that patterns of inter-household relations shape settlement structure and mobility patterns, whose comprehension would otherwise be inaccessible to an external eye. This suggests that we can understand governance structures in the informal context by exploring a settlement's spatial growth patterns. Second, integration denotes that the interconnections between formal and informal are not opposed or parallel systems but rather integrated in a unique process of urban development. Doherty and Silva (2011), for example, observe close relationships between urbanization and informal urbanism, defining "informal" as "a construction of the formal." Pinches (1994) similarly contends, based on his work in Manila, that slums are integral to modernization and commodification processes.

Concepts of "order," "rhizomic growth," and "integration" challenge the *formal/informal* dichotomy to emphasize the complexity of dynamics and networks operating in contemporary cities. Yet, to explain how different forms of spatial control work and are related, more knowledge is needed about specific socio-spatial order and growth patterns. Furthermore, while attention typically focuses on informal settlements as residential communities, their connections with markets are often overlooked. As the paper shows below, revealing socio-spatial dynamics and structures in informal markets provides opportunities for understanding the functioning of informal settlements at large.

3 Methods Used to Understand Informality in Nairobi

This paper explains governance and spatial organization and control dynamics in Nairobi's Toi Market, through a longitudinal case study conducted between 2004 and 2016. The single case study strategy (Stake 1994; Yin 2003) allows for understanding a phenomenon occurring in Nairobi's context, within its own complexity and different scales (Contandriopoulos 1990). Processes of spatial transformation are thus studied in their socio-economic and political context through a multi-scale analysis (Arefi 2011).

Empirical evidence was obtained through 58 semi-structured interviews with market traders; 5 informal conversations; 3 focus groups with market leaders; 1 meeting with 43 market leaders and section representatives; informal conversations with 3 Nairobi City County officials; conversations with representatives of 2 community-based institutions (Muungano Support Trust and Jamii Bora, a micro-finance association that supported the Toi Market reconstruction in 2008); analysis of statutory documentation from the Nairobi City County, grey literature, and documents collected at the Toi Market office; and drawings and mapping-at-scale exercises.

Interviews and conversations, typically an hour in length, aimed at capturing the point of view of different actors dealing with informal urbanism. Interviews targeted men and women mostly aged between 30 and 50 years old who belong to different tribes and market sections, as well as current and past community leaders, and a mix of tenants, owners, and hawkers. Drawing and mapping-to-scale exercises were conducted over different fieldwork periods to analyze the market's physical features, spatial configurations, and building typologies. Interviews and mapping were carried out with a team formed by four architecture students from the Jomo Kenyatta University of Agriculture and Technology and one former member of Architecture Open Circuit onlus, who has volunteered to collect information and prepare documents in the Toi Market since 2007.

Data analysis included triangulation of first-hand information, examination of information from field notes and residents' narratives, and drawing inferences from local facts (Flyvbjerg 2006). This was later compared with theoretical categories drawn from literature. Triangulation of data (Stake 1994) obtained from interviews, direct observation, and focus groups with legal, contractual, and published documents confirmed the relevance of personal perspectives and narratives. Grey literature helped validate or challenge collected information. Analyzed data, such a graphics and maps, were also validated with market representatives during meetings held by the research team. Whenever differences emerged, these were made evident in findings explanation. Results allowed for drawing analytical generalizations (Yin 2013) about structures and dynamics of space occupation and organization.

4 The Impact of Migration on Informal Settlement Formation

Kenya is a rapidly urbanizing country, with an annual urban population growth averaging 5.4% since the 1970s (FAO 2012). According to the International Organization for Migration (IOM 2015), the Kenyan population growth is mainly due to natural increase[1], but decades of migration, especially internal displacement, has fueled slum development. Despite the hostile conditions of informal settlements, migrants flock to slums in search of work opportunities and proximity to services (Beguy et al. 2010). Internal displacement is often due to environmental degradation, forced evictions, and violence caused by political conflicts, such as the ones that occurred after 1992, 1997, 2002, and 2007 presidential elections. The 2007 conflict led to internal displacement of an estimated 650,000 people (IOM 2015). Currently, the country is affected by new social tensions following the August 2017 election result deletion by Kenya's Supreme Court.

However, internal displacement is also linked to historical-political causes and regional economic disparities. Anyamba (2011) identifies three waves of informal settlement proliferation linked to population displacement. First, Nairobi's foundation led to the expropriation of large rural areas by European settlers, and the formation of the first generation of slums by Africans, who, excluded from the city, concentrated in peripheral unplanned areas (Anyamba 2011). After the First World War, informal settlements grew due to the arrival of young migrants looking for employment in the city. A third wave occurred after the Second World War and is mostly ascribed to people fleeing rural poverty (Anyamba 2011); poor living conditions combined with natural hazards like drought, flood, and famine drove people to search for livelihood opportunities, better health care, and education in urban areas (Adepoju 1995; Oucho 1998; in Beguy et al. 2010). Additionally, depauperation of agricultural land due to intense, mechanized practices has meant rural-urban migration becomes a sur-

[1]Natural increase is births minus deaths, which excludes population increase by in-migration and out-migration.

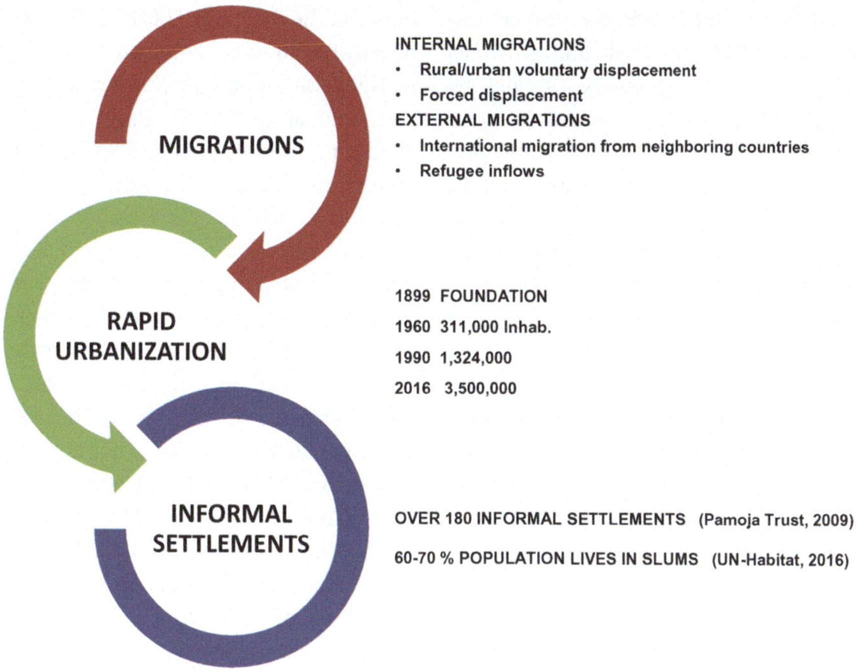

Fig. 4 Relationships between migration, urbanization, and informal settlement formation in Nairobi. Source: G. Cardosi, 2017. (Nairobi, the Kenya capital city is only 117 years old and counts 3.5 million inhabitants. Its expansion has been largely afftected by internal and external migrations, which have also contributed to the formation of informal settlements.)

vival mechanism for thousands rural families who depend on remittances. Intra-urban movements are also significant in Kenya as over a third of migrants move between different slum areas. Finally, although international migration accounts for only 2% of the entire population in 2013, Kenya accommodates one of the largest refugee populations in Africa, mainly from Somalia (Muindi et al. 2009, in Beguy et al. 2010); refugee arrivals have contributed to the recent formation of some of Nairobi's largest informal settlements (Rahbaran and Herz 2014) (Fig. 4).

4.1 Living Conditions and Ethnicity in Informal Settlements

With 7% population growth per year, Nairobi is one of the fastest growing African cities. Contrary to national data provided by IOM, Nairobi's population increase is mainly due to rural-urban migration (World Bank 2009). According to Beguy et al. (2010), about 80% of Nairobi residents aged 25–59 are migrants. It is also estimated that 70% of Nairobi's population lives in over 180 slums (MuuST, SDI, UC Berkeley, UoN, & Pamoja Trust 2011) and that 52% live in conditions of poverty (ADBG 2014).

These slums, or "informal" settlements, are the most affordable solution for millions of urban poor and migrants, while the "informal sector" and "informal markets" lead to job creation in Kenya. The so-called "informal sector" share of total employment increased to 82.7% in 2014 (KNBS 2015). The *formal-informal* dichotomy used here represents a purely Western view of urban development, according to Anyamba (2011) and other Kenyan critics; informal urban processes, instead, embody mitigation strategies adopted by people to adapt urban spaces to their needs. Since the country's dominant planning framework (derived from the British colonizers) has failed to meet needs for housing and services, most informal practices in

Kenya should be viewed as adaptive rather than illegal (Anyamba 2011; Mitullah and Kibwana 1998). Moreover, legal frameworks should be based on real urban practices instead of oversimplistic dichotomies. For instance, since the majority of Kenyan citizens are slum dwellers, treating slums as illegal settlements strains the credibility of the legal system. Concern for the dualism emerging from the formal-informal contrast as an artificial distinction has been expressed by non-Kenyan critics as well; in particular, conceptualization of informality as a contextualized process emphasizes that activities largely accepted in a society are considered informal typically when the legal system is in the process of reformulation and no longer corresponds to the real socioeconomic interactions (Bornstein 1992).

Although Kenya has recently experienced important economic growth (GDP growth of 5.3% in 2014) (KNBS 2015), access to housing, social services, and land titles is still denied to millions of urban dwellers. Poverty remains a structural feature that, combined with high unemployment rates and political instability, hinders socio-economic development. Security of tenure remains a critical challenge in informal settlements (Pamoja Trust 2011). Slum eviction and demolition, largely adopted in the 1960s, is still carried out in Kenya, notably in preparation for infrastructure and housing projects and urban redevelopment initiatives (Otiso 2002). Pamoja Trust (2009) estimates that 65,200 households in the Nairobi's slums of Kibera, Korogocho, Embakasi, and Dagoretti were affected by eviction in 2004. These conditions lay the ground for ethnic conflicts, such as the ones that erupted in 1992, 1997, and 2008 (BAD et al. 2015). Informal settlements are particularly vulnerable to conflicts. In fact, ethnic tensions commonly erupt in informal settlements, and ethnic groups become targets for political manipulation (Oyugi 2000); during election periods in particular, politicians often take advantage of slum dwellers' vulnerable conditions and fuel ethnic tensions to gain political capital.

Ethnic tensions can be traced back to Kenya's independence from British rule, which implied major redistribution of land and economic opportunities (Oyugi 2000). IOM explains that political turmoil is combined with ethnic divides in Kenya and consequently presents geographic variations according to ethnic affiliations (IOM 2015, p. 71). Geographical distribution by ethnicity is also characteristic of Nairobi's slum population. Kibera, for instance, is historically home to Nubians, who settled in Kibera's land during the first wave of migration, in 1913, to work as soldiers for the British Army. Mango (1980) explains that British authorities allowed Nubians to settle in Kibera without land titles until independence in 1963. The new Kenyan government then allowed other tribes to settle in Kibera, and a change occurred in both ethnic and administrative control of the land (Mango 1980). Landlordism and tenure insecurity increased with the settlement's densification and paved the way to ethnic conflicts related to land occupation. For instance, one century later, thousands of Nubians still claim land property in Kibera. In 2008, as political and ethnic conflict sparked in the slum, the Luo and Nubian supporters of the opposition party, who were settled in the area, provoked forced eviction of hundreds of Kikuyu residents.

4.2 Urban Governance and Planning in Nairobi

Over the 1990s and 2000s, the Kenyan Federation of slum dwellers, Muungano Wa Wanavijiji, and other civil society representatives conducted a negotiation process with Kenyan authorities for the development of inclusive legal and political frameworks. Relevant progress can be seen in the 2010 National Constitution's establishment of a National Land Commission (Government of Kenya 2010); adoption of the National Eviction and Resettlement Guidelines that aim to protect, prevent, and compensate for forced eviction (Government of Kenya 2012); and government reorganization, as described below. The political measures, finalized after the 2008 crisis, are to foster social stability and reduce ethnic divides.

Before 2010, local planning and land policies were under the control of the central government, acting through the provincial administration, and the Ministry of Local Government. The 2010 Constitution, Devolved Governments Act, the Urban Areas and Cities Act (Government of Kenya 2011), and other political tools made two important changes to the governance of Nairobi's slum. First, in an effort to improve policy implementation and monitoring at the local level, the Constitution replaced provinces with a new intermediate structure, the county government. Government bodies for Nairobi's informal settlements now act at the county level and through sub-county divisions. Second, the Nairobi City County (NCC) replaced the City Council of Nairobi. The NCC is in charge of planning, public health, social services, and housing, as well as public services hitherto provided by the national government. Seemingly, the absence of an obligation to serve informal settlements (Taylor 2005) has been continued. Neither change eliminates the need for multilevel coordination. Since several informal settlements are on government lands, the central government remains involved (Fig. 5). And while the NCC has responsibility for housing and city planning, its activities in the slum are done in collaboration with civil society. For example, the NCC can designate informal settlements as "Special Planning Areas" and allocate public land for community-led slum upgrading projects by NGOs and CBOs (Pamoja Trust 2011).

5 Toi Market

Toi Market was formed in the late 1980s on the northern edge of Kibera (Fig. 6) and is now the second largest informal market in Nairobi (Cardosi et al. 2016). Currently, it accommodates about 2400 stalls and provides employment to over 3000 traders and their families (Interview with market member 2011). Toi Market offers a variety of goods, food, and services to Kibera and other formal and informal settlements (Cardosi et al. 2016). Interviews showed that most vendors are internal migrants coming from rural areas in search of jobs and other economic opportunities. Many of them support relatives in rural villages, where they typically travel to over the course of

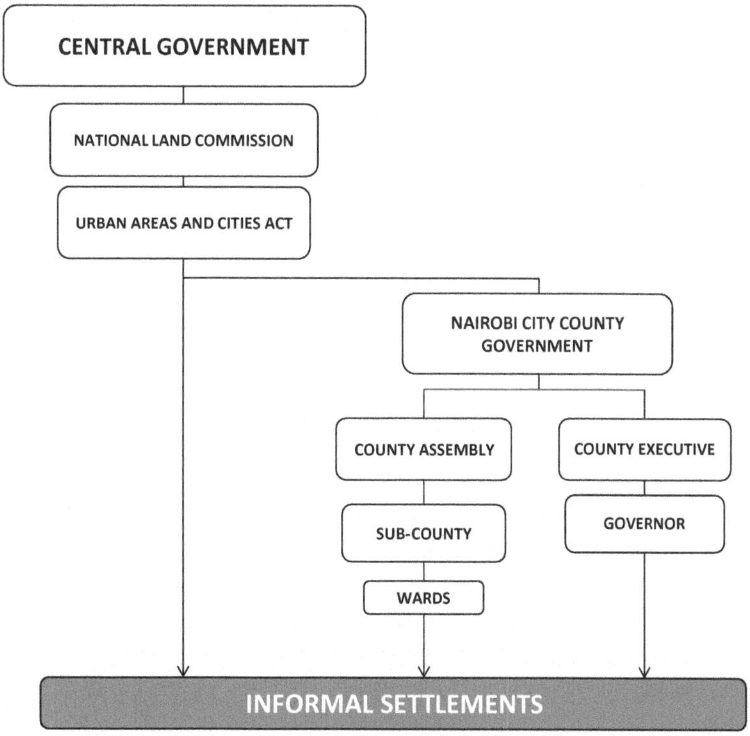

Fig. 5 Central government and county government control planning and land policies. Source: G. Cardosi, 2017

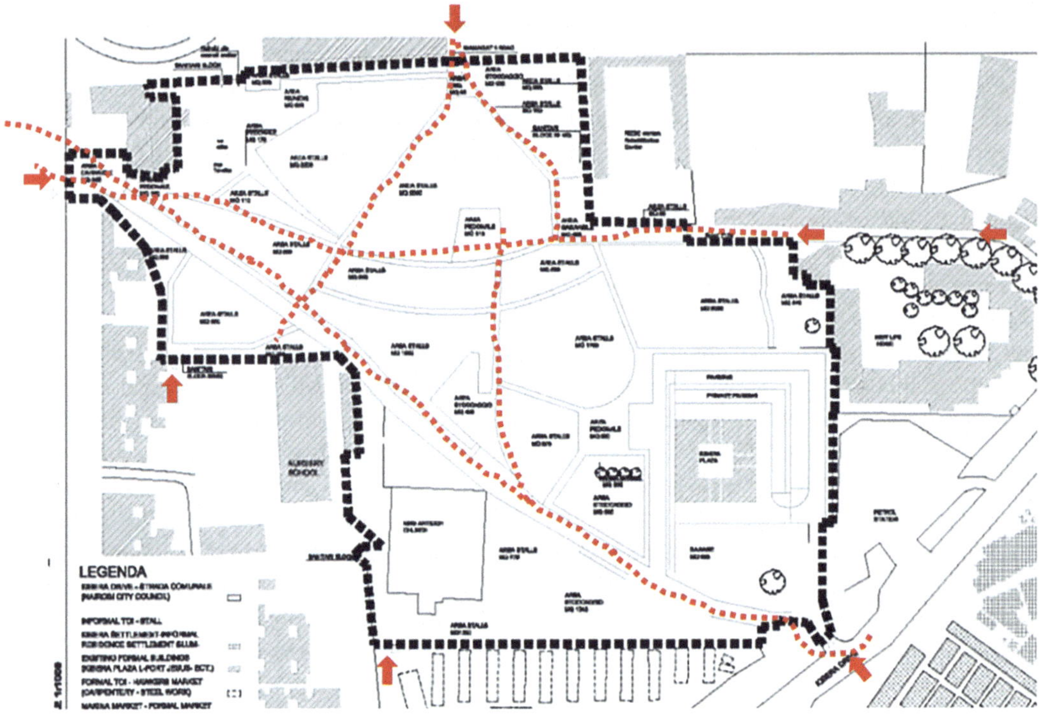

Fig. 6 Level A of informality: Toi Market path system (in red) in the early 1980s. Source: G. Cardosi, 2011

the year. The majority of interviewees live in Kibera and other slums; proximity is the main reason for them to choose to operate in Toi Market. Half of the interviewees have been working in Toi Market since before 2000 and 15 of them since the early 1990s. Since its formation, the market has peacefully accommodated tens of tribes: Luo, Kikuyu, Luhya, Kamba, and Nubians, among others. Yet, ethnic tensions have increased since 2008. Several interviewees used the word "tribalism" while talking about the market governance and decision-making. Corruption has also increased according to them. Generally, fear of internal and national conflicts is stronger now, due to the impact that the 2008 conflict had on the market community.

The market land is highly contested, and tenure security conditions are very complex. The NCC considers the market as an "open-air market" that is unauthorized, hence temporary, although it collects market fees from traders. In particular, respondents from the NCC section responsible for markets said that urban renewal projects may threaten the market's existence in future. At the same time, interviewees at the sub-county level said traders could not be evicted, as the Limitation of Actions Act (Government of Kenya 2012) protects those who occupy land for 12 years or more without litigation among the parties. Traders who have been operating there for years consider the market permanent, though they fear eviction and are ready to use all legal means to fight it. However, the situation is complicated by allocations, in the 1990s, of some plots within the market area to powerful individuals who, when later claiming the properties, prompted market traders to contest their claims in court. Typical dichotomies between authorized and unauthorized activities would be difficult to apply to the market.

Tenure insecurity makes the market existence precarious, confirming that informality has an uncertain future. Risks of eviction and environmental degradation threaten it. And a multiplicity of actors with different interests pave the way for internal division, controversies, and conflicts. In 2008, the settlement was hit by post-election conflict, razed by a fire caused by youth gangs, and reconstructed, in the same year, by a task force. The Community Organization Practitioners Association of Kenya (COPA 2008) explains that the market invasion was organized by a group of former traders who teamed up with Nubian youth gangs from Kibera. The Nubians wanted to get back what they claimed to be their land, while the Gatwikira groups intended to grab spaces belonging to other tribes, especially Kikuyu. In addition, since the market land lies between Kibera and middle- and high-income areas destined for densification, it attracts many investors.

6 Multilevel Informality

Informality in the Toi Market is structured through a system of alliances, established among diverse stakeholders, that give shape to processes of formation, transformation, and eventually consolidation of the market settlement on contested land. This system is comprised of multilevel arrangements affecting the community's physical and sociopolitical configuration, temporarily and more "permanently." Six main levels of informality and one sub-level can be identified, in which both community associations and public institution members operate. Infringement of laws, rules, codes, and practices operate at multiple levels (see Table 1).

Level A of informality (Fig. 6) is what we define as a *community-led initiative,* adopting an expression used by market members at the High Court. This level of action begins through spontaneous incremental formation of the settlement and contravenes urban planning rules dictated by local authorities. The market land was originally governmental and planned to accommodate public services (Ministry of Lands and Settlement 1994). These were never realized, so, in the late 1980s, groups of women coming from Nyanza province established premises along the Kibera Drive to sell vegetables (Cardosi and Lizarralde 2014). Gradually, other vendors set up temporary structures along the pedestrian paths there connecting residential areas (Kimani 2004); by 1990, the area was full. Land occupation was done through unfixed, implicit, and ad hoc planning, wherein patterns of local circulation and location of main accesses guided spatial configurations (Fig. 6). Activities were distributed mainly by business type, separating, for example, second-hand clothes from food kiosks (Interview with market member 2016).

Until the first eviction in 1996, traders simply enjoyed de facto tenure security without worrying about land registration or designation as a market (Kimotho 2013). It was during years under risk of eviction that the market members started organizing in groups and eventually developed a sense of community, to fight against eviction. In the meantime, the market area was densified and developed by vendors of various tribes (Luo, Luhya, Kamba, and Kikuyu, among others). As yet we have no data concerning ethnic distribution in the market area in this early Level A phase, but informal conversations with vendors confirm that ethnic groups lived together peacefully. In this Level A of informality, a hierarchical though flexible governance structure prevailed, where the market management office and other parties had a certain freedom to interact and develop other levels of informality. Power was also shared; market members confirmed that Muungano Wa Wanavijiji (the slum dwellers' federation) has decisional power on project initiatives equal to, or more than, the market committee, in part because the federation's organizational capacity allows it access to small project funding.

Level B of informality (Fig. 7) or *politicized gang-led action* is how we describe the violent invasion of a consolidated informal community by foreigners. In this case, actors are Nubians and Luo youth gangs from Kibera, who were politicized to protest against the victory of the new Kikuyu president in 2008. Most of them reside in Kibera and claim ownership of the Toi Market land. This initiative is informal within Level A, in that it breaks consolidated rules established by the existing informal community. Level B activities created important losses for the vendors. For example, about 800 Kikuyu members had to leave the market area (Kimotho 2013) to settle in another underused space with enormous economic and community losses. As

Table 1 Levels of informality, stakeholders and actions, and types of order infringement. Source: G. Cardosi, 2017

Level	Actions/actors	Type of order infringement and creation of a new order
A *Community-led initiative*	Toi Market incremental growth by informal vendors	Unauthorized (but not explicitly forbidden) gradual occupation of public land planned to accommodate public utilities Establishment of internal rules for land use and management De facto tenure regime
B *Politicized gang- led action*	Toi Market invasion and destruction by outsiders – Nubians and leader-residents of Kibera-Makina	Infringement of consolidated community rules Disruption of the livelihood system Disruption of the social system Destruction of the market's physical structures
C *Multiparty led initiative for community interest*	Toi Market reconstruction by the Task Force (MoU 2008)	Peacekeeping Establishment of new governance system and new rules (new market constitution) Reconstruction process implemented without formal approval or explicit prohibition (or ban) by authorities Ad hoc planning New plans and new spatial configuration
D *Independent group-led initiative*	Transformation of stalls into housing by some structure owners	Land use transformation within the market, not considered by the market constitution and without consent by the market committee
D.1	Reconstruction by structure owners of houses demolished by the market committee	New infringement of community rules and market constitution
E *Market committee-led action*	Demolition of housing structures by the market committee Intimidation and eviction of vendors by market officials Gate construction in two access points to the market to improve security	Infringement of governmental and constitutional rules allowing only the state police to demolish structures after a court order and upon adequate notice to affected residents Infringement of the principle of bona fide tenants/ structure owners and of principles of the market constitution Initiative taken without permit by the NCC
F *Neighborhoods-led initiative*	Gate construction by neighbors at Fort Jesus entrance	Infringement of the Toi Market and of NCC's rules; construction of gates on the southeast access to Toi Market, done by Fort Jesus estate residents

many interviewees noted, tribalism is now stronger in the market.

Level C of informality is defined as a *multiple-party-led initiative for community interest.* The Toi Market reconstruction process in January 2008 is illustrative. This initiative is informal in that it was planned and implemented without official government approval; however, it was not opposed by any party. This initiative consisted in peacekeeping, coordination, fundraising, and reconstruction. The local micro-finance institution Jami Bora took charge of the coordination and included the newcomers – Nubian and Luo residents of Kibera's Makina villages – in the process, a central strategy for restoring peace in the area (Interview with Jamii Bora 2016). The second strategy was to form a task force composed of former market members, outsiders, and representatives of public authorities. Level C is characterized by rational planning outside the government system. Here it was dictated by the urgency to rebuild the market as an essential source of livelihood and food for the market population and Kibera's residents. Planning aimed at redistributing space among traders and rebuilding stalls and other structures. Land was subdivided and allocated to a list of beneficiaries. The resulting market form has a more rational, stable, and rigid configuration.

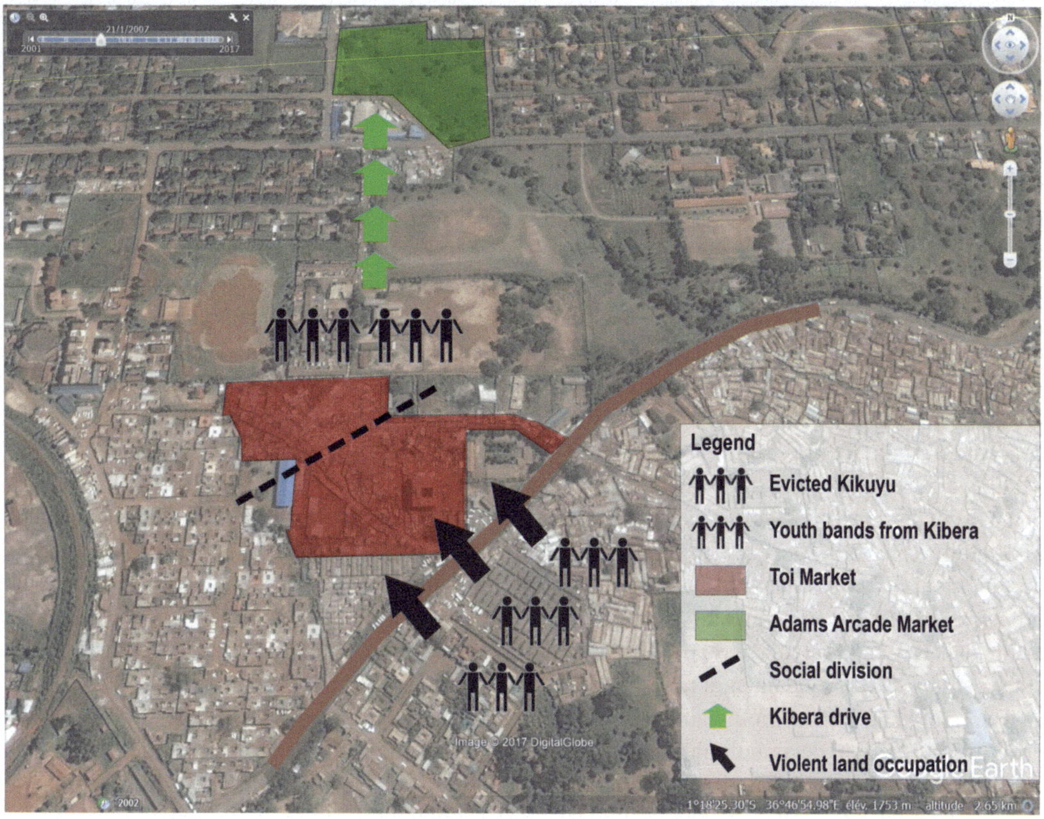

Fig. 7 Level B of informality. Drawing on Google Earth. Source: G. Cardosi, 2017

Level D of informality is defined as an *independent group-led initiative.* In the market, it is seen in the transformation of stalls into housing by some structure owners. Stall owners wish to change land uses when located in areas where clients cannot efficiently circulate and sales are consequently limited. However, land use transformation within the market is not an option listed in the market's constitution. As such, housing construction was implemented without any assembly or consent of the market committee. Within this level, a sub-level is recognized, Sub-level D.1, which is the reconstruction by structure owners of the houses demolished by the market committee. Here, structure owners repeatedly disobeyed the community rules and market constitution.

Level E *of informality*, or *market committee-led action*, consists of unsanctioned actions undertaken by a sanctioned actor, in this case the market committee. Examples include (1) demolition of "unauthorized" housing structures (Level D), (2) vendors' eviction, and (3) gate construction in two market access points to improve security. Action 1 represents a transgression of governmental and constitutional rules by which only the state police can demolish structures after explicit order by the High Court of Kenya and only upon adequate notice to residents (Interview 2016). Action 2 contravenes the principle of bona fide tenants and structure owners as well as the market constitution. Action 3 has been undertaken without a permit from the NCC, done by the market committee, allegedly with funding from a small group of structure owners.

Finally, *Level F of informality*, named *neighborhood-led initiative*, is illustrated by the establishment of physical structures by those in neighborhoods adjacent to the informal market. For example, gates on the southeast access to Toi Market were constructed by Fort Jesus Estate residents, an infringement of both Toi Market and NCC's rules.

7 Multilevel Informality and Settlement Consolidation

The identified levels of informality underline a multilevel organizational structure that is both hierarchical and horizontal, as it connects different governance groups inside and outside the market. Formal organizations, such as government and civil society groups, constitute this multilevel structure. Formal decision-making structures are highly dynamic and interact with informality, challenging the kinetic/static conceptual dichotomy (Mehrotra 2005). Multilevel informality, and the messiness of non-dichotomous relations, confirms that urban informality cannot be opposed to formal urban development but is integral to a unique urbanization process implemented by numerous actors, where – at least in the case of the market – ethnic groups play a fundamental role.

The Toi Market is both authorized and unauthorized. The traders and public institutions consider it an informal market, whereas the Kenyan government sees the "essence of informal settlements" as an absence of tenure security and planning (Government of Kenya 2009). Being informal has not impeded Toi from functioning for almost four decades and becoming Nairobi's second largest market. Its existence and success challenges stereotypes of slums as temporary, unplanned, and based on short-term planning vision. The market reconstruction relied on rational layout that, on the one hand, facilitates more rigid governance – and spatial control – by ethnic groups who participated in reconstruction planning and, on the other hand, does not guarantee land rights to traders, especially the most vulnerable. Indeed, many vendors lost their stalls under the new power structure.

The planned/unplanned dichotomy cannot be applied to the Toi Market, which has grown through both ad hoc incremental and planned processes. Local knowledge and ad hoc planning have contributed yet concepts such as "long-term/short-term," "chaotic," or "dynamic" poorly fit in reference to spatial configurations. Certainly, mobility of people, adaptability of function, and flexibility of location are patterns that future design and urban planning should address. Likewise, the role of context-specific governance issues, such as ethnicity and migration in the case of Nairobi, must be considered.

The Toi Market also represents a new category of settlement. Pamoja Trust's (2011) Policy Brief on land regularization in Kenya identifies four categories of informal settlements, namely, informal settlements on public land, on private land, on private land with absentee landlords, and on indigenous freehold land. The Toi Market lies partially on public land and partially on private land with absentee landlords. It still lacks public services and infrastructure, but there is no program to upgrade or redevelop it in the near future. Tenure insecurity and lack of financial resources typically discourage individual vendors from improving the built environment and investing in permanent structures. Nevertheless, slow incremental upgrading has been undertaken by groups who manage saving schemes or get external funds for small projects. The construction of such structures as permanent toilet blocks and group halls has contributed to settlement consolidation over the years. Sadly, the risk of divisions within the market community again is high. And, although the new constitution protects slum dwellers from eviction under certain conditions, eviction is a real risk for the Toi Market traders due to real estate market pressure in Kibera. The Toi Market traders faced eviction and demolition in 1983, 1996, 1999, 2005, and 2014 and had often tenuous relations with the Nairobi City Council. While the prognosis for the future of the market is uncertain, the multilevel informality operating there, and documented here, helps us to understand complex processes of informal settlements' spatial organization, transformation, and consolidation.

This paper started by presenting concepts of order, growth, and integration in informal settlements. The need to go beyond dichotomies was emphasized to better understand how informality works and interacts with different structures to define urbanization processes. In particular, Dovey and King's (2011) representation of informality as a rhizomic order, horizontally infiltrated in formal structures, is an important element in defining a new analytical framework toward informal urbanism. Our case study of the Toi Market demonstrated the difficulties in applying dichotomies and that the need to deeply explore the governance dynamics ruling spatial organization and control. Informality, understood as a system of alliances among politicized, administrative, and civil society stakeholders, exists in the Toi Market in at least six different levels; both traders and representatives of external and internal structures contribute to that multilevel informality. Informality can be visualized as made up of both horizontal and vertical structures, shaping a unique process of urban development, where government-led and community-led processes are not opposed. This implies that future studies should further analyze informal multilevel structures, including context-specific cultural-ethnic mechanisms that contribute to shaping informality. Finally, the results show that multilevel informality is both a factor of sociopolitical fragmentation, which puts the Toi Market community at risk of becoming divided, and *an adaptive mechanism* that enhances the settlement consolidation in the urban context. Although further investigation is needed, plans and interventions can benefit from considering *multilevel informality*.

Acknowledgments We wish to thank Œuvre Durable (the Canadian Disaster Resilience and Sustainable Reconstruction Research Alliance) for support to the 2016 fieldwork in Nairobi. We particularly wish to thank the Toi Market members who have welcomed the research team and participated in this research project; Patrizia Piras, former member of AOC onlus (Architecture Open Circuit), for volunteering in this case study since 2007 and the School of Architecture and Building Sciences of the Jomo Kenyatta University of Agriculture and Technology, which contributed to the 2016 fieldwork with a team of four hardworking and passionate students.

References

ADBG (2014). Kenya. Country strategy paper 2014–2018: African Development Bank Group.

Adepoju, A. (1995). Emigration dynamics in sub-Saharan Africa. *International Migration, 33*(3-4), 313–390. https://doi.org/10.1111/j.1468-2435.1995.tb00032.x.

Anyamba, T. (2011). Informal urbanism in Nairobi. *Built Environment, 37*(1), 57–77.

Arefi, M. (2011). Order in informal settlements: A case study of pinar, Istanbul. *Built Environment, 37*(1), 42–56.

BAD et al. (2015), Perspectives économiques en Afrique 2015 : Développement territorial et inclusion spatiale, Éditions OCDE, Paris, http://dx.doi.org/10.1787/aeo-2015-fr.

Beguy, D., Bocquier, P., & Zulu, E. M. (2010). Circular migration patterns and determinants in Nairobi slum settlements. *Demographic Research, 23*, 549–585.

Bornstein, L. (1992). Introduction to special section on the informal sector. *Berkeley Planning Journal, 7*(1), 121–123.

Cardosi, G., & Lizarralde, G. (2014). Understanding urban form and space production in informal settlements: The Toi Market in Nairobi, Kenya. In A. Osman, G. Bruyns, & C. Aigbavboa (Eds.), *Conference proceedings, XXV World Congress of Architecture "Architecture Otherwhere Resilience – Ecology – Values"* (pp. 3–7). Durban, South-Africa.

Cardosi, G., Lizarralde, L., & Labbé, D. (2016). *Autonomous design and planning initiatives in informal markets: The case of Nairobi´s Toi Market. Sfc 2016 – Sustainable futures conference.* Nairobi: Architecture and construction in the global south. Conference proceedings.

Contandriopoulos, A. P. (1990). *Savoir préparer une recherche: la définir, la structurer, la financer*. Presses de l'Université de Montréal.

COPA, C.O.P.A.o.K (2008). An update on the post-election violence in Nairobi's informal settlements.

Crawford, M. (2005). *Everyday urbanism: Margaret crawford vs. Michael speaks*. Ann Arbor, MI: University of Michigan, A. Alfred Taubman College of Architecture.

Deleuze, G., & Guattari, F. (1980). *Mille plateaux: Capitalisme et schizophrénie ii*. Paris: Minuit.

Doherty, G., & Silva, M. L. (2011). Formally informal: daily life and the shock of order in a Brazilian favela. In P. Hall & D. Banister (Eds.), *Built Environment* (Vol. 37, pp. 30–41). Oxon: Alexandrine Press.

Dovey, K., & King, R. (2011). *Forms of informality: Morphology and visibility of informal settlements. Built Environment 37*(1), 11–29. Oxon: Alexandrine Press

Elsheshtawy, Y. (2011). The informal turn. *Built Environment, 37*(1), 5–10.

FAO (2012). *Growing greener cities in Africa. First status report on urban and peri-urban horticulture in Africa* Rome, Food and Agriculture Organisation of the United Nations.

Fayazi, M., & Lizarralde, G. (2013). The role of low-cost housing in the path from vulnerability to resilience. *ArchNet-IJAR, 7*(3), 146–167.

Flyvbjerg, B. (2006). Five misunderstandings about case-study research. *Qualitative Inquiry, 12*(2), 219–245. https://doi.org/10.1177/1077800405284363.

Government of Kenya. (2009). *Ministry of lands. Sessional paper no. 3 of 2009 on National land policy*. Nairobi: Government Printer http://www1.uneca.org/Portals/lpi/CrossArticle/1/Land%20Policy%20Documents/Sessional-paperon-Kenya-National-Land-Policy.pdf.

Government of Kenya. (2010). *The Constitution of Kenya, 2010*. Nairobi: National Council for Law Reporting with the Authority of the Attorney-General.
Government of Kenya. (2011). Urban areas and cities Act NO. 13 of 2011. Nairobi: National Council for Law Reporting with the Authority of the Attorney-General.
Government of Kenya. (2012). *Limitation of Actions Act. Chapter 22*. Nairobi: National Council for Law Reporting with the Authority of the Attorney-General http://kenyalaw.org/lex/rest//db/kenyalex/Kenya/Legislation/English/Acts%20and%20Regulations/L/Limitation%20of%20Actions%20Act%20Cap.%2022%20-%20No.%2021%20of%201968/docs/LimitationofActionsAct21of1968.pdf.
International Organization for Migration (2015). Migration in Kenya. A Country Profile, 2015. Nairobi: International Organization for Migration (IOM).
Interview with Jami Bora. (2016). Nairobi, Jami Bora office. October 2016.
Interview with market member. (2011). Nairobi, Toi Market. May 2011.
Interview with market member. (2016). Nairobi, Toi Market. October 2016.
Kimani, B. (2004). *Re-development of Toi Open Air Market*. Kibera: Creating Legible Environments for the Informal Sector Unpublished document.
Kimotho, R. (2013). Undp/danida legal empowerment study Kenya report (Draft 2: June 30, 2013).
KNBS (2015). Economic survey 2015. Kenya National Bureau of Statistics.
Lefebvre, H. (2000). *La production de l'espace*. In Anthropos (Ed.), (4th ed.). Paris: Anthropos.
Mandelbrot, B. (1983). *The Fractal Geometry of Nature*, revised ed. New York: W.H. Freeman.
Mango, C. (1980). The growth of Kibera: From Nubian enclave to squatter real estate. UoN Digital Archive.
Mehrotra, R. (2005). *Everyday urbanism : Margaret crawford vs. Michael speaks*. Ann Arbor, Mich.: University of Michigan, A. Alfred Taubman College of Architecture.
Ministry of Lands and Settlement, Department of physical Planning (Cartographer). (1994). Kibera Development Plan. REF. NO. 42.32.94.9.
Mitullah, W.V., & Kibwana, K. (1998). A tale of two cities: Policy, law and illegal settlements in Kenya. *Illegal Cities Law and Urban Change in Developing Countries*, Zed, London.
Muindi, K., Zulu, E., Beguy, D., Mudege, N., & Batten, L. (2009). Characteristics of recent in-migrants in the Nairobi Urban Health Demographic Surveillance System. Paper presented at the XXVI IUSSP Conference. Marrakech, Morocco, September 27 – October 2, 2009.
Muungano Support Trust, Slum Dwellers International, University of Nairobi, & University of California Berkeley. (2012). *Mathare Zonal Plan*. Nairobi, Kenya: Collaborative plan for informal settlement upgrading.
Norris, F. H., Stevens, S. P., Pfefferbaum, B., Wyche, K. F., & Pfefferbaum, R. L. (2008). Community resilience as a metaphor, theory, set of capacities, and strategy for disaster readiness. *American Journal of Community Psychology, 41*(1–2), 127–150.
Otiso, K. M. (2002). Forced evictions in Kenyan cities. *Singapore Journal of Tropical Geography, 23*(3), 252–267.
Oucho, J. O. (1998). Recent internal migration processes in sub-Saharan Africa: Determinants, Consequences and Data Adequacy Issues. In R. E. Bilsborrow (Ed.), *Migration, urbanization, and development: New directions and issues* (pp. 89–120). Norwell: UNFPA-Kluwer Academic Publishers.
Oyugi, W. O. (2000). Politicized ethnic conflict in Kenya: A periodic phenomenon. In *African conflicts: Their management, resolution, and post-conflict reconstruction*. Addis Ababa: DPMF/OSSREA.
Pamoja Trust. (2009). An Inventory of the Slums in Nairobi. Unpublished report.
Pamoja Trust. (2011). Securing land tenure for the urban poor in Kenya. A policy brief. Nairobi Environmental Outlook, Nairobi: Government printers, February 2011.
Pinches, M. (1994). Modernisation and the Quest for Modernity: Architectural Form, Squatter Settlements and the New Society in Manila. In M. Askew, & W. Logan (Eds.), *Cultural Identity and Urban Change in Southeast Asia: Interpretative Essays* (First ed., pp. 13–42). Geelong: Deakin University Press.
Rahbaran, S., & Herz, M. (2014). Nairobi, Kenya. Migration Shaping the City. In J. Herzog & P. de Meuron (Eds.), *ETH studio base*. Zurich: Lars Muller Publishers.
Stake, E. R. (1994). Case studies. In K. Norman Denzin & S. Yvonna Lincoln (Eds.), *Handbook of qualitative research*. London: SAGE Publications.
Task force for the Toi Market Reconstruction. (2008). *Memorandum of understanding*. 13 January 2008. Unpublished document.
Taylor, W. (2005). Communities and communication in Nairobi. In Patrick Wakely and Elizabeth Riley (Ed.), *Communities and Communication: Building urban partnerships* (pp. 64-84). Rugby, England: ITDG Publishing.
Turner, J. (1972). Housing as a verb. In C. Macmillan (Ed.), *Freedom to build: Dweller control of the housing process* (pp. 148–175). The Macmillan Company, New York.
UN-Habitat. (2016). *Slum almanac 2015/2016. Tracking improvement in the lives of slum dwellers*. Nairobi: UNON, Publishing Services Section.
World Bank (2009). Kenya – Poverty and inequality assessment : Executive summary and synthesis report. Washington, DC.
Yin, R. K. (2003). *Case study research: Design and methods* (3rd ed.). London: SAGE Publications.
Yin, R. K. (2013). *Case study research: Design and methods*. SAGE Publications.

Part II

Resettlement Challenges of Disaster Displaced Populations

The Impact of Post-disaster Housing Reconstruction Policies on Different Beneficiary Groups: The Case of Bam, Iran

Mahmood Fayazi and Gonzalo Lizarralde

Abstract

Disaster management studies have demonstrated that housing reconstruction programs often lead to different levels of community recovery. Yet, insufficient knowledge still exists about how reconstruction policies and decisions produce diverse impacts among different social groups. The purpose of this paper is to explore *why* and *how* housing reconstruction policies impact households in different ways. This research focuses on low-income housing programs implemented in response to disasters. It examines post-disaster reconstruction policies through the lenses of a comprehensive body of knowledge about the evolution of housing policy in developing countries over the last seven decades. Using a set of indicators from pre- and post-disaster conditions among six household categories, the qualitative enquiry examines the housing reconstruction program conducted after the 2003 earthquake in Bam. Empirical results show that the scant attention to different categories of tenancy, families' socioeconomic conditions, and demographic changes (before and after the disaster) led authorities to adopt housing reconstruction policies that benefited some groups of households, while having the opposite effect on others. *Single-family house-owners*, for instance, rebuilt their permanent houses quickly and resumed normal activities in a relatively short period of time. *Members of extended families* – who before the disaster relied on a complex social fabric based on proximity – were instead adversely affected by policies that allocated them a unit in a residential complex located in the city outskirts. Results reveal the inefficiency of the one-policy-for-all approach in housing reconstruction. The coexistence of a multiplicity of measures and programs can allow households to choose the solution that best fits their needs, conditions, and expectations. Findings also highlight a gap between general housing and housing reconstruction policies in developing countries. Pre-disaster policies must be constantly assessed to identify and understand their effectiveness and drawbacks in reducing vulnerabilities. Post-disaster reconstruction brings an opportunity to do this and ensure a sustainable development based on resilience enhancement and disaster risk reduction.

Keywords

Housing reconstruction · Recovery policy · Households' recovery · Bam earthquake

M. Fayazi (✉)
The Institute of Disaster Managment and Reconstruction, Polytechnic University of Hong Kong and Sichuan University, Chengdu, China
e-mail: mahmood.fayazi@umontreal.ca

G. Lizarralde
The Royal Society of Canada, Professeur École d'architecture, Université de Montréal, Titulaire de la Chaire Fayolle-Magil Construction en architecture, bâtiment et durabilité, Directeur du Groupe de recherche IF (grif) et OEuvre durable, Montreal, Canada
e-mail: gonzalo.lizarralde@umontreal.ca

A. Asgary (ed.), *Resettlement Challenges for Displaced Populations and Refugees*, Sustainable Development Goals Series, https://doi.org/10.1007/978-3-319-92498-4_9

1 Introduction

In the last few decades, disaster- and reconstruction-related studies have made considerable endeavors to determine the variables behind the failure and success of housing reconstruction programs. Numerous studies have examined the short- and long-term impacts of interventions to pave the way for improving reconstruction policy (Alexander 2008; Barenstein 2006; Comerio 1998). Duyne-Barenstein (2006), for instance, explores how housing reconstruction after the 2004 tsunami in Tamil Nadu paid inadequate attention to the social-cultural and environmental conditions of the local population, thereby affecting peoples' cultural identity and livelihood resources. Other authors have found that reconstruction policies adopted after disasters often neglect the variety of beneficiaries and the diversity of their needs and desires (Aysan and Oliver 1987) and fail to consider how they affect communities differently (Aldrich 2012; Davidson et al. 2007). Despite the existence of consensus over the uniqueness of every disaster and the need for the adoption of an appropriate reconstruction policy, little is still known about *how* and *why* policy causes different levels of recovery among affected families.

This study aims at bridging this gap. It is based on a detailed, qualitative case study of the permanent housing reconstruction program conducted after the devastating earthquake that struck the historic city of Bam in Iran on December 26, 2003. This paper is divided into four sections. First, the main approaches to housing policy in developing countries, reconstruction policy, and the importance of housing reconstruction policy in the process of recovery are reviewed. A section about the methods used to examine the evolution of pre- and post-disaster conditions among different household types affected by the disaster follows. The third section reviews the adopted policies in Bam and describes how the policies impacted different categories of households, confirming that pre-disaster vulnerabilities and conditions (that vary significantly among household groups) largely determine the success or failure of policies. Finally, the discussion and conclusion sections further elaborate on the theoretical and practical implications of these findings.

2 Housing Policy in Developing Countries: From Turnkey Projects to Habitat III

A better understanding of housing reconstruction policies and their impact requires first to review the comprehensive body of knowledge about housing policy evolution in developing countries, which, as we shall see, is deliberately associated with international policy reforms.

From the end of World War II to 1972, policy largely prioritized the state's role as the provider of public housing, which often took the form of standardized residential complexes. Relying on research in Latin America conducted by Turner (1967) and Mangin (1967), the World Bank formulated in 1972 the *self-help policy*, which is now considered the first generation of international housing policy. It emphasized the positive effect of self-help activities without direction from centralized bureaucracies (World-Bank 1974). According to Pugh (1994), this phase of the Bank's policy expressed the intention of achieving affordability, the use of budget limits to define feasible standards, and resource allocation for land acquisition and infrastructure. However, this policy was seen by many as simplistic and narrow in its understanding of the relationship between the state, markets, and households' roles in housing, thus insufficient to tackle the underlying qualitative and quantitative housing deficits that affect developing countries in general (Burgess 1978; Lizarralde 2015).

The second-generation policy (1983–1989) went beyond the narrow emphasis on housing and brought shelter into a closer relationship with macroeconomic and development policy (Pugh 1992). Conditionality clauses were attached to international loan agreements with governments, aiming at stabilizing macro- and microeconomic indicators. Their influence, according to (Lizarralde 2015), resulted in the formulation of neoliberal policies, pushing governments to reduce their involvement in housing and to transfer increased responsibilities to municipalities, which often lacked adequate financial mechanisms and administrative structures. The second generation of policy – *market-enabling policy* – was soon accused of creating "adjustment poverty," putting more than 30% of the urban population in developing countries in poverty (Pugh 1992). In response, the third generation of World Bank housing policy – *well-functioning housing policy* – was pursued from 1992 to the 2000s. This policy emphasized the growth and development of the whole housing sector in its urban and national context (World-Bank 1992). It extended housing development to social objectives, including poverty and health alleviation (Jenkins et al. 2006). The overall package of reforms required strict administration and complex systems of cooperation. For the majority of developing countries, however, the achievement of this comprehensively envisaged reform was rather unlikely (Zanetta 2004).

International policy reforms continued in the late twentieth century. World conferences such as the Second United Nations Conference on Human Settlements – Habitat II in 1996 (Turkey), the United Nations Convention on Climate Change in Kyoto in 1997 (Japan), and the Millennium Summit Conference in 2000 (United States) broadened the scope of international policy, promoted sustainable development objectives, warned about climate change, and established Millennium Development Goals to be achieved by 2015 (Jenkins et al. 2006). More recently, the Third United Nations Conference on Human Settlements – Habitat III in Quito (Equator) in 2016 reinvigorates the global commitment to housing and sustainable urbanization, assesses accomplishments, and identifies emerging challenges toward sustainable development. Regarding current challenges such as population growth, urban sprawl and informal settlements, and unprecedented displaced populations, Habitat III establishes new urban agendas to reaffirm global commitments to sustainable development. It emphasizes, for instance, enhancing the public supply of land for affordable and sustainable housing in central and consolidated areas of cities, strengthening municipal finance and local fiscal systems, and encouraging mixed-income development to promote social inclusion and cohesion.

3 Reconstruction Policy in General

The increased frequency and severity of natural disasters is a radical challenge to the sustainable development of human settlements (CRED 2016). The World Bank considers reconstruction policy as a source of reform in the power relationship or allocation of resources within society (Pugh 1995; Zanetta 2004). When disasters occur, reconstruction policy lays out the rules for recovery (programs and projects), defining how different actors will coordinate, provide various forms of support, and adopt risk reduction measures against future disasters. Academics and practitioners in the disaster management field commonly believe that a holistic policy must address institutional and financial strategies; the role of stakeholders; mechanisms of coordination and financing; modes of communication; infrastructure reconstruction; training; environmental management; land use planning; and economic development.

The emergence of reconstruction policy began in parallel with the appearance of self-help housing policy in developing countries in the 1970s. Davis (1977,1978) recognized housing reconstruction as "a human social process as much as a technical one" and considered survivors as "active participants" instead of passive "victims." This insight grounded a move from top-down toward people-centered and participatory approaches. About two decades later, in 1994, the World Conference on Natural Disaster Reduction, in Yokohama (Japan) recognized community involvement and participation as a necessary component in every effective recovery program. Since then, numerous studies and world conferences have encouraged the development of comprehensive reconstruction policy and implementation. In the early years of the twenty-first century, the adoption of the resilience approach was consolidated. This approach integrated the holistic consideration of the natural, built, and social environments, stressing how society can deal with (and adapt to) disturbances caused by extreme events. Bosher (2008), for instance, encouraged a shift from "resistance" to "resilience" and emphasized the need to recover rapidly but sustainably. The World Conference on Disaster Risk Reduction in Hyogo in 2005 (Japan) provided critical guidance to building the resilience of nations and communities to disasters (UNISDR 2005). More recently, resilience policy has also been encouraged by other international programs such as Rockefeller Resilient City Program (Rockefeller Foundation 2016) and the UNISDR program of Making Cities Resilient (UNISDR 2012).

The trend of increased emphasis on reconstruction policy was consistent in the late 2000s. Lyons et al. (2010) and Lizarralde et al. (2010) recognized reconstruction as an opportunity to reduce vulnerabilities and enhance sustainable development, in what is now called the owner-driven housing reconstruction policy. According to them it is an efficient and ethically appropriate approach that can help to reduce risks, enhance preparedness, and build back better (Lizarralde et al. 2010 and Lyons et al. 2010). Lately, the Third United Nations World Conference on Disaster Risk Reduction in Sendai in 2015 (Japan) argued for a better integration of research findings into policies, plans, and programs, to help governments and relevant stakeholders identify risks and invest properly in improving resilience. See more about the evolution of reconstruction and housing policies in Table 1.

4 Housing Reconstruction Policy in Particular

Housing reconstruction can be singled out as an important, if not the most important, ingredient in the recovery of households after disasters (Alexander 1993; Quarantelli 1999). Housing reconstruction policy evolved under the influence of reforms in reconstruction and housing policies in developing countries (see Fig. 1). Post-disaster reconstruction literature often recognizes the following approaches in housing reconstruction policy.

Procurement and Turnkey Policy It typically refers to replacing damaged houses with houses provided by governments or reconstruction agencies. This approach relies on standardization, technology-oriented solutions, and the use of professional developers to increase the speed of reconstruction, keep costs down, and ensure the quality of final products. However, final products are often the repetition of a basic module that takes little account of the beneficiaries' capacities and skills, tends to neglect cultural and local conditions, and disregards users' real needs (Barenstein 2006). Dikmen et al. (2012) studied the reconstruction program after the Dino earthquake in 1995 (Turkey) and explained how an inadequate fit between beneficiaries' way of life and the basic model houses caused users' dissatisfaction, which eventually led to leaving houses vacant. Housing reconstruction using the turnkey policy caused similar outcomes in Bou'in-Zahra, Iran, after the 1962 earthquake (Fallahi 1993), Yemen after the 1982 earthquake (Barakat 1993), and Abruzzo, Italy, after the 2009 earthquake (Bologna 2010).

Community-Driven Reconstruction (CDR) This policy gives increased rights to affected communities to play active roles in planning the reconstruction processes and take crucial decisions in resource investment (Maskrey 1989). Jha et al. (2010) explain that the role of a community in reconstruction may vary considerably. Reconstruction after the 1983 earthquake in Popayán, Colombia, was one of the earliest and successful examples in which community members played leading roles. Communities were organized in groups of 15–20 families, under the direction of a locally elected coordinator, a treasurer, and a secretary. In this case, around 87% of participants occupied the houses they had built, and roughly 84% were satisfied with the size and design of their houses and the

Table 1 Housing and post-disaster reconstruction policies since the 1950s

Housing policies in developing countries	Post-disaster reconstruction policies	
	Literature (selected literature)	World conferences
Procurement and turnkey policy (1950–1972) Provision of public housing in the form of standardized residential complexes *Self-help policy (1972–1983)* Self-help activities with minimum direction from centralized bureaucracies *"Market-enabling" policy (1983–1989)* Housing and macroeconomic and development policy and the emergence of neoliberal policies *"Well-functioning housing" policy (1992–2000s)* Development of the whole housing sector in its urban and national context *Second United Nations Conference on Human Settlements – Habitat II (1996)* Improvement of human settlements on sustainable basis, economic reforms, social investment, improvements to the environment, and democratic governance *United Nations Convention on Climate Change in Kyoto (1997)* Emission reduction *Millennium Summit Conference (2000)* Millennium goals *Third United Nations Conference on Human Settlements – Habitat III (2016)* (a) Public supply of land for affordable and sustainable housing in central and consolidated areas of cities, (b) strength of municipal finance and local fiscal systems, (c) mixed-income development to promote social inclusion, etc.	*Emergency Shelter* (Davis 1977) *Shelter after Disaster: Guidelines for assistance* – (UNDRO 1982; Davis 1978) (a) Local authorities are the most capable to manage reconstruction programs, (b) surviving families have motivations for the reconstruction of their houses, (c) relocation is rarely feasible, (d) reconstruction is an opportunity for disaster risk reduction, and (e) avoid foreign ill-adapted solutions *Hazards and the Built Environment: Attaining Built-in Resilience* (Bosher 2008) A shift from *resistance* to *resilience*, being capable to both resist *and* recover rapidly *Building Back Better – delivering people-centered housing reconstruction at scale* (Lyons et al. 2009) (a) Reconstruction is an opportunity to reduce vulnerabilities and reach development, (b) people-centered housing (owner-driven in particular) is efficient and ethically appropriate *Rebuilding After Disasters: From Emergency to Sustainability* (Lizarralde et al. 2010) (a) Problems have to be tackled within their real complexity, (b) a system approach is necessary to understand complexities and to develop an appropriate organization, (c) owner-driven reconstruction approach is efficient and ethically appropriate, (d) shift from tactical to strategic planning *Safer Homes, Stronger Communities – A handbook for reconstruction after natural disasters* (Jha et al. 2010) (a) Emphasis on civil society and private sector, (b) assessment and monitoring can improve reconstruction outcomes, (c) community members should be partners in policy making and leaders of local implementations, and (d) sustainable reconstruction ensures long-term development	*World Conference on Natural Disaster Reduction, Yokohama, Japan* (1994) (a) Disaster prevention, mitigation, preparedness, and relief into development plans, (b) international cooperation in technology transfer and information sharing, (c) appropriate technology and data, (d) community involvement and participation *Second World Conference on Disaster Risk Reduction, Hyogo, Japan* (2005) (a) Strengthen institutions; (b) monitor disaster risks and enhance early warning; (c) knowledge, innovation, and education to build a culture of safety and resilience; (d) reduce the underlying risk factors; (e) strengthen disaster preparedness *Third World Conference on Disaster Risk Reduction, Sendai, Japan* (2015) (a) Understand disaster risk, (b) strengthen disaster risk governance, (c) invest in disaster risk reduction for resilience, (d) enhance disaster preparedness and "build back better"

building materials used (Barakat 2003). Despite its strengths, the CDR policy typically opens the door for different interpretations of the degree of community participation. According to Barenstein (2010, p. 98), the CDR also permits that agencies consult only with community elites and leaders "whose views do not reflect those of community." Similarly, agencies can impose restricting standards and regulations, limiting families' preferences and ignoring their expectations (Hidellage and Usoof 2010; Karunasena and Rameezdeen 2010).

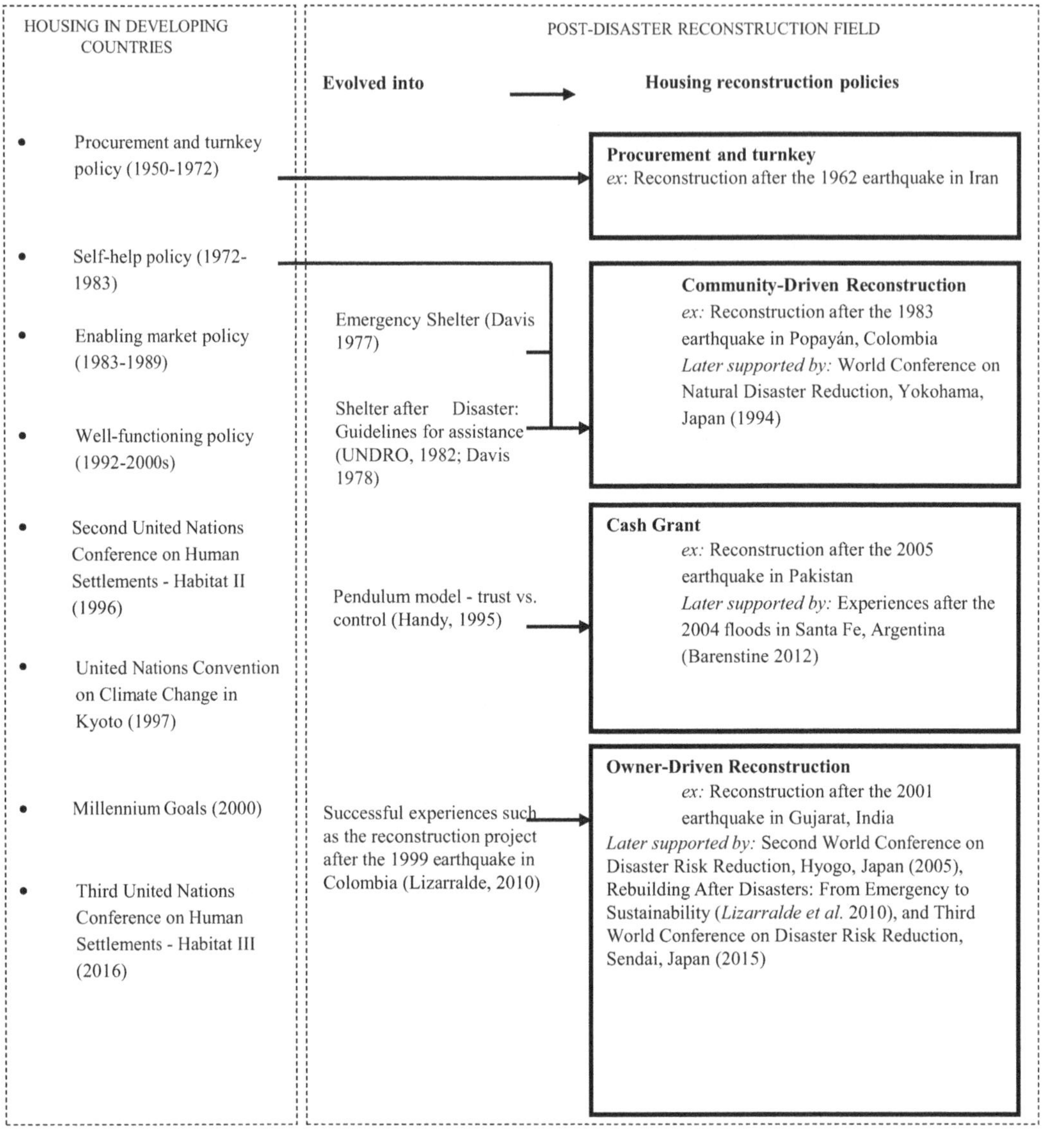

Fig. 1 Links between housing reconstruction policies and reforms in both reconstruction and housing policies in developing countries

Cash Grant Policy This policy stands on the belief that the distribution of cash (as a replacement for "in-kind" assistance) helps empower beneficiaries, decreases dependency on reconstruction assistants, and gives residents additional freedom to make decisions (Davis and Alexander 2015). The cash approach gives beneficiaries the choice to use the assistance based on their priorities, which may not necessarily be housing. The pendulum (trust vs. control) model explains the fact that if leaders "place more trust in their workforce, fewer controls are needed" (Handy 1995). Drawing on the pendulum model, Davis and Alexander (2015, p.105) explain how disaster managers may assume that "beneficiaries will use the money prudently and not on the purchase of non-essential goods." In Pakistan after the 2005 earthquake, cash grants acted as leverage, ensuring the compliance of new buildings with building standards and introducing families to banking – an essential entry point to further economic development (Davis and Alexander 2015). On the other hand, many researchers warn some potential risks such as corruption, reproduction of pre-disaster vulnerabilities, the use of financial assistance for not housing-related requirements, and the lack of opportunity for labor training and introducing better techniques and materials (Barenstein and Iyengar 2010; Karunasena and Rameezdeen 2010; Pugh 1995).

Owner-Driven Reconstruction' (ODR) policy This approach consists on providing conditional financial assistance under strict control and supervision (Davis and Alexander 2015). In an ODR program, families are expected to reconstruct their houses and manage the process of reconstruction using a combination of financial and technical assistance. The ODR policy first appeared during post-disaster reconstruction in the 1999 earthquake in Colombia (Lizarralde 2010) and the 2001 earthquake in Gujarat (Barenstein and Iyengar 2010). The ODR policy aims at enabling families to return to normal life faster and helping people who have been through a trauma restore their sense of pride and well-being. Barenstein and Iyengar (2010) explain that the adoption of ODR policy in Gujarat led to empowering disadvantaged and marginalized communities, upgrading technical solutions, and improving sociocultural practices and local self-governance. They emphasize that the success of ODR relies on providing an enabling environment that can be created by subsidizing the price of, and strengthening access to, key building materials, ensuring administrative support to the most vulnerable, developing relevant technical guidelines, and facilitating technical assistance and training (Barenstein and Iyengar 2010, p. 165).

The ODR policy is often conceived as "the most empowering and dignified approach" to household's recovery (Barenstein 2010, p. 95). However, given the potentially dissimilar impact of disasters on households and their unequal capacity to recover, *how* can housing reconstruction policies, and the ODR policy in particular, cater to the needs of different social groups and promote social equity after disasters?

5 Methodology

In order to answer this question, this study adopts a deductive reasoning strategy, which, according to Babbie (2012), includes *pattern matching*, that is, a research proposition that might be logically or theoretically expected and observations that examine whether the expected pattern actually occurs. Conducting a case study, as part of the experimental research, is an appropriate method to confirm or contrast the theoretical expectation (hypothesis) and produce research findings (Creswell 2013). This study is based on a detailed, longitudinal, qualitative-quantitative case study of the housing reconstruction program conducted after the earthquake that struck the city of Bam in Iran on December 26, 2003. The case of Bam was selected for its diversity of housing reconstruction policies and the different levels of recovery among various groups of households, so as to observe the impacts of distinct policies – from owner-driven to subsidized programs – adopted.

In the first steps of the study, a detailed and extensive review of housing policies in developing countries, and post-disaster housing reconstruction policies in particular, led to the formulation of a hypothetical proposition. The main preposition (predicted pattern) is that the adoption of a single policy is hardly likely to lead to the recovery of all variously affected households. This recovery instead is largely influenced by a number of variables such as ownership rights, social connections, lifestyle, and livelihoods. In a second stage, empirical findings from the case of Bam were matched with the predicted patterns described above. The results highlight the effects of housing policies on different categories of households and validated (but also nuanced) the theoretical proposition. The findings then contribute to theory building, or what Yin (2003, p. 33) describes as an "analytical generalization," that is, a generalization to "the theoretical propositions rather than to populations or universes."

Data collection occurred over five separate field trips to Bam: (1) July–August 2004, (2) November 2008, (3) January 2011, (4) March 2012, and (5) June–July 2014. Data was collected using 12 interviews with officers and authorities in Bam and Tehran and 70 interviews with residents in Bam. Interviews with households took about 45 min, and those with officers and authorities, regarding their responsibilities in the reconstruction program, varied from 30 min to 2 h.

Secondary data was also collected from more than 32 reports and 6 policy documents, including minutes of project meetings, press releases, and construction documents, and the 11 thematic reports of the Bam Reconstruction Documentation Project (BRDP) conducted by the Housing Foundation of Islamic Republic (HFIR).

The use of multiple sources of evidence helps support facts and hypotheses. In order to verify and corroborate information collected from the interviews, we reviewed reports and policy documents and conducted some field visits and direct observations. This triangulation of data decreased the risk of personal interpretation and interviewees' distorted memories, which eventually minimize the danger of incomplete and conflicting reports. In other words, the trinangulation of data "converge lines of inquiry," Yin (2008, p. 98). Using the triangulation of collected data and methods, we could create a coherent narrative of the events and decisions made in the reconstruction project during 10 years. This permitted to follow up not only the implementation of different policies over time but also the effects of these policies over a 10-year period, providing unique information about the reconstruction process.

6 Results

6.1 The 2003 Bam Earthquake in Iran and First Response

On December 26, 2003, a 6.7 magnitude earthquake severely struck the city of Bam, Iran (Ghafory-Ashtiany and Hosseini 2008). Because of the intensity of the earthquake, the time of occurrence, and the instability of traditional mud-straw houses, the event led to a high rate of casualties and damages: approximately 22,400 people died, more than 75,000 residents were left homeless, and nearly 93% of urban buildings were destroyed (Statistic Center of Iran 2003). Immediately after the earthquake, the Iranian government set up the Reconstruction Supervision and Policymaking Association (RSPA), an interministry organization headed by the Minister of Housing and Urban Development with extensive power paralleling that of the president's cabinet (Fallahi 2007; Fayazi 2012; Fayazi and Lizarralde 2013a). The RSPA made all decisions related to the recovery and reconstruction phases. Setting recovery objectives and defining reconstruction policies were among these decisions. The Housing Foundation of Islamic Republic (HFIR) was quickly identified as the sole housing reconstruction executor. To prevent the emergence of parallel organizations and excessive bureaucracy, the RSPA designated the HFIR to coordinate the relationship between contractors, banks, affected families, and the municipality.

6.2 Household Types

For the purpose of this study, affected households were classified according to a set of three indicators that address their most significant pre-disaster conditions, one indicator of impacts of the disaster on households, and one indicator about the way the program responded to these conditions. The five indicators are (1) pre-disaster ownership rights, (2) lifestyle, (3) quality and location of houses, (4) casualties and emotional impacts of the disaster, and (5) type and quality of temporary houses received after the disaster. By a combination of all variables, beneficiaries could be categorized in more than 100 types. However, the detailed observation of social constructs on the ground, the experience of the main researcher with Bam social groups, the identification of the most significant communities, the responses of interviewees regarding social groups, and the analysis of demographic data (notably the social groups identified by the Statistics Center of Iran) revealed that six main

Table 2 Categories of beneficiaries according to the set of five indicators chosen for the study

		Household categories					
Variables		1	2	3	4	5	6
Pre-disaster ownership status	House-owners	X					
	Apartment owners				X		
	House tenants		X			X	
	Apartment tenants						
	Without legal ownership rights			X			X
Pre-disaster lifestyle	Single families	X			X	X	X
	Extended families	X	X				X
Pre-disaster location	Inner city (downtown)	X	X	X	X	X	
	Affluent neighborhood	X	X	X		X	
	Low-income neighborhoods	X	X	X	X	X	
	Informal settlements (in city outskirts)						X
Loss of bodies and emotional impacts of the disaster	Minor impacts (no loss of family members)	X	X	X	X	X	X
	Major impacts (loss of family members and/or depression)	X	X			X	X
Temporary solution offered after the disaster	Temporary units in the yard of destroyed houses	X	X	X		X	X
	Camps in the outskirt of the city or inside the city			X	X	X	X

types of beneficiaries can be considered socially representative in the city (Table 2).

The main six categories of beneficiaries are (1) pre-disaster house-owners (about 42% of the population), (2) members of extended families (17%), (3) young couples who married after the disaster (3%), (4) pre-disaster apartment owners (10%), (5) pre-disaster tenants (20%), and (6) informal settlers (3%).

6.3 Reconstruction Policy in Bam

The recovery program assumed that enabling people to have a leading role in the reconstruction of their houses and supporting them with assistance would lead to overall recovery. In practice, however, the recovery of heterogeneously affected households needed the adoption of three sets of policies in different periods of time, leading to varied levels of recovery.

6.4 The Initial Policy: The Provision of a Financial and Technical Aid Package for House-Owners

Relying on the learned lessons from previous experiences in Iran, the RSPA adopted the ODR policy about 2 months after the earthquake, recognizing house-owners as managers in the reconstruction process of their houses and enabling them to recover faster. Equal compensation and

Fig. 2 An example of the ODR reconstructed houses (Source: author 2014)

distribution of the same resources (5% interest loans of about $10.750 US) were provided to all affected house-owners. Families were responsible for consulting with designers to choose a plan, supplying the needed materials, inspecting the construction, optimizing the construction materials usage, and cooperating with the inspection authorities throughout the different reconstruction phases (Omidvar et al. 2010).

Two years after the disaster, investigations showed that only 44% of the population was able to reconstruct their houses (Fig. 2), and about a third of pre-disaster house-owners had not reconstructed their houses. Before the disaster, members of extended families or renters were sharing dwellings with parents or landlords. The provision of equal compensation regarding the number of destroyed houses led to the replacement of big houses with small ones that were designed for single families. This change in size left pre-disaster renters and members of extended families in either temporary housing camps or in temporary units located in the yards of reconstructed houses. According to Tafti and Tomlinson (2013), families who started building more than one unit (one for their housing and another for rent or family members) were less likely to finish the rebuilding of their own houses.

The initially adopted policy could only help single-family house-owners (about 32% of the population) start reconstruction of their houses quickly, participate actively in design and implementation, and receive sufficient training about safe construction technologies and the use of new materials. These residents also had a chance to stay in touch with their neighbors and friends and take care of their date palm groves, which often provided their primary source of livelihood (Rafieian and Asgary 2013).

After the emergency phase, the adverse conditions (including harsh climate conditions) and the expectation of a long process of reconstruction forced authorities to move affected families to temporary units until permanent reconstruction could be completed. Around 37,900 units were ultimately built by adopting distinctive strategies, including temporary housing camps and separated units on the yard of destroyed houses (Fayazi and Lizarralde 2013). The quality of temporary houses that were developed was critical in the recovery. "Temporary" units made of masonry (with safe construction techniques) eventually provided an extra bedroom or an area for conducting domestic businesses after the reconstruction of permanent houses on the same lot. Prefabricated units, instead, did not last long and

Fig. 3 Prefabricated temporary units (left). Temporary units made of masonry materials (right) (Source: author 2014)

became rapidly dilapidated and of little use for residential purposes (see Fig. 3).

Some owners found additional struggles in reconstructing their homes. The loss of family members, and heads of families in particular, caused long delays, additional costs, and other difficulties in the reconstruction process. The value of financial aid provided by the government decreased because of rising inflation at the time, which then eroded the households' capacity to reconstruct their houses. Thus, about 3% of house-owners were never able to reconstruct their houses and eventually remained in the temporary camps almost 2 years after the disaster. Abdolreza was a house-owner who never reconstructs his house. He expressed "We lost everything; my two daughters, brother, and parents in law. My wife and I were so depressed, felt lonely, and didn't want to continue [our life]. Zahra [his wife] didn't want to go out of our tiny cabin [temporary housing unit] for days and talk to nobody, even me. We had tough days for almost eighteen months. When we came to reconstruct our house, everything was too expensive, and the financial aid was hardly enough for the construction of a 40 m^2 unit. We thought if we wait longer, prices will return to normal conditions; but that never happened."

Likewise, the vast majority of pre-disaster apartment owners lived in the temporary camps for more than 2 years. They met significant technical, logistic, and legal challenges in the reconstruction of their buildings – such as the complexity of dealing with inheritance laws when neighbors died. One of the pre-disaster apartment owners said "... two out of our five neighbors died, and their children inherited their apartments. I begged them to reconstruct the building, but they were reluctant. They live in Kerman and Tehran and have no intention to reconstruct that building." A revision of policy was then necessary (see Table 3).

6.5 The Revised Policy: Ownership Rights

Two years after the earthquake, a grant of $10,750 US was provided to tenants, members of extended families, and young couples who married after the disaster to start the reconstruction of their houses. The prerequisite for receiving the grant was to own land in the city or to be able to use a plot of land with the agreement of its owner (NDTF 2014; Tafti and Tomlinson 2013). While a total of 4950 residential units were built using this grant, it insufficiently addressed the most vulnerable households, low-income tenants in particular. Those who received this assistance were among wealthier and middle-class tenants, apartment owners, and extended family members.

Before the disaster, hundreds of young couples lived in their parents' houses. Even though many stayed with their parents in temporary shelters after the disaster, some were able to reconstruct new houses in the same yard (about 12% of the population). Many of young couples

Table 3 The impact of the housing reconstruction policies on different beneficiary groups and the estimated demographic distribution in each category

Beneficiary groups	The initial policy	The second policy	The third policy
Pre-disaster house-owners	Single-family house-owners reconstructed their houses (≈ 32%)	–	–
	Extended family house-owners reconstructed small houses (≈ 12%)	–	–
	Stayed in temporary shelters (≈ 3%)	Stayed in temporary shelters (≈ 3%)	Received an apartment (≈ 3%)
Members of extended families	Stayed in temporary shelters (≈ 17%)	Split land and reconstructed in the yard of parents' houses (≈ 12%)	–
Young couples who married after the disaster	Not considered as beneficiaries of aid (≈ 3%)	Purchased a piece of land and built a new house (≈ 5%)	–
		Stayed in temporary shelters (≈ 3%)	Received an apartment (≈ 3%)
Apartment owners	Lived in temporary camps for more than 2 years (≈ 10%)	Purchased a piece of land and built a new house (≈ 8%)	–
		Stayed in temporary housing camps (≈ 2%)	Received an apartment (≈ 2%)
Pre-disaster tenants	Stayed in temporary housing camps (≈ 20%)	Purchased a piece of land and built a new house (≈ 13%)	–
		Stayed in temporary housing camps (≈ 5%)	Received an apartment (≈ 5%)
		Migrated from the city (≈ 1%)	–
		Formed a new informal settlement (≈ 1%)	–
Informal settlers	Not considered as beneficiaries of aid (≈ 3%)	Not considered as beneficiaries of aid (≈ 3%)	Received an apartment (≈ 3%)

from wealthier families were able to buy a piece of land and build a new house, while others received their parents' agreement to split their lands and construct new houses. Instead, those who could neither afford land nor split their parents' land had to stay in temporary shelters (about 5% of the population). According to one resident: "…we [my parents, my wife, children and I] have to live together, it's our lifestyle. My mother takes care of our kids when we [my wife and me] work on our inherited date palm groves. Using the financial aid and our savings, we could only afford the reconstruction of a small house [about 60 m^2] for our parents. Many families could reconstruct an extra house in the same yard, but our yard is too small and doesn't let us build anything more. My wife and I still sleep in the cabin [temporary housing units in the yard] and wish to sleep in our real house one night" (see Fig. 4).

The second policy failed to reach the majority of tenants (Tafti and Tomlinson 2013). While few better-off tenants could shift their tenure to ownership, others stayed in the temporary housing camps, or migrated from the city, or to new informal settlements. In fact, with hopes of receiving funds to build new houses, some pre-disaster tenants rushed to buy land at low prices in the outskirts of the city, in Janbazan Town. However, the municipality did not give them permission to construct new houses in this area. "Nobody told us that this is a forbidden land. I gave all I saved

Fig. 4 Members of extended families who could not afford land and could not split their parents' land to build their own houses were still living in these temporary units (Photo by Fayazi 2008)

Fig. 5 Informal settlement (Janbazan) in the outskirts of the city (Photo by Fayazi 2014)

during the last ten years, and now this is all I have: living in a Cabin [temporary unit] afraid of eviction" said one inhabitant of Janbazan Town in 2014. It is estimated that more than 100 affected families carried their prefabricated temporary housing units from the camps to Janbazan and formed an informal settlement (Fig. 5).

The impacts of this second-generation policy for pre-disaster apartment owners varied. The affluent ones could purchase a land and construct their house, while lower-income apartment owners could not receive the grant. Given the rising inflation at that time, and the fact that households had to pay for purchasing land as well, this policy only helped wealthier tenants, part of extended families' members, and the apartment owners who could provide land. It confined lower-income tenants, informal settlers, and much of new couples to camps and temporary housing units. Please see Table 3.

6.6 Third-Generation Policy: Residential Complexes Located in the Periphery of the City

Three years after the disaster, a significant number of the most vulnerable, low-income households were still living in temporary housing camps (about 15% of the population). The first ODR policy and the subsequent modification could not reach them. The following policy was an agency-driven reconstruction plan in a relocated site. The HFIR in collaboration with the Ministry of Housing and Urbanism built about

Fig. 6 Residential complexes located in the periphery of the city (Razmandegan Town) (Source: author 2014)

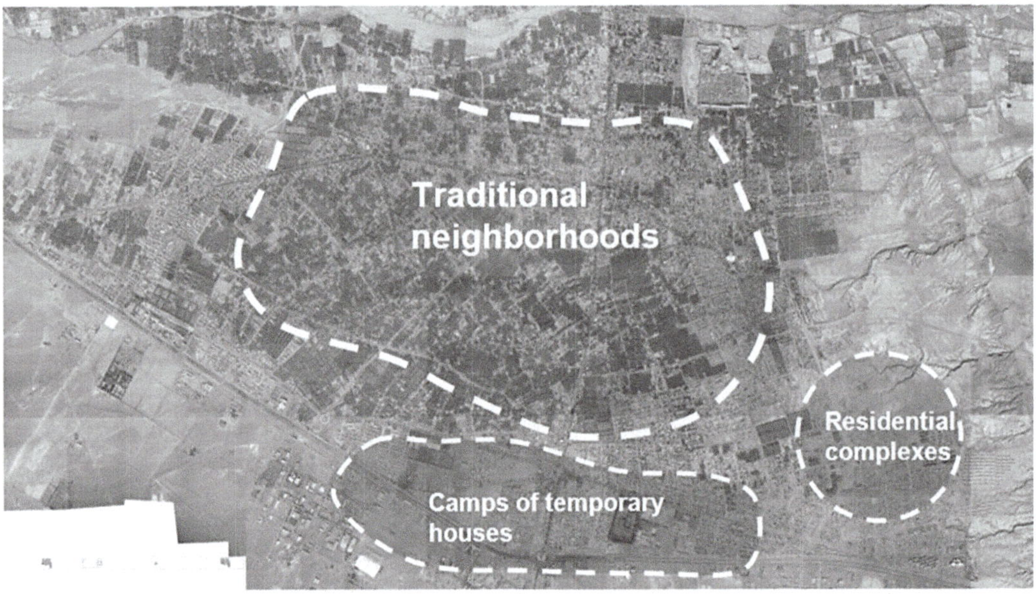

Fig. 7 Location of the main areas (Adapted from Fayazi et al. (2015))

4300 apartment units located in 50 multistory residential complexes on the eastern side of the city called Razmandegan Town (Figs. 6 and 7). Almost all the units were completed 7 years after the disaster. During the construction process of the residential complexes, the households had to register for the program by providing about $6650 USD. This down payment was affordable for many, and families could pay it from the grant provided by the government. The families also had to pay $150 USD per month for 10 years. In total, every apartment unit cost $24,650 USD that must be paid in 10 years. However, the units were generally more expensive than the owner-driven reconstructed houses, and the relocation imposed increased transportation costs for residents (see Fig. 7).

This policy considered informal settlers equal to the rest of vulnerable families in Bam. Before the disaster, a still unknown number of families lived in an informal settlement on the eastern side of the city. They became eligible to financial aid and an apartment in the residential complexes. The government also gave them equal access to basic services, reducing their pre-disaster socio-economic vulnerabilities (see Table 3).

Though pre-disaster informal settlers enjoyed feeling the sense of ownership of their new apartment, pre-disaster tenants and apartment owners, who could not afford a piece of land and were economically, socially, and emotionally dependent on their neighbors, extended family, and communities, were adversely affected when they moved into the new residential complexes (Fallahi 2007; Fayazi 2012). They addressed this issue several times during the interviews. For instance, one of the pre-disaster apartment owners said: "we live on an island, disconnected from our community. We finally received a roof, but lost our friends, families, and neighbors."

Different pre-disaster conditions, priorities, and needs caused various levels of recovery among households. While the acquisition of a new apartment decreased tenants' vulnerabilities in different dimensions, for many of them, living in residential complexes meant being excluded from their communities. Many of them deplored the increase in transportation costs, loss of social networks, and limited capacity to receive support from family and friends.

7 Discussion

The importance of sheltering and housing reconstruction in the recovery of affected families after disasters is well recognized in hazard-related literature. An appropriate sheltering and housing program can stimulate risk reduction, foster social structures, and achieve development in the broader sense. In practice, however, housing reconstruction programs seldom help all different beneficiary groups recover in the same way and result in unexpected heterogeneities. Given various pre-disaster vulnerabilities and resilience levels, every disturbing event impacts different groups of families in distinct ways. Low-income families and informal settlers are typically more fragile than affluent ones who live in less vulnerable areas and more resistant buildings. Loss of family members (and heads of families, in particular) causes serious problems and radically hampers recovery. Seemingly chaotic conditions after disasters often involve supposedly similar families losing their loved ones and looking for help, but survivors are always very different and have diverse needs, desires, and expectations for recovery. Numerous studies have tried to understand the exact impacts of undesirable events on society and find solutions for addressing the diversity of needs and expectations in the reconstruction of houses (Barenstein 2008; Bolin 1982; Caporale 1989); however, such an understanding still remains inadequate.

Our empirical results show that the initially adopted ODR policy in Bam helped merely single-family house-owners and the core of extended families to reconstruct their houses. The modification of land issues in the revised ODR policy failed to cover all different groups of households and excluded the most vulnerable families: tenants, apartment owners, and informal settlers. Encouraging families to buy land and rebuild their houses generated a new informal settlement, exacerbated social gaps, and increased vulnerabilities and inequalities. The third policy was a retreat to the procurement and turnkey approach, which provided apartments for those who remained in transitional sheltering camps 4 years after the disaster. However, residential complexes in the outskirts of the city increased transportation costs, expelled families from their communities, and limited their capacity to receive support from family and friends.

The main challenge for policy makers in Bam was the recognition of households' diversity and the adoption of appropriate housing policies to address the specific conditions of every group of households. The housing reconstruction program in Bam overlooked the diversity of affected households regarding their pre-disaster status and the impact of the disaster on them. In addition, the intervals between different policies in Bam, which were at least 1 year, caused insecurity, frustration, and a sense of "being excluded" from the housing reconstruction program among some households. In the period of time between policies coming to effect, disappointed families took irrevocable decisions, permanent migration, or unsafe reconstruction in vulnerable areas, which ultimately affected their recovery. The Bam case shows us how crucial it is to develop effective housing reconstruction policies that respect diversities and heterogeneities, allowing households to choose a set of solutions that fit their conditions, priorities, and needs.

There are important theoretical implications of these results. First, they highlight the fact that the one-policy-for-all approach cannot effectively lead to the recovery of affected families. Reconstruction policy needs to include all types of households, respecting their contextual and specific conditions. Second, results show the gap between pre-disaster housing and post-disaster housing reconstruction policies. National governments and international agencies sometimes establish housing reconstruction policies based on insufficient knowledge about both pre-disaster conditions and the disaster's impact on families. There is often a need for a better integration of housing reconstruction and general housing policies, ensuring equal recovery and preventing the reproduction of vulnerabilities. In fact, housing reconstruction policies can ensure sustainable development, resilience enhancement, and disaster risk reduction by addressing pre-disaster deficits and vulnerabilities, such as providing ownership rights, increasing access to services in informal settlements, reducing social gaps, and mitigating disaster risks.

There are also relevant practical implications of these results. First, authorities can make better decisions if they reassess pre-disaster policies, evaluate their impacts on households' vulnerability, and conceive housing reconstruction programs as an opportunity to address pre-disaster deficits. Second, policy makers can issue sets of policies and initiatives at the same time to help different categories of households choose the most suitable solution to their conditions. These results, however, have to be taken with prudence, given that this research experienced several limitations, including scarce information about general housing and urban policy before the disaster in Iran. Pre-disaster policies were not examined, and results cannot explain how the housing reconstruction policies could tackle pre-disaster barriers and deficits. More research is still needed to bridge this methodological gap.

8 Conclusion

Post-disaster reconstruction experiences show varied levels of recovery among different groups of households, which sometimes exacerbate pre-disaster social conditions such as poverty, social exclusion, and marginalization. By studying the housing reconstruction experience after the 2003 earthquake in Iran, this article reveals how housing reconstruction policy often oversimplifies pre-disaster conditions and overlooks diversity of households after disasters. Scant attention to families' socioeconomic conditions and demographic changes before, and after, the disaster led authorities to adopt housing reconstruction policies that benefited some groups of households while having the opposite effect on others. Theoretical implications of these findings point to the inefficiency of the one-policy-for-all approach in housing reconstruction after disasters. Results indicate how crucial is to issue multi-initiatives and solutions at the same time, letting households choose the most appropriate solution. The results also call for the integration of general housing and housing reconstruction policies to address causal factors of pre-disaster vulnerabilities, reveal households' diversities, and ensure long-term recovery, sustainable development, and disaster risk reduction. At the practical level, results call for the reassessment of pre-disaster conditions before the adoption of reconstruction policies, notably to prevent the preservation of vulnerabilities and inequalities. Further studies must be conducted in the housing, urban development, and post-disaster housing reconstruction fields to bridge the gaps between pre- and post-disaster policies.

References

Aldrich, D. P. (2012). *Building resilience: Social capital in post-disaster recovery*. Chicago: University of Chicago Press.

Alexander, D. E. (1993). *Natural disasters*. The Netherlands: Springer.

Alexander, D. (2008). Mainstreaming disaster risk management. In *Hazards and the built environment: Attaining built-in resilience* (pp. 20–36). Londres: Taylor and Francis.

Aysan, Y., & Oliver, P. (1987). *Housing and culture after earthquakes: A guide for future policy making on housing in seismic areas*. Oxford: Oxford Polytechnic Press.

Babbie, E. R. (2012). *The basics of social research*. Belmont, USA: Cengage Learning.

Barakat, S. (1993). *Rebuilding and resettlement, 9 years later; a case-study of the contractor built reconstruction in Yemen, following the 1982 Dhamar earthquake*. IAAS. Post-war reconstruction and development unit.

Barakat, S. (2003). Housing reconstruction after conflict and disaster. *Humanitarian Policy Group, Network Papers, 43*, 1–40.

Barenstein, J. (2006). Challenges and risks in post-tsunami housing reconstruction in Tamil Nadu. *Humanitarian Exchange, 33*, 38–39.

Barenstein, J. (2008). *From Gujarat to Tamil Nadu: Owner-driven vs. contractor-driven housing reconstruction in India*. Paper presented at the 4th international i-Rec conference: building resilience.

Barenstein, J. (2010). Reconstruction approaches. In A. K. Jha, J. D. Barenstein, P. M. Phelps, D. Pittet, & S. Sena (Eds.), *safer homes, stronger communities – A handbook for reconstructing after natural disasters* (pp. 93–107). Washington, DC: The International Bank for Reconstruction and Development / The World Bank.

Barenstein, J., & Iyengar, S. (2010). India: From a culture of housing to a philosophy of reconstruction. In M. Lyons, T. schilderman, & C. Boano (Eds.), *Building Back Better: Delivering people-centered housing reconstruction at scale* (Vol. 163). Warwickshire, UK: Practical Action Publishing Ltd.

Bolin, R. C. (1982). *Long-term family recovery from disaster*. Colorado, U.S: Institute of Behavioral Science, University of Colorado.

Bologna, R. (2010). *Planning and designing post disaster transitional housing: The case of the Abruzzo (Italy) earthquake*. Paper presented at the Proceedings TG63–Special Track 18th CIB World Building Congress, Salford, UK.

Bosher, L. (2008). *Hazards and the built environment: Attaining built-in resilience*. London: Taylor and Francis.

Burgess, R. (1978). Petty commodity housing or dweller control? A critique of John Turner's views on housing policy. *World Development, 6*(9), 1105–1133.

Caporale, R. (1989). Toward a cultural-systemic theory of natural disasters: the case of the Italian earthquakes of 1968 in Belize, 1976 in Friuli and 1980 in Campania-Basilicata. In Quarantelli & Peland (Eds.), *Preparations for, Response to, and Recovery from Major Community Disasters* (pp. 253–271). Newark, DE: University of Delaware.

Comerio, M. C. (1998). *Disaster hits home: New policy for urban housing recovery*. Berkeley, CA: University of California Press.

CRED (2016). *2015 disasters in number*. Retrieved from Belgium http://www.unisdr.org/files/47804_2015disastertrendsinfographic.pdf

Creswell, J. W. (2013). *Research design: Qualitative, quantitative, and mixed methods approach*. London: Sage.

Davidson, C. H., Johnson, C., Lizarralde, G., Dikmen, N., & Sliwinski, A. (2007). Truths and myths about community participation in post-disaster housing projects. *Habitat International, 31*(1), 100–115.

Davis, I. (1977). Emergency shelter. *Disasters, 1*(1), 23–39.

Davis, I. (1978). *Shelter after disaster*. Oxford: Oxford Polytechnic.

Davis, I., & Alexander, D. (2015). *Recovery from disaster*. Abingdon, New York: Routledge.

Dikmen, N., Elias-Ozkan, S. T., & Davidson, C. (2012). Comparison of Post-Disaster Housing Procurement Methods in Rural Areas of Turkey. *Open House International, 37*(1), 28–39.

Fallahi, A. (1993). Learning from the lessons of the post-earthquake reconstruction projects in Iran: An introduction. *Macedon Digest, The, 8*(4), 19.

Fallahi, A. (2007). Lessons learned from the housing reconstruction following the Bam earthquake in Iran. *The Australian Journal of Emergency Management, 22*(1), 26.

Fayazi, M. (2012). *Temporary housing experience after Bam Earthquake 2003*. Tehran: Housing Foundation of Islamic Republic.

Fayazi, M., & Lizarralde, G. (2013). The role of low-cost housing in the path from vulnerability to resilience. *International Journal of Architectural Research, 7*(3), 146–167. Retrieved from http://www.archnet-ijar.net/index.php/IJAR/article/view/56

Fayazi, M., & Lizarralde, G. (2013a). Temporary housing: an obstacle or an opportunity for enhancing community resilience in Bam, Iran In G. Lizarralde, J. Barenstein, G. Cardosi, & A. Oliver *the 6th international i-Rec conference* Ascona, Switzerland L'Observatoire universitaire de la vulnérabilité and la reconstruction durable (Oeuvre durable); pp. 92-107 http://www.grif.umontreal.ca/i-Rec2013/conferenceIREC2013papers.htm.

Fayazi, M., Kikano, F., & Lizarralde, G. (2015). The Impacts of Post-Disaster Reconstruction Policies on Different Categories of Households in Bam, Iran. Paper presented at t*he 7th International i-Rec conference*, July 6-8 in the Bartlett Development Planning Unit, University College London.

Ghafory-Ashtiany, M., & Hosseini, M. (2008). Post-Bam earthquake: Recovery and reconstruction. *Natural Hazards, 44*(2), 229–241.

Handy, C. (1995). Trust and the virtual organization. *Harvard Business Review, 73*(3), 132–142.

Hidellage, V., & Usoof, A. (2010). Scaling-up people-centered reconstruction: Lessons from Sri Lanka's post-tsunami owner-driven programme. In M. Lyons, T. Schilderman, & C. Boano (Eds.), *Building Back Better: Delivering people-centered housing reconstruction at scale* (pp. 77–112). Warwickshire, UK: Practical Action Publishing Ltd.

Jenkins, P., Smith, H., & Wang, Y. P. (2006). *Planning and housing in the rapidly urbanising world*. London, UK: Routledge.

Jha, A. K., Barenstein, J. D., Phelps, P. M., Pittet, D., & Sena, S. (2010). *Safer homes, stronger communities. A handbook for reconstruction after natural disasters*. Washington, DC: The World Bank.

Karunasena, G., & Rameezdeen, R. (2010). Post-disaster housing reconstruction: Comparative study of donor vs owner-driven approaches. *International Journal of Disaster Resilience in the Built Environment, 1*(2), 173–191.

Lizarralde, G. (2010). Decentralizing (re) construction: Agriculture cooperatives as a vehicle for reconstruction in Colombia. In M. Lyons, T. Schilderman, & C. Boano (Eds.), *Building Back Better: Building Back Better: Delivering people-centered housing reconstruction at scale* (pp. 191–214). Warwickshire, UK: Practical Action Publishing Ltd.

Lizarralde, G. (2015). *The invisible houses: Rethinking and designing low-cost housing in developing countries*. London: Routledge.

Lizarralde, G., Johnson, C., & Davidson, C. H. (2010). *Rebuilding after disasters: From emergency to sustainability*. London and New York: Spon Press.

Lyons, M., Schilderman, T., & Boano, C. (2009). *Building Back Better: Delivering people-centered housing reconstruction at scale*. Warwickshire, UK: Practical Action Publishing Ltd.

Lyons, M., Schilderman, T., & Boano, C. (2010). *Building back better: Delivering people-centred housing reconstruction at scale*. Warwickshire, UK: Practical Action Publishing Ltd.

Mangin, W. (1967). Latin American squatter settlements: A problem and a solution. *Latin American Research Review, 2*(3), 65–98.

Maskrey, A. (1989). Disaster mitigation: A community based approach.

NDTF. (2014). *The minutes of the National Disaster Task Force (NDTF) meetings on 03 April. 2004*. Tehran: Ministry of Interior Affairs.

Omidvar, B., Zafari, H., & Derakhshan, S. (2010). Reconstruction management policies in residential and commercial sectors after the 2003 bam earthquake in Iran. *Natural Hazards, 54*(2), 289–306.

Pugh, C. (1992). International finance and housing policies in developing countries. *Cities, 9*(2), 117–137.

Pugh, C. (1994). Housing policy development in developing countries: The World Bank and internationalization, 1972–1993. *Cities, 11*(3), 159–180.

Pugh, C. (1995). *The role of the World Bank in housing*. London: Zed Books.

Quarantelli, E. L. (1999). The disaster recovery process: What we know and do not know from research.

Rafieian, M., & Asgary, A. (2013). Impacts of temporary housing on housing reconstruction after the Bam earthquake. *Disaster Prevention and Management, 22*(1), 63–74.

Rockefeller Foundation (2016). 100 resilient cities. Retrieved from http://www.100resilientcities.org/resilience#/-_

Statistic Center of Iran. (2003). *Summary of Bam statistic*. Tehran: Statistic Center of Iran.

Tafti, M. T., & Tomlinson, R. (2013). The role of post-disaster public policy responses in housing recovery of tenants. *Habitat International, 40*, 218–224.

Turner, J. C. (1967). Barriers and channels for housing development in modernizing countries. *Journal of the American Institute of Planners, 33*(3), 167–181.

UNDRO. (1982). *Shelter After Disaster: Guidelines for Assistance*. New York: United Nations Publication.

UNISDR (2005). *Hyogo framework for action 2005–2015: Building the resilience of nations and communities to disasters*. Paper presented at the Extract from the final report of the World Conference on Disaster Reduction (A/CONF. 206/6).

UNISDR. (2012). *How to make cities more resilient: A handbook for mayors and local government leaders*. Geneva: UNISDR.

World-Bank. (1974). *Housing*. Washington: World Bank.

World-Bank. (1992). *Housing: Enabling markets to work*. Washington: World Bank.

Yin, R. (2003). *Case study research: Design and methods* (Vol. 5). California: Sage Publications Inc.

Yin, R. K. (2008). *Case study research: Design and methods* (Vol. 5). Thousand Oaks, CA: SAGE Publications, Incorporated.

Zanetta, C. (2004). *The influence of the world bank on national housing and urban Policies: The case of Mexico and Argentina during the 1990's*. London: Ashgate Publishing, Ltd.

Resettlement Challenges for Children After Disasters (Case Study): Bam City

Hedyeh Gamini and Hamid Amouzad Khalili

Abstract

The Bam earthquake catastrophe has had many negative effects on children. The purpose of establishing post-disaster "child-friendly spaces" is to provide an opportunity for children by designing safe spaces, in addition to creating good physical and psychological conditions that will be effective in helping children to rehabilitate faster after disasters. Bam city, after the horrible earthquake in 2003, was an example of a situation in which a child-friendly approach was considered, and governmental and international organizations and NGOs became involved there and constructed several child-friendly spaces. It seems that, 10 years after the earthquake, an assessment of child-friendly spaces' impacts is particularly important. Methodology included qualitative assessment-based approaches, and the content analysis method was adopted. In this study, interviews and group meetings were conducted, including presence of children who became juvenile and adolescent during the 10 years after the earthquake, along with parents and educators working in child-friendly spaces; and thus comments of children in relation to child-friendly spaces were collected and analyzed. Results indicate that despite most children being completely satisfied with these spaces, there are still challenges in the optimal planning and design of such spaces. For example, there are proposed strategies including localization of activities, considering the effect of climate on design and also the use of indigenous architectural knowledge, as well as paying attention to secure pathways for children. Providing furniture, interior design, and suitable equipment for children, separating health services, the use of resistant and waterproof tents and colorful ones with age-appropriate and happy schemes, and also increasing green space in addition to the closed spaces were requested by children as feedback on the post-disaster child-friendly space design of Bam.

Keywords

Post-disaster planning and design · Child-friendly spaces · Bam earthquake · Content analysis

1 Introduction

Building and operating "child-friendly spaces" (CFS) has an important role in post-disaster reconstruction programs under rehabilitation

H. Gamini (✉)
Shahid Beheshti University, Tehran, Iran

H. A. Khalili
University of Melbourne, Melbourne, Australia
e-mail: hamouzad@student.unimelb.edu.au

A. Asgary (ed.), *Resettlement Challenges for Displaced Populations and Refugees*, Sustainable Development Goals Series, https://doi.org/10.1007/978-3-319-92498-4_10

programs and psychological support frameworks for survivors. The purpose of establishing "child-friendly spaces" is creating safe places where children can participate in organized activities for gaming, social behaviors, and education and to introduce themselves. "Child-friendly spaces" is one of the planned interventions to protect children from physical trauma and injuries that help them in training and development, immediately after emergencies (UNICEF, A Practical Guide for Developing Child Friendly Spaces 2008, 7). In these spaces, various activities such as games, rehabilitation, education, health, and social and psychological works are performed. Generally, concepts of CFS are defined as activities of post-catastrophe in short or medium time that are done usually in tents or even open spaces without much planning or forethought to conditions. These spaces are usually provided by NGOs or local governments, and the UNICEF has also put them into practice and has assisted in construction of CFSs and coordination and supplying standards for evaluating them (Child Protection Working Group 2012).

On the 5th of January, 2003, an earthquake struck Bam city in Kerman province. Many children affected by the earthquake were harmed and faced with various psychological, physical, economic, social, and legal problems. To overcome the severe mental pressures, they needed immediate psychosocial support, including safe place for their leisure time. After the earthquake, governmental organizations and NGOs and international organizations supporting children responded in some places immediately in order to deal with these problems, and in cooperation with the organizations concerned with children's well-being, they planned and executed some projects (UNICEF 2004). These places were known in Bam as "child-friendly spaces." Creating the physical infrastructure and activities of child-friendly spaces was addressed by various authorities from the first days after the earthquake, but due to the lack of documented experiences in the country, these activities were carried out through trial and error. Ignoring the concept of efficiency as well as short-term, medium-term, and long-term logistics and difficulties of planning and providing activities and infrastructure to such spaces can lead to premature abandonment of these programs. Architectural design of these spaces was not highly organized and purposeful, due to most NGOs not being familiar enough with the crisis and crisis management concepts and processes, children's rights in critical conditions, and also tools for equipping and designing architectural spaces according to the needs of children. This study tries to investigate "child-friendly spaces" after the 2003 earthquake in Bam, to explore visible and hidden issues in planning, design, and construction of such spaces, and to analyze and explain results of the process, and thus to help develop programs, rules, and regulations for planning and design of post-disaster child-friendly spaces for authorities. The present study was conducted aimed to answer key questions: First, what were the planning, designs, and construction processes of child-friendly spaces in Bam after the 2003 earthquake? What were the short- and long-term effects on children? Then it evaluates the efforts by measuring success rates of child-friendly spaces from children's, parents', and educators' viewpoints. Accordingly, in this study and in the section reviewing the theoretical background, concepts related to child-friendly spaces and fundamental principles are described, and in the research methodology section, the qualitative approach and tools that were employed in the study are explained. In recognition of the research area, this planning process and physical progression of child-friendly spaces in Bam were described. Then, based on the analytical framework of the research, interviewees' important words were surveyed to extract related issues from them. After explaining concepts, summing up and results of the content of the interviews were inserted in table. To achieve this, first, key statements were extracted from individuals' talks to identify issues related to design and plan such spaces. Finally, conclusions are presented.

2 Research Background

Childhood is a very important part of human growth. Paying attention to physical and mental health of children in these critical years not only prevents occurrence of a lot of damage but also provides a happy and healthy life in adulthood (Save the children 2011). Given that in disaster and crisis situations, children's life and health would be threatened, it is essential that necessary measures are taken to protect them. Generally, children's rights are ignored after catastrophes, and this is more pronounced in issues related to participation, protection, and development. Under these conditions governments would be ineffective, and families often make the necessary efforts. Therefore, protection and safekeeping of children in this period of time are critical (Child Protection Working Group 2012). If children are not supported well in times of crisis, a direct consequence will impact their lives in adulthood. Catastrophe has traumatic and destructive physical, mental, social effects on children that can be alleviated by design and planning of a cooperative atmosphere and child-friendly spaces (Global Protection Cluster et al. 2011). "Child-friendly spaces" are considered as one of the planned interventions in order to protect children from physical trauma and injuries, as well as to aid to train and development, immediately after an emergency (UNICEF 2008, 7). Thus, two main objectives of "child-friendly spaces" are creating opportunities for children in order to enhance development, education, play, and create or strengthen resistance after a state of emergency or during prolonged state of crisis and finding out and identifying ways to respond to threats particular to all children or a particular group of them, including vulnerable groups after a state of emergency, crisis, or during prolonged crisis.

In fact, CFS approach is supporting children's rights programs, and their welfare in times of crisis is the first concern in child-friendly spaces program. This concept has been widely used since 1999. Child-friendly spaces in places where trauma has occurred, or military conflicts which threaten children, take a participatory approach that seeks to help children and protect people's safety. Such spaces have been proposed in specialized texts as a new idea, and it is only a short time that this idea has attracted the attention of experts in times of crisis and catastrophe situations (UNICEF 2008, 16). It is important to secure child-friendly places. If children feel secure and safe in an environment, they will be able to move easily in such environments. This movement not only will be beneficial for their physical growth but also enables them to express and release their internal energies. Even with tumbling, children can become familiar with their environments and acquire experiences (Kostenly 2008).

Provided facilities are tailored to the needs of children in child-friendly spaces. Such facilities will aid physical, social, and intellectual development of children. Generally, equipment in such spaces must be appropriate for children's physical dimensions (and the changing of these dimensions with their growth) as well as their needs at the time. To maintain and repair equipment used in such spaces, preventative measures to ensure the safety of children should be in place in these spaces. Moreover, taking precautionary measures in such spaces increases the sense of security for children and enjoyment of children using the equipment. If CFS is a physical space newly constructed, a clean, safe, and healthy environment can be prepared for children.

The concept of child-friendly space, due to the use of the word space, could create a false impression that the purpose of CFS is limited to adjusting relief efforts and infrastructure to the body dimensions and physical aspects of children. While addressing these things when planning CFS is always very important (UNICEF 2008, 15–16), there is more in creating CFS than engineering and procuring child-friendly equipment. Hence, CFS programs are implemented in five basic steps that are also partially separated and constitute CFS program's overall structure. All active employees and officials working with CFS should be familiar with the principles and rules governing child-friendly spaces and recognize standards, child protection policies, general laws, and regulations related to CFS facilities. Children and their input shall be included in the development of these rules. Activities of all those involved in CFS should be continuously monitored, and any person involved should have an equal voice in the process. It is very essential at child-friendly spaces that minimum standards be considered for safety and quality of environment to increase effectiveness. All stakeholders should be familiar with these standards and follow them in order to apply the CFS design and implementation process properly (UNICEF 2008).

3 Methodology

The data this study used as primary data was from structured and semi-structured interviews, in groups and individually, as well as interviews with key informants. Data was also obtained using collaborative tools such as questionnaires. Meanwhile secondary data was used which has been collected through a series of meetings related to the " role of NGOs (SAMAN) in Bam" at Tehran mosque Ershad[1], and a review of reports and studies was conducted. Since various issues are likely to be addressed by interviewees, the issues which are most common among individuals' opinions would be considered most relevant. Considering the difficulty of participating in an environment and objective observation, the effect of post-disaster child-friendly spaces was evaluated by collecting observations spanning over 10 years. The research carried out can be considered to be field-based. The study was qualitative and exploratory in kind, exemplified by content analysis and group discussion methodology used by children, educators, and parents.

In this study, the notion of space its features and properties such as its psychological impacts, practicality, and environmental qualities has been examined. On another level, the essay attempts to redefine the fundamental definitions, indicators, and parameters related to child-friendly spaces. Accordingly, significant cases are identified and coded. After undertaking the data collection process, the content and collecting data have been categorized and classified based on the utilized codes. During the research process, selected indicators help researchers in a way to conclude and thus equip accurate identification of topic with a logical bridge to be inferred. After inference, an analysis step has been carried out, as a matter of fact; this step suggests or describes results of content analysis in this research.

4 Analysis of Case Study

4.1 Planning and Design Process of Child-Friendly Spaces in Bam

After the Bam earthquake, almost all governmental agencies, international organizations, and NGOs of children have child-friendly spaces set up, in their programs, and in these spaces NGOs' role was very significant. This project was designed to provide opportunities for children; so children in child-friendly spaces can spend their leisure time in a useful way and raise their living conditions in a positive and creative way. Other objectives of this project were to help them while simultaneously returning to a normal life, to be useful to the community and to take practical steps, and develop positive attitudes to the future and learning life skills, as well as processing the experience of the disaster in a healthy way. Child-friendly spaces built in Bam not only want to provide a safe space for children but also trained and supported them psychologically in these spaces. Generally, before organizations' presence in the region, an internal meeting consulting with community members and stakeholders to anticipate their activities and needs takes place, after which a specific schedule is adopted.

After the arrival of the aid organizations in Bam, the process of identifying eligible children was begun by conducting a census. On the strength of the observation and interviews, it appeared that the problems faced by the people in Bam did not originate from the earthquake only but indeed also from internal societal problems that the earthquake revealed to them. Being faced

[1]Coincident with the tenth anniversary of the earthquake in Bam on January 5, 2003, a series of meetings on "The Role of NGOs (SAMANs) in Bam" was held in the library rooms of Ershad mosque by Khrasanchy. In calling upon the NGOs somewhat participated in the earthquake were invited in the opportunity to present their studies, activities, and proposals. In three consecutive terms in the first week of January, February, and March 2014, the meetings were held.

with these problems led to the conclusion that although planning is essential before coming to the region, recognition that a regional plan makes such a broad vision must lead to a review of programs and activities at the local level and, cannot, to some extent, be predetermined. So after this stage, discussions pointed out to assessment and identification of deficiencies. According to the executive agencies, authorities who are responsible for child-friendly spaces, the most important requirement for people was to provide a happy atmosphere in the region, because one of the most important issues is the psychological effects of the earthquake on children of this time of sorrow dominating the region. Disengagement and unhappiness in children lead to feelings of depression, and this is why the need was felt for assistance and training alongside other fun programs in the affected regions. Over time, when families became familiar with activities carried out in child-friendly spaces and children tended to learn, people were getting more and more interested in participating in these spaces.

4.2 Physical Infrastructure Progression Process of Child-Friendly Spaces in Bam

4.2.1 Tent

At first, child-friendly activities took place under tents. Tents were erected in parks and public spaces, along with other locations. Over time and based on the needs of children being in each region, tents were added. Some tents remained in the same places where they were originally stationed, but some others according to their needs and increase in number of participating children changed location (for instance, the tent of Child Protection Community of work children moved from park to a nearby location). Over time, the activities and projects which were defined for children were also changed. Some of the projects required larger areas than tent space (e.g., workshops, sports classes, etc.). Yet at that time, activities such as collective games or tournaments were held in open spaces. In such conditions, women were not willing to participate in activities. So in some cases, former tents were replaced with larger ones, or an even larger tent was built next to a small one.

In some months of the year, strong winds are blowing in Bam. Winds ruined child-friendly tents which did not have enough strength and disrupted activities. However, officials repaired the damaged tents and continued the activities. Hot weather also created problems. To deal with such problems, air conditioners were used in tents and trailers. In addition to such facilities, native shelters (traditional) were used to deal with intense heat (climate problems in detail are given in Sect. 5).

4.2.2 Conex

By distributing conexes over time in some areas, tents were replaced by child-friendly conexes. Conexes at that time were a great blessing. In places that required more spaces, these conexes were placed next to tents. After receiving conexes, there were painted using child-friendly paintings and colors that helped to change the mood.

4.2.3 Kawar[2]

Traditional species is defined as shelters that were built using local materials and common construction technique by local craftsmen (Fallahi 2009, 16). These are often set up quickly. With hot climate in Bam and possibility of children being impacted by the heat, parents and local people suggested making Kawar (native structures belonging to the region) to the authorities, and after they were faced with disinterest, they decided to build them by themselves. Kawar were used as shelter and as space for holding classes in hot summer conditions.

4.2.4 Shed

Some of organizations with aims at providing services as a long-term goal chose some lands to continue their activities. It should be noted that

[2]Kawar, species of native architecture of Bam that can be built with a palm tree filament; in summer the weather is very hot and the native's use of these structures is temporary. That makes it doubly cool splashing water on.

supplying land has always been one of fundamental problems of child-friendly spaces. These organizations, along with conexes or tents available to them, also built sheds. These spaces have been very suitable for activities with more people participating, and authorities' long-term goals would have been met.

5 Sample Selection and Data Collection Process in Bam

All the people in the samples who attended in-depth interviews (individually and in groups), as well as the authorities, were among those who attended child-friendly spaces or were related as executive officials to these places. What has been important over this research is to achieve a deeper understanding of the subject under the study. So inevitably, data collection is limited to a relatively small sample of the study population. Being a small size of the sample in this study is not a disadvantage, because smaller sample provides an opportunity for a deeper level of investigation. In addition to that depth of interview, in this study methods including narration of stories, photos, and videos were used as a supplementary materials alongside the main research method.

Thirteen people were interviewed in person in Bam in the time of research. Among them ten participants were children at the time of earthquake and were present in one of the constructed child-friendly spaces; two participants were mothers, one of whom currently is in charge of one of the child-friendly spaces; and a person who was responsible for teaching music to children. Another 12 persons were interviewed by phone about various characteristics of child-friendly spaces, and their comments were analyzed and evaluated. Among them there were five parents (three fathers and two mothers), five trainers from the time of earthquake, and two people who were children at that time and attended in spaces.

After studying related documents and combining data collection methods, an innovative method has been taken with regard to the research position and existing restrictions.[3] The method first provides information about research and activities of desired activity to people. Thereafter making sure of absence of ambiguity, children were asked to provide their demographic information[4] to researchers. First, some questions were raised as initiator[5] between researcher and interviewees; these questions are as follows: How long did everyone spend in child-friendly spaces? What activities have they participated in? What are lasting and memorable mementos for each of them from child-friendly spaces? Additionally, a set of questions which was already designed was provided to persons to answer.[6]

After gathering all the questionnaires, as it is shown in Fig. 1, participants were given blank paper to write or draw graphically what they recalled from child-friendly spaces and to note their views without any fear. Most of the focus during this part of the interview is put on activities that had been applied in child-friendly spaces, and satisfaction of people was measured. After the end of this section, and a break for study participants, time was given to each participant to express their information as a group collectively, and a facilitator noted information as data in a table and comments to be sorted out. Followed by group discussions, a sheet was handed out entitled "benefits and problems" to any person able to write. Individuals' views were collected on this sheet (separately). After its completion,

[3] Meeting (a group meeting) took 6 h (14–9) within the conex

[4] These specialized texts, as well as international experience, show that so far only two child-friendly spaces detailed assessments were conducted in Uganda and Ethiopia. The sample after 6 months of construction of child-friendly spaces is embodied; therefore, the present study was after 10 years; an attempt was made to recover the most information that could be collected after 10 years had elapsed.

[5] Information and basic.

[6] Due to the specialized literature, information was collected by attending in meetings "NGOs in Bam earthquake"; the questions were designed and after assuring the validity of the questions, according to experts' viewpoint; the question before the presence in the region and experimentally were laid to five individuals who are introduced to child-friendly spaces.

Fig. 1 A group meeting with study participants in Bam

the sheets were presented to other group members for them to agree or disagree with issues by placing ✓ and ✗. This process was carried out for each sheet completed by participants. The main points were extracted from the comments of the participants, and its accuracy was evaluated by sharing it in a social context.

6 Data Analysis

To evaluate child-friendly spaces in Bam after 10 years and determine better architectural design criteria of child-friendly spaces, some indicators had been chosen from specialized texts and also considering views of interviewees and listing items as criteria which garnered more attention and emphasis. Because authors were trying to create a proper relationship between parameters extracted and child-friendly space conditions in Bam, subjects were classified into the following fields:

1. Quality of the environment characteristics such as security and protection, relationship between green space and artificial space, health and hygiene, space visual aspects, climate conditions, ease of access and traffic
2. Physical environment characteristics such as dimension, choice of materials, tools and equipment, and fences
3. Sociopsychological characteristics such as partnership, relationship between caregiver and child, quality of teaching activities, and separating in terms of ages and genders

Based on the extracted criteria and in order to more accurately analyze those cases which interviewees pointed out, the desired issues were rearranged and classified in tables (Tables 1, 2, and 3).

Analyzing different views of interviewees in relation to characteristics of child-friendly spaces indicated that different features should be considered in formation of child-friendly spaces, according to their views. From the perspective of the studied group, security includes not only the physical environment but also the psychological security and the performance of the system. Due to restrictions in Bam at that time (limited number of instructors and space limitations), age separation were not possible for all child-friendly spaces programs. The nature of some programs

Table 1 Indicators of environmental quality

Topic	Issues	Results	
Security and protection	Parents and educators to be ensured of the safety of the environment	Increasing the feeling of comfort for the parents	
		Positive change in mothers' mood	
Relationship between green space and the artificial environment	Holding some outdoor class	Increasing child-friendly spaces, stability	
	Children are more interested in outdoor classes	Preventing the accumulation of children in a particular space	
		Creating opportunities for mobility	
		Increased sense of responsibility to the environment	
		Mental conditions improvement and increasing socialization	
		Increasing social skills	
Health and grooming	Not separating WC for children and adults	Putting additional pressure on the children in delaying fulfillment of biological needs	
	Not separating WC for boys and girls	The risk of disease outbreaks, especially in girls	
	Lack of sufficient WC	Microbe contagion	
	Being insecure WC	Refusing to use WC	
	Stench from WC	Physical complications in children	
	Contaminated WC	Increased insect around WC	
	Not possible handwashing with soap	Insecurity feelings	
	Being tents of plastic for WC	Preferring dilapidated houses to use WC	
	Placement conditions of WC		
How to access and transit	Possible entrance of strangers into the WC	Increased insecurity	
	Failure to monitor the adults' traffic	Increased feelings of fear to the children	
	Degradation of pathways	Pathways are unsafe for children	
	The authorities' lack of action to quickly rebuild paths	Children depend on their parents to attend	
	Insecurity of Bam		
	Lack of public transportation		
Visual aspects of space	Stained tent with the participation of children	Dissatisfaction with dark tents	
	Colorful tent of pre-designed	Increased the participation of children when staining tent	
		Increased sense of belonging	
		Increased happiness	
		Increased activeness	
Climate	Instability of the tents against strong winds of Bam	The use of indigenous knowledge	Using Kawar, a native structure
	The inability of women and children in the setting up of tents for exercising	The use of local resources	
	Exceed air heat due to accumulation of children in a space	Increase parent involvement	
	Lack of heating and cooling equipment in child-friendly spaces	Increased social skills	
	Children gathering in spaces equipped with cooler	Accelerated the rehabilitation process to the native	
		Increasing child-friendly spaces stability	
		Preventing the accumulation of children in a particular space	

Source: Authors (Adapted from Metzler et al. (2013a, b), Kamelnia and Haghir (2009))

Table 2 Indicators of physical environment quality

Topic	Issues	Results
	Ironically small size	Lowering quality of education and the individuals' concentration
	Unfavorable conditions	Fatigue trainer
	Unexpected high number of children	Warming too much the space in the warm seasons
		Energy depletion of children
		Extreme disorder in the classroom
Dimension		No motion for children
Choice of materials	Infiltration of rainwater into the tent	Lowering quality of education and individuals' concentration
	Influence of exceed heat and cold inside the tent	Difficulties to the class
	Unstable tents	Warming too much the space in the warm season
		Bad smell in class
Tools and equipment	Lack of suitable tools and equipment	Not wanting the children to read books
	Not using the equipment according to age groups	Educating superficially and fleeting
	Lack of appropriate equipment for the disabled	Children's unwillingness to continue classes offered
	Low quality and low number of children's books	Wasting time and energy to planning in the region
Fence	The use of metal barbed wire	Increasing barbed wire and then feeling of being surrounded
	Painting on the walls (in some cases)	Painting the fence to improving the visual aspect of
		Painting on the wall space and then increased sense of ownership

Source: Authors (Adapted from Metzler et al. (2013a, b), Kamelnia and Haghir (2009))

was such that age separation does not cause problems (such as storytelling and poetry), and children did not express dissatisfaction due to multi-age classrooms. However, failure to age-segregate in other programs, such as exercise classes, language training, and life skills training, could be unpleasant for children and created problems.

As shown in Table 1, another thing that endangered the safety of children is how to navigate traffic to CFSs. At that time, with the degradation of pathways and also authorities' inaction in rebuilding them quickly, routes to CFSs were unsafe, and there was danger for children trying to reach them. The lack of functioning public transportation made travel to CFSs difficult for children. According to all of the above data and general conditions of insecurity prevailing at the time in Bam City, children were dependent on elders to accompany them to attend in child-friendly spaces, thus limiting their ability to make use of these spaces.

One of the problems which girls mentioned about lack of separation in classes in terms of gender and age was that toilets were inappropriate. The toilets were not constructed separately in child-friendly spaces in Bam, and children were forced to use the public WCs. This puts additional pressure on them, and in many cases they suppressed the need to go to lavatory out of fear or discomfort until nighttime or at home, and this obviously caused negative physical impacts for children.

According to Table 2, from the issues that were addressed in relation to the construction of child-friendly spaces, the greatest was the lack of tents' strength against wind in Bam. The tents of child-friendly spaces were erected in the face of such strong winds that they did not have enough strength and were simply blown away. According to one official, "we had lunch one day in a tent when wind was blowing, then a layer of soil covered on our food and our tent was blown away after that." In such conditions it would be impossible to continue activities in tents, and it would disrupt the program development process. Tents were blown away by strong winds, and it took a

Table 3 Indicators of sociopsychological environment

Topic		Issues	Results
		Children's participation in painting tents and conex	Positive effect on the morale of children
			Increased satisfaction of space
			Creating a sense of belonging
			Using valuable ideas for children
	Children		Meeting children together
		Adolescent participation as assistant trainer	Strengthening relationships between children and adolescents through collaboration
Partnership		Participation in executive programs	Improving mental condition
		Participation in training programs	Increased self-confidence
	Juvenile		Learning of teaching skills and classroom management along with trainer
			Better understanding of the past learning by teaching it
			Preparing for the future life
			Making the relationship strength between assistant adolescents
		Outdoor setting in order to offer some suggestions	Creating lively in a damp atmosphere of Bam and boosting enthusiasm for mothers
		Parental participation in community activities	Meeting and co-talking of parents
		Parental involvement in educational activities as trainer	Sympathy during participation
		Participation in constructing the aboriginal structures, Kawar	Forgetting loneliness and sadness
	Parents	Increasing parents' tendency for participation in the program over passage of time	Increased sense of calmness due to being close to the children space
			Increased satisfaction sense
			Exiting doubt and passivity
			Improved financial conditions for receiving wage
			Using indigenous knowledge
		Lack of physical punishment of children	Enhancing the intimate relationship between children and educators
		Lack of attention to children's self and nobody is very spoiled	Increased sense of security
Relationship between trainer and child		Trainers' intimate behavior with children	No fear of educators
		Sympathy of trainers with parents and children	Learn faster content with fun
		Trainers' respectful behavior by adolescents	Children are happy and relaxed
		Aboriginal educators need to assistance training	Successful trainers in gaining trust of the children and parents
			Children relieved by the trainers' warm embrace
		Offering a number of programs	Enhancing hope and motivation and becoming children stronger to interact with the community
		Engaging children's mind in the things other than sorrow	Increased self-confidence and satisfaction feelings

(continued)

Table 3 (continued)

Topic	Issues	Results
	Revealing the feelings, thoughts, wishes, needs, and internal conflicts	Appearing fears
Quality of teaching	Expressing through by paintings ideas	
	Understanding with their experiences and cope with them	
	Liberation from the constraints of the environment	
Segregation of age and gender	The lack of gender segregation in educational programs	Reducing the feeling of security in the classes related to physical abilities
	Lack of different educational programs for age segregation	Unequal conclusions between boys and girls from classes
	Differing abilities of girls and boys	The girls did not participate in some programs
		Shy girls from attending classes
		More young children do not play more sport programs and competitions

Source: Authors (Adapted from Metzler et al. (2013a, b), Kamelnia and Haghir (2009))

lot of time to reconstruct them, because children were often present in such spaces and many trainers were women who were faced with significant difficulties in reerecting the tents.

7 Conclusion

Establishment of child-friendly spaces after earthquake in Bam, although successful in many cases, and strengthening relationships and friendships, increasing creativity and life skills, and enhancing self-confidence in children, in some cases, failed because of lack of experienced executive organizations and institutes and unsuccessfully conducted case studies in Iran. Significant problems have been encountered regarding planning and architectural design of these spaces.

One of the problems faced is that in addition to inappropriately small-sized tents or conex for the purpose of establishing the classes and programs of child-friendly spaces, undesirably poor environmental conditions of these spaces caused dissatisfaction among many people. In many cases, infiltration of rainwater into tents and conex made difficult to continue classes in such damp conditions. In some cases, undesirable odor flowing into tents and conex from poorly designed and sited latrines led classes to be cancelled early. In these situations, people also offered suggestions to use waterproof material on tents and conex and recommended to manufacture of child-friendly tents and conex from waterproof material when it is more likely to get serious rainfall in a region. If tents produced by these facilities are costly or not possible at that time, it is recommended to design a separate waterproof layer for tents and conex to be treated with a waterproof coating so as to prevent penetration of water, in the event that this becomes necessary.

For children, it should also be noted that visual aspects of a tent ought to be considered, and tents with bright colors or designs can help in making child-friendly spaces fit and positively reinforce children's moods. It is better when tents have sufficient quality in terms of visual aspects as well as being appropriate to the physical needs of children. Children were unhappy because of the dark color of tents they were allotted. It should also be noted that the main target of such tents and conex are traumatized and injured children, so taking measures in order to improve physical conditions should be such that they have an overall positive impact on children both psychologically and mentally. To achieve this, attention must be paid carefully to the visual characteristics of the space and coating such that special, cheerful effects and colors can have a positive impact on children who likely have a depressed mood. In this context, children also presented suggestions, and also they very much welcome ideas for designs on simple tents. Therefore, it is advisable pre-designed tents be considered for child-friendly spaces. Tents which are used for other purposes will be insufficient because visual features of such tents have a measureable negative effect on the children. If a predesigned and nice looking tent for children is not available, it is recommended that children are allowed to draw on tents and design them with their own creativity. Another frequently cited concern is that children be separated by age and gender in classes. Most girls agreed to an age separation program. Both categories of girls and boys also tended to report on separated classes positively. Because of differing abilities and temperaments of girls and boys, separation of their classes can make conditions for them easier. Girls who were too shy to attend classes with boys and could not show off all their abilities fully in mixed gender classes were not able to participate in these classes as dynamically. The children also pointed out the need to improve WC services as part of child-friendly spaces. Being insanitary, unsafe, and terribly stinky was so disturbing as to make public WCs unusable for children; so if WC services of child-friendly spaces could be separated from other spaces of camp, the WC service's management and security will be much stronger. Also, it should be noted that as supply of tents, conex and other equipment at the time of the catastrophe is important, preparation of books and other cultural materials for children should be initiated quickly and yet retain quality carefully. These materials can be very effective in the recovery process of traumatized children and help them return to a normal life. Children, in this regard, suggested that to improve the quality of classes, more equipment and a higher quality of teaching aids should be used.

So the above issues associated with the planning and design of post-disaster child-friendly spaces, especially in projects formed with the aim of meeting the needs of children and their participation as well as other local resources, can also accelerate construction of these spaces after a catastrophe and increase their quality, provide an environment that matches the needs of children, and increase stability of these spaces, by creating opportunities for participation of children and also local people as staff.

References

Child Protection Working Group. (2012). *Minimum Standards for Child Protection in Humanitarian Response*. Geneva: CPWG.

Fallahi, A. (2009). *Disaster assessment, risk-taking, vulnerability and damage*. Tehran: Scientific-Applied High Education Institute of Crescent.

Global Protection Cluster, Global Educational Cluster, INEE, IASC. (2011). Guidlines for child freindly spaces in emergencies.

Kamelnia, H., & Haghir, S. (2009). Landscaping design patterns in child friendly city (case study: Child-friendly city of bam). *Quarterly Garden, 6*, 77–88.

Kostenly, K. (2008). *Starting up child centered spaces in emergencies: A field manual*. Richmond: Christian Children's Fund.

Metzler, J., et al. (2013a). *Evaluation of child friendly spaces: Uganda field study summary report*. New York: World Vision International & Columbia University Mailman School of Public Health.

Metzler, J., Savage, K., Vojta, M., Yamano, M., Schafer, A., & Ager, A. (2013b). *Evaluation of child friendly spaces: Ethiopia field study summary report*. World Vision International & Columbia University Mailman School of Public Health.

Save the Children. (2011). Community based child protection emergency project: Internal evaluation report.

UNICEF. (2008). A practical guid for developing child friendly spaces. 7–16.

UNICEF. (2004). Minimum standards for child friendly spaces and children's centres darfur states- FINAL. October.

Sheltering Status a Year After the Multiple Disaster in Fukushima

Kenji Koshiyama

Abstract

Severe damage to houses was caused by Great East Japan Earthquake in 2011. Furthermore, people had to evacuate the area in which the nuclear power plant accident occurred. Victims were given long-term sheltering over a vast area. Six years after the disaster, resettlement is still a challenge. The construction of prefabricated housing by the public sector near the damaged area was the main method of providing temporary housing after the disaster. However, the difficulty of constructing prefabricated housing because of a lack of space and capacity to supply all with homes resulted in the provision of new temporary housing, 'Private Rental Housing As Temporary Public Housing'. In this study, the problems of evacuees' circumstances and the uncertainty of recovery are examined. First, the temporary housing provision plan and its characteristics are explained. Second, the results of a questionnaire survey completed by 10,082 families are presented. Special attention is paid to the status of the evacuees 1 year after the disaster by analysing their housing, level of damage, attitude towards nuclear damage, change of work and economic conditions, transformation of family forms and outlook for the future. The results show that evacuees living in private rental housings were in better conditions compared to those living in prefabricated housings. However, many of the evacuees have lost or changed their employment in the new settlement due to moving to long distance from their previous locations. Furthermore, they were forced to live in a new environment and had to adapt to it. Thus, this sheltering process may not be aligned with the Sustainable Development Goals (SDGs). It is important that a provision plan should include life recovery processes and assist their adaptation to a new environment.

Keywords

Nuclear accident · Long-distance evacuation · The East Japan Earthquake disaster · Temporary shelter

1 Introduction

The physical destruction caused by the Great East Japan Earthquake in 2011 and the radioactive damage that was brought about from the ensuing nuclear power plant accident resulted in extensive loss not only of housing, food supplies and infrastructure but also of community func-

K. Koshiyama (✉)
Faculty of Societal Safety Sciences,
Kansai University, Osaka, Japan
e-mail: k-koshi@kansai-u.ac.jp

A. Asgary (ed.), *Resettlement Challenges for Displaced Populations and Refugees*,
Sustainable Development Goals Series, https://doi.org/10.1007/978-3-319-92498-4_11

tions, thus forcing most of the survivors into widespread, long-term evacuation. Lack of experience with evacuation from a nuclear accident (as opposed to a natural disaster) and the ensuing temporary resettlement left many disaster victims very confused and in a state of chaos, both during the evacuation and, subsequently, in their lives away from home. While a number of studies such as Imae (2011a, 2011b, 2012) and Institute of Disaster Reconstruction Research, Fukushima University (2012) have since been conducted on the process of evacuation from the nuclear crisis and the evacuees' temporary living situation, the actual conditions of the disaster victims' housing and daily lives have yet to be examined, 6 years after the incident.

In this paper, the aspects that characterised the lives of evacuees from the nuclear disaster, a phenomenon that is distinctly different from a natural disaster, are examined. The data obtained from a survey questionnaire that focussed on the evacuees' financial conditions were analysed.

The analysis was based on data from the 'Survey on Evacuation Conditions Associated with the Tokyo Electric Power Company (TEPCO) Fukushima Nuclear Accident', which was conducted in 2012 by the Nuclear Damage Claim Dispute Resolution Centre; the latter is an organisation operated by the Ministry of Education, Culture, Sports, Science and Technology (MEXT). While a large number of survey topics were designed for the survey, the focus of this paper is on the analysis and emanating discussion of the actual conditions of the evacuees' daily lives and changes in their spending.

2 Outline of Evacuation Status of Fukushima

According to the MEXT report, there were approximately 150,000 evacuees 6 months after the earthquake. Of these, 95,000 people stayed inside Fukushima Prefecture with the exception of the area designated as the evacuation zone, and 55,000 moved to outside Fukushima. Among them, 36,000 were referred to as 'voluntary evacuees' distinct from 'designated evacuees' because although they lived in Fukushima, they did not stay inside the evacuation zone. They were regarded as evacuees by choice so as to escape from the effects of radioactivity. This phenomenon characterised the nuclear accident. The consequence thereof was multiple sheltering for the Fukushima evacuees.

There are two methods of providing temporary housing for those who have lost their houses in Japan after a disaster. Prefabricated housing is new housing provided by the public sector, and private rental housing is used as temporary public housing. Many private rental housings in Japan were arranged for the evacuees of the East Japan Earthquake because it was difficult to build many new prefabricated houses in the damaged area. Furthermore, several evacuees availed themselves to move to other areas all over Japan, even outlying areas.

To generalise, prefabricated housing is not perceived as being very comfortable for evacuees in modern-day Japan. This was defined by the Disaster Relief Act; it was pointed that there were several problems with prefabricated housing; in particular, from the viewpoint of housing specialists, this form of housing cannot meet the modern lifestyle of the Japanese. As noted previously, private rental housing is an alternative method employed by Japan of providing temporary housing after disasters.

3 Results of Analysis

3.1 Survey Overview and Characteristics of Respondents

The details of the survey are described in the 'Report on the Results of a Survey Concerning Settlement Mediation in Claims for Compensation for Nuclear Damage from the 2011 TEPCO Fukushima Nuclear Power Plant Accident'. A brief overview of the survey is presented in Table 1. The report cited above was a preliminary version; however, an analysis using the final data, including additional responses to the question-

Table 1 Outline of Questionnaire

Period: 1–19 March, 2012
Respondent: Representative of household
Target: 41,754 householders
Residents in municipalities including 13 areas designed as the restricted zone, the planned evacuation zone, the emergency evacuation zone and the specific spots recommended for evacuation
In addition, residents registered in the system of the evacuation information in MIC
Method: Mail
Number (%) of respondents: 10,082 (24.1%)
Respondents sex: Male 61.5%; female 38.0%
Respondents age: Under 29, 5.7%; 30–49 years, 27.4%; 50–69 years, 44.3%; over 70, 22.1%

Source: Report on the research of the mediation for compensation of the damage by the nuclear accident at the Fukushima Daiichi power plant of TEPCO (2014)

Table 2 Current residential place and time moved in

Place (n = 10,082)	No	%	Time (n = 10,082)	No	%
Hokkaido	81	0.8	March 2011	1848	18.3
Tohoku (except Fukushima)	822	8.2	April	1601	15.9
Fukushima	5883	58.4	May	1008	10.0
Kanto	2231	22.1	June	1023	10.1
Middle, West Japan	829	8.3	July	1160	11.5
No answer	236	2.3	August	918	9.1
			September	659	6.5
			After October	983	9.8
			No answer	882	8.7

Table 3 Reason for selection of place

Reason (n = 10,082)	No	%
Near the workplace of a household of a family member	2614	25.9
Long distance to avoid radioactive effect	2243	22.2
Existence of hospital, welfare institute	905	9.0
My children and relatives lived in the area	3184	31.6
My friends and acquaintances lived in the area	1474	14.6
Accompanying neighbour	443	4.4
Leading by local government	1328	13.2
Availability of an assistance programme	2046	20.3
Select location	2169	21.5
Fewer earthquakes	264	2.6
Others	1990	19.7

naire, is presented in this paper. The characteristics of the survey respondents as revealed by their responses to the questions concerning the location of their residence in March 2011 and 1 year after the disaster, the timing of relocation and the reason for the choice of location are shown in Table 2. One year after the disaster, 58.4% lived in Fukushima, and 22.1% lived in the Kanto region. It is significant that 9.0% lived in regions other than Tohoku and Kanto; furthermore, this is also indicative of the reality of the expansiveness of the evacuation that was captured through the survey.

With reference to the timing of the evacuees' relocation a year after the earthquake, 65.9% relocated between March and July; however, March was the most commonly accepted response. The most common reasons for choice of location were as follows: 'my children and relatives lived in the area' and 'my friends and acquaintances lived in the area', thus, showing the respondents' reliance on relatives and personal ties in the community (Table 3). Another commonly selected reason was 'near the workplace of a householder or a family member'. On the other hand, 22.2% selected 'long distance to avoid radioactive effect', which was one of the distinguishing characteristics of the nuclear disaster evacuation. This clearly shows that, in addition to the problems associated with rebuilding their lives and securing housing, the evacuees also had to evacuate from the ongoing impact of the disaster. In addition, 20.3% chose 'availability of an assistance programme', another distinguishing characteristic of disaster evacuation. This indicates that the availability of information on assistance was an important factor in determining evacuation destination.

The relationship between the designation of the home area at the time of the earthquake (as perceived by the respondents) and the place of residence 1 year later is exhibited in Table 4. Among the group of respondents who responded 'unknown' or 'others (including voluntary evacuation)' to the question related to the type of disaster risk designation their area received, the

Table 4 Designed zone by residential place after 1 year

Zone (n = 10,082)	No(%)	Percentage by region				
		Hokkaido	Tohoku	Fukushima	Kanto	Middle West
Restricted	6256(62.0)	0.2	5.0	65.1	21.9	5.8
Planned evacuation	869(8.6)	0.5	3.2	76.1	13.8	3.6
Emergency evacuation	1590(15.7)	0.5	14.8	54.0	18.7	8.4
Specific spots recommendation for evacuation	46(0.4)	0.0	15.2	41.3	23.9	17.4
Unknown	706(7.0)	4.7	20.0	17.0	32.3	23.9
Others (including voluntary evacuation)	254(2.5)	3.5	15.4	6.7	46.1	26.8
Total	10,082(100)	0.8	8.2	58.4	22.1	8.3

percentage of those who stayed in Fukushima Prefecture was very low. However, the percentage of those who evacuated to a distant region was high. This shows that their evacuation destinations encompassed a wide geographical area. It is likely that the uncertainty of information concerning the risk levels of radiation contamination and fluctuation in risk awareness had a substantial impact on the choice of evacuation destination. Furthermore, this evacuation pattern can be interpreted as one of the phenomena caused by radiation-related information.

3.2 Housing Types 1 Year Later

When answering the question about the ownership type of their home 1 year after the disaster, 71.7% of the respondents chose 'temporary housing units' and 'private rental housing deemed as temporary housing units'. This revealed that many households took advantage of the public housing assistance scheme. What is characteristic about the nuclear disaster in question, however, is that 'deemed temporary housing' was the main type of housing selected by the evacuees. The 'deemed temporary housing' programme provides multiple housing options to the evacuees while affording many households with the opportunity to evacuate to a distant area. At the same time, however, a certain number of households had temporary housing arrangements that are associated with mental and financial stress; 8.7% lived with relatives and people in the community, and 8.8% rented a house at their own expense or resided in private rental housing.

When asked about rent and gratuities, many indicated '0 yen' or '50,000 to 100,000 yen'. '0 yen' accounted for the highest proportion of 'temporary housing units' and '50,000 to 100,000 yen' for 'rental house/private rental housing'. However, even for 'deemed temporary housing', whether rent was fully subsidised or not was dependent on the size, location and property. Furthermore, over 30% of those living in 'public housing' or 'with children, relatives or acquaintances' paid 50,000–100,000 yen. This made it difficult to establish what factors determined rent. Furthermore, an examination of designated area at the time of the earthquake did not reveal any major patterns.

Regarding the housing type before the disaster, 82.6% (n = 8329) chose 'a detached house' and 11.1% (n = 1124) 'flat'. The residential type shown in Table 5 is related to 'flat', 'temporary building' or 'room sharing'. Because of the evacuation, many householders were forced to change their type of housing, which resulted in lifestyle changes.

3.3 Separation of Family Members

An examination of the number of family members who were living together at the time of the earthquake and in their new housing 1 year after the disaster showed that the single-member households doubled and two-member households increased 1.5 times. On the other hand, households with six or more members decreased to one-third. These results reveal that a large pro-

Table 5 Residential type and rent amount

Residential type (n = 10,082)	No	%	Rent amount (yen) (n = 10,082)	No	%
Temporary housing (prefab)	1548	15.4	Free	1390	13.8
Private rent housing as temporary housing	5673	56.3	Under 10,000 yen	553	5.5
Available public housing	64	0.6	10,000–30,000	542	5.4
House of children, relative, friend	875	8.7	30,000–50,000	651	6.5
Private rental house	892	8.8	50,000–100,000	1379	13.7
Residence for employees	273	2.7	Over 100,000	320	3.2
Others	452	4.5			
No answer	305	3.0	No answer	5247	52.0

Table 6 Change in family make-up

Selection (n = 10,082)	No	%
No change	3895	38.6
Someone has lived in another place temporarily	350	3.5
Someone has been living in another place	3829	38.0
Living with another relative who lived in another place	642	6.4
Living with friend	76	0.8
Others	491	4.9

Table 7 Reasons for change in family make-up

Reason (n = 5356)	No	%
Family's work	2472	50.5
Child's school	977	20.0
Child's minor	475	9.7
Going to a hospital or nurse care	647	13.2
Spaciousness of rooms for householders	1522	31.1
Fear of effect of radiation	1426	29.1
Difference of the way of thinking of effect of radiation	318	6.5
Others	670	13.7

portion of the changes led to multi-member households becoming two-member households and large households becoming smaller households. This pattern revealed that the evacuation process caused a decrease in the number of household members; in other words, family members were separated.

With respect to changes in family make-up, 38.6%, the largest proportion, stated 'no change'. In contrast, 38% responded, 'family members that lived together before the earthquake are now living separately'. Thus, almost 40% of the respondents experienced separation from their family members. When asked about the reason for separation, the response most given was 'work situation' (50.5%). This was followed by the 'inability to secure housing with enough space in the resettled area' and 'concerns over the impact of radiation'; approximately 30% forwarded this as a reason. The latter reason, in particular, is irrelevant in natural disasters. The results of the survey indicated this as the main reason for separation of families after the nuclear power plant disaster. Moreover, a significant number of households were affected.

When reasons for the change in family make-up were examined according to designation area, distinct patterns emerged from the responses related to 'concerns over the impact of radiation'. The more ambiguous the risk-related information was, the higher the rate of selection of the latter response. The responses were as follows: 'warning zone' (19.8%, $N = 2979$), 'emergency evacuation preparation zone' (38.3%, $N = 731$) and 'unknown' (60.4%, $N = 404$). These results demonstrate that confusion over risk-related information was a driving factor behind family separation. Combined with the earlier analysis on area-wide evacuation, the fact that information concerning radiation contamination has an effect on the break-up of families, the basic units of everyday life, highlights the urgency of establishing assistance measures and raises an important aspect for studying future measures that need to be discussed (Tables 6 and 7).

Table 8 Householders' jobs before the disaster and 1 year after the disaster

		Current job (%)				
Type ($n = 10, 082$)	No (%)	No change	Part-time	Unemployment	On leave	Job change
Self-employed	1531(15.2)	24.7	5.4	49.3	3.3	5.7
Employee	4457(44.2)	59.8	1.2	17.4	6.9	8.3
Pensioner (unemployed)	1481(14.7)	–	–	–	–	–

3.4 Changes in Income and Work

With reference to the status of employment of family breadwinners at the time of the earthquake, 44.2% were full-time company employees and 15.2% were self-employed. The result does not provide strong evidence of characteristics commonly seen among disaster victims such as a high level of social vulnerability (Table 8). This indicates that the degree of impact from the nuclear crisis was determined by geographical characteristics rather than social class and, consequently, there was a wide variation in the degree of disaster impact and life changes following the disaster.

The results of a cross-tabulation between the disaster victims' work situation at the time of the earthquake and their work situation after 1 year show that 47.2% indicated, 'no change since before the earthquake', thus, demonstrating that the majority of breadwinners' work situation changed. The major changes in their living environment as well as in their employment/work situation as a result of the evacuation are distinctive characteristics of this process of life reconstruction. The situation surrounding the self-employed was particularly difficult because 45.6% of the unemployed did not have unemployment insurance. Because the prospect of the reconstruction process in areas affected by the nuclear crisis was even more unclear than that of the areas with serious tsunami damage, there is no doubt that the outlook of reconstruction was completely unclear for the self-employed after 1 year. On the other hand, while company employees tended to have relatively stable employment at the time of the disaster, their employment has not been sufficiently stable since; this is shown in Table 8. The damage at the nuclear power plant is possibly the primary cause. However, a sudden change in the employment environment as a consequence of a complete shutdown of the core industry because of a disaster is an issue that should also be addressed in other areas, not just in affected areas where the nuclear power generation industry has played a central role. This implies that, in addition to the damage suffered in one's home areas, there is a risk of the impact of another very large disaster.

In Table 9, the aggregate changes in income and spending by income level before the earthquake are displayed. Approximately 48.8% of the households experienced a decrease in income, while 14.1% had no income. Combined with the fact that 73.6% experienced an increase in spending, these results reveal the financial difficulties the evacuees experienced 1 year after the disaster.

In addition, a decrease in income was common among those who were middle- and high-income earners at the time of the earthquake; this could possibly be attributed to employment change and family separation. Compared with past natural disasters, the nuclear crisis can be described as having caused a drastic change in the economic environment of even middle- and high-income earners. Accordingly, the establishment of a diverse range of assistance schemes for disaster victims is of paramount importance.

3.5 Changes in Spending

In Table 10, the responses to the question concerned with increases in the evacuees' monthly household spending 1 year after the disaster are displayed. More than 50% of the respondents indicated an increase in spending for costs related to food, utilities, transportation and communication. An examination of increases in monthly

Table 9 Family income before the disaster and current income and expenses

Annual income (thousand yen) ($n = 10{,}082$)	No (%)	Change of income				Change of expenses		
		Increased	Unchanged	Decreased	No income	Increased	Unchanged	Decreased
0–2000	1699(16.9)	6.9	41.0	41.0	25.1	70.7	22.2	7.1
2000–4000	2973(29.5)	5.0	36.9	36.9	14.7	73.3	19.3	7.4
4000–6000	2428(24.1)	3.5	29.4	29.4	10.9	74.5	17.2	8.4
6000–8000	1026(10.2)	2.7	26.0	26.0	7.6	77.9	14.4	7.8
8000–10,000	638(6.3)	2.6	22.5	22.5	9.2	75.0	14.8	10.2
Over 10,000	462(4.6)	3.6	23.0	23.0	12.3	69.5	17.0	13.5

Table 10 Items of expenses and increases

Items	No(%)	Up to expenses (%)			
		Under 10,000	10,000–50,000	50,000–100,000	Over 100,000
Housing	3046(30.2)	12.6	36.0	28.7	22.7
Food	5300(52.6)	8.7	51.2	27.0	13.1
Water, electronic	5164(51.2)	15.1	75.3	8.1	1.5
Clothing	3340(33.1)	25.8	63.5	6.7	4.0
Health care	1401(13.9)	47.0	47.4	3.1	2.4
Education	886(8.8)	25.5	60.0	10.2	4.3
Entertainment	2016(20.0)	12.0	69.4	11.8	6.9
Transportation	5332(52.9)	12.8	71.5	11.6	4.1
Communication	5156(51.1)	28.7	66.4	3.7	1.1
Others	995(9.9)	30.4	48.7	11.3	9.6

expenditure revealed that spending rose by a wide margin in many households, particularly expenses for housing and food. The increase in cost of living is probably the result of many evacuees relocating from rural areas to cities. Large increases in housing and food costs, which are the main sources of daily spending, have a major impact on daily life. The latter is a critical issue that needs to be addressed when rebuilding life after widespread evacuation. The increases in communication and transportation costs probably reflect the reality of family separation and wide-scale evacuation. The items outlined involve regular spending, and when the evacuees' daily needs are considered, there are no possibilities to limit expenditure for these items, thus indicating the burden essentials place on household finances.

In this survey, respondents were also asked about special income. In the case of disaster-related special income, 72.5% responded that they received public donations and allowances. Results showed that the largest proportion of respondents (43.5%) received special income that totalled between 1 million and 10 million yen. However, this survey did not assess whether this level of special income was for evacuation and life assistance or compensation for property damage. However, the largest amounts were from 'insurance payments and others'. In addition, whether or not one had an insurance policy before the crisis had a decisive impact on the evacuees' living circumstances. It is evident that those who had access to insurance payments have been able to reduce the level of hardships while living as evacuees. In this survey, however, no questions were asked about the amount of compensation received from Tokyo Electric Power Company. Furthermore, responses to questions concerning monetary amounts, including those described previously, may be less reliable than others. Therefore, analysis and discussion of economic circumstances based on monetary amounts must be studied in further detail (Tables 11 and 12).

Table 11 Special income (multi-answer)

Items	No	%
Contribution from public sectors	7306	72.5
Aid and gifts from personal	3599	35.7
Insurance	2342	23.2
Others	550	5.5
Consolation from TEPCO	3896	38.6
No special income	488	4.8

3.6 Spending Related to Evacuation

The most common response to the question on disaster-related spending other than monthly household spending was 'expenses for the purchase of new furniture and household appliances' (71.9%), followed by 'moving costs' and 'costs for travelling to city halls, etc. to gather information'. Since evacuees from the nuclear crisis had to leave home without any belongings once the evacuation order was issued and residents in some areas are still not able to return to their

Table 12 Temporary expenses for evacuation and sheltering (multi-answers)

Items	No	%
Moving housing	4878	48.4
Purchase of furniture, electronic appliances	7249	71.9
Purchase of new equipment for child's change of school	1771	17.6
Cost of seeking new house	3255	32.3
Transportation costs for gathering information and contacts of public sectors	4221	41.9
Transportation costs for public procedures	3103	30.8
Transportation costs for TEPCO compensation	2901	28.8
Cost of aid for other people who were evacuated	775	7.7
Others	2111	20.9

homes, there was a great chance that evacuees needed to procure all daily necessities for their life away from home. Furthermore, those who chose to evacuate at their own expense have suffered a much greater financial burden than those living in temporary housing units or private rental housing deemed temporary housing units, where private assistance programmes such as a starter kit that consists of six household appliances have been made available. Such assistance programmes are not available for all the diverse forms of evacuation. In addition, economic status is not a determinant of assistance. This suggests those families that chose to rely on their own resources and take action independently to avoid risks instead of relying on public assistance schemes may have suffered a financial burden.

In widespread evacuation, evacuees have inevitably faced difficulty obtaining information from local governments. Results of the survey show that evacuees had no choice but to find their own way to obtain information about subsidies and support programmes as well as forms and documents and information related to recovery efforts. Future measures should also focus on the phenomenon of an increase in spending in order to acquire disaster-related information. Another question pertained to expenses for dealing with radiation. The response rate for 'expenses for purchasing and renting measuring equipment' was 15.0% and that for 'expenses for gaining knowledge related to the impact of radiation' was 16.3%. While these expenses are characteristic of this particular evacuation, the survey did not explore the relationship between these expenses and other types of spending or their impact on daily life. It is worth noting, however, that the cost of monitoring radiation damage is added to the initial household spending required to achieve stability in a living situation that is normally incurred in the post-disaster recovery process. Consequently, the radiation-related expenses are likely to have had an enormous impact.

4 Conclusion

An examination of the living environment and economic situation of evacuees from nuclear disaster zones not only revealed the various difficulties they face but also reconfirmed the need to distinguish those issues that are the same as those in natural disasters from those that are specific to the evacuation pattern associated with a nuclear crisis. Issues such as separation of family members, the quality of housing associated with extensive evacuation, spending accrued as a result of changes in the living environment and community issues have been raised in past post-disaster housing reconstruction processes. However, these issues are likely to be more pronounced in a nuclear crisis when combined with factors such as 'private rental housing deemed temporary housing units' and 'concerns over radiation contamination' that may magnify these issues. However, because of the focus on tsunami victims because of the seriousness of the damage in the tsunami-stricken areas, the issue of accountability for radiation contamination and the spatial heterogeneity in areas surrounding the nuclear power plant following the disaster, no attention has been paid to the conditions of the evacuees from the nuclear disaster. Six years have passed without fully addressing these issues. Six years may be long enough for people to adjust to difficult circumstances, and the impact on the evacuees may slowly dwindle with time. Feelings of guilt over evacuation, the desire to remain anonymous and privacy issues may have contributed to avoiding the issue.

Two issues identified by the analysis in this paper are related to disaster information. One issue relates to risk-related information available to evacuees and victims during the evacuation. The authorities' selection of certain types of risk-related information over other information affected evacuees' actions and their lives and resulted in lifestyle changes that naturally lasted for an extended period of time. It is important to examine and organise issues pertaining to the long-term nature of this impact. The second aspect is related to information about the recovery process. It is necessary to address the issue of an increase in spending incurred by disaster victims as they evacuate and relocate to other regions and attempt to rebuild their lives in the face of uncertainty. Compared with the rapid daily flow of information associated with the progress of technology, efforts to provide all families with access to information at the time of the disaster and the use of digital technology for accessing information are outdated. Furthermore, it is important to recognise that the disaster victims face an enormous financial burden as a result thereof.

The Fukushima disaster has been creating various new problems though Japan has the advanced urbanisation system. Especially, this result points that the migration for safety leads to the distress of life in Japan. Even if disaster, even if voluntary evacuees, it must be necessary to sustain the environment for daily healthy life.

References

Imai, A. (2011a). The primary survey of inhabitants who were evacuated from the nuclear power plant disaster. Monthly papers of the Japan research Institute for Local Government, No.393, 1–37. (Japanese).

Imai, A. (2011b). The second primary survey of inhabitants who were evacuated from the nuclear power plant disaster. Monthly papers of The Japan Research Institute for Local Government, No.398, 17–41. (Japanese).

Imai, A. (2012). The third primary survey of inhabitants who were evacuated from the nuclear power plant disaster. Monthly papers of the Japan research Institute for Local Government, No.402, 24–56 (Japanese).

Institute of Disaster Reconstruction Research, Fukushima University. (2012). Report on the research of the status for the disaster recovery in 8 damaged municipalities in Fukushima. (Japanese).

The Center of Alternative dispute resolution for the nuclear damage. (2014). Report on the research of the mediation for compensation of the damage by the nuclear accident at the Fukushima Dai-ichi power plant of TEPCO (Japanese).

Urbanism of Emergency: Use and Adaptation of Public Open Spaces in Disaster-Induced Resettlement Sites

Sandra Carrasco and David O'Brien

Abstract

Post-disaster reconstruction projects in areas with high pre-disaster vulnerabilities can lead to both temporary and permanent resettlement outcomes. During this complex process, efforts are centred to the provision of housing and basic infrastructure. In general, the limited land available restricts opportunities for high-quality spatial design within the settlements to optimize the land and provide the maximum number of houses possible. Open spaces in these settlements are rarely considered during the planning and implementation of the projects, and consequently there is little scholarship of community space. However, the importance of open spaces to the community and the liveability of these new settlements depend on the interactions which spontaneously take place in these areas. This study explores the different uses and characteristics of open spaces in temporary and permanent resettlement sites established in the aftermath of major disasters in Japan and the Philippines. This study addresses the limitations of the immediate post-disaster urban planning, compared to the longer-term urban space that emerges spontaneously in an attempt by the residents to improve their community. It was observed that the planning of open and public spaces was not a priority for the governments and the other stakeholders involved in settlement design. Despite these compromises, this study reveals that resettled communities appreciate any available open spaces and transform these spaces to accommodate their needs and cultural preferences reflecting their vital role as urban elements for convergence of the multiple sociocultural activities towards the recovery and revitalization of the disaster-affected communities.

Keywords

Public spaces · Disaster-induced resettlement · Community interaction · Adaptation

1 Introduction

The physical reconstruction of the built environment in the aftermath of disasters usually prioritizes the recovery of roads, the urban infrastructure and the construction of temporary and permanent housing for the victims. The large numbers of homeless people force governments to allocate people in emergency and temporary accommodations while their former homes are repaired or

S. Carrasco (✉) · D. O'Brien
University of Melbourne, Melbourne, Australia
e-mail: sandra.carrasco@unimelb.edu.au; djobrien@unimelb.edu.au

A. Asgary (ed.), *Resettlement Challenges for Displaced Populations and Refugees*, Sustainable Development Goals Series, https://doi.org/10.1007/978-3-319-92498-4_12

rebuilt. However, in cases where there are potential future risks for the communities that are directly related to the site where their pre-disaster settlements are located, it is necessary to resettle the communities in safer areas (Jha et al. 2010), which could be either temporary or permanent housing. Resettlement is considered to be the less desirable scenario for housing reconstruction because this not only means the removal of a community from its place, but it creates a cultural and physical crisis with profound dimensions (Oliver-Smith 1991). Jha et al. (2010) claim that resettlement may lead to the loss of livelihoods, sense of community and social capital and poverty and force people to leave the new sites and return to their former settlements. Affected people consider resettlement as a disaster (Partridge 1989), and involuntary resettlement has to be avoided and/or minimized (Oliver-Smith 1991; Jha et al. 2010) in order not to add more grief to the already traumatized communities.

The emotional stresses and trauma to which disaster-affected people are exposed can produce extreme situations such as the rise in suicide rates (Krug et al. 1998) and increased vulnerabilities for groups such as the elderly and people with disabilities (Sakamoto 2014) as these people are unable to establish links within their new community.

After disasters the planning of temporary and permanent new settlements focusses on the construction of a large number of housing units, and generally this occurs on limited plots of available land. The pre-disaster spatial characteristics are rarely able to be replicated or considered in the layout of the new settlements (Zetter and Boano 2010). The importance of the public open space is often ignored although studies demonstrate that these spaces can influence recovery and adaptive behaviour towards recovery (Allan and Bryant 2010) as they are considered spaces for social cohesion. Carr (1992) argues that public spaces generate links that connect the sense of place with a person's associated culture, economics and politics. Hence, these places are where social networks are created and contribute to the revitalization of a disaster-affected community and its support networks.

The importance of community-building in the alleviation of victims' stresses was evidenced in Japan after the Hanshin-Awaji Earthquake (January 17 1995) (Sakamoto 2014). This experience has been fundamental for researchers investigating the psychological needs and care for survivors and the importance of community and social spaces in the spatial planning of temporary housing in Japan (Maly 2014).

This paper examines the planning limitations concerning public and community spaces in the settlement layouts designed for temporary housing in Japan after the 2011 Great East Japan Earthquake and Tsunami and the permanent resettlement sites after the 2012 Typhoon Washi in the Philippines. The purpose of this study is to analyse how residents of these settlements adapt the open public spaces in their neighbourhood and around their homes and how this affects their social interaction in an attempt to improve their living conditions in the community. The analysis of the case studies is done based on the information obtained during field visits to the disaster-affected sites and temporary relocation sites in Japan between September and December 2011, and the second visit in July 2012, and visit to the permanent resettlement sites in the city of Cagayan de Oro, Philippines, between July and August 2014. The data for this paper was obtained through government reports and review of relevant bibliography. For primary information, we considered meetings with government officials, interviews with local residents and community leaders as well as direct observation of the residents' social behaviour and the use of the open spaces in their new settlements.

2 Spatial Planning for Post-disaster Reconstruction

After a disaster, a highly complex and often chaotic process of coordination and decision-making is necessary between multiple stakeholders involved, each of them with diverse but specific capacities that contribute to the whole process of planning and implementation of the reconstruction projects (Da Silva 2010; Davidson 2010).

This rapid, intricate, and usually ad hoc process is similarly present in long-term solutions or permanent housing projects (O'Brien and Ahmed 2011; Carrasco et al. 2015) and in the middle- or short-term solutions or temporary housing projects (Johnson 2010; Felix et al. 2013). After a disaster, recovery is focused on different sectors, where housing becomes a most pressing issue (O'Brien and Ahmed 2011). Housing is considered essential for the well-being and the development of the communities (Barakat 2003), and it is fundamental to establish some sense of normality in the life of the affected people (Felix et al. 2013). Despite these people-centred considerations, the housing dimension in the built environment of the victims is often reduced to conceive housing as isolated physical elements instead of considering them as human settlements. It is crucial to recognize the importance of the relationship between the house as a material commodity and its spatial and social importance (Zetter and Boano 2010). Therefore, the understanding of the human and collective dimension of community is compatible with the desirable development of long-term sustainable communities and urban built environments (Kondo and Karatani 2016).

In the aftermath of a disaster, housing occurs in different stages, which may overlap, and not all the victims will pass through each stage (Johnson 2010). The time people live in temporary housing, it is expected that they gradually resume their daily activities and routines, while permanent houses are built. Permanent housing is considered the last stage towards the long-term recovery and development of disaster-affected communities (Quarantelli 1995; UNOCHA and IFRCC 2015). Temporary housing is intended to accommodate affected people and allow them to perform their daily activities. These buildings are built with prefabricated or non-durable materials and are intended to last from 6 months to 3 years (Felix et al. 2013), although this timeframe can be extended to more than 5 years, such as the case in Japan after the 2011 Great East Tsunami. This might increase the risk to become permanent or semi-permanent, not guarantying the quality of housing and living conditions. In contrast, permanent housing represents a long-term solution for affected communities towards their complete recovery and further community revitalization and socio-economic development. Permanent housing implies both returns to rebuilt houses or moves to new resettlement sites. The process of construction of permanent housing may take several months or several years.

3 Public Spaces, Community Identity and Place Attachments in Disaster-Induced Resettled Communities

The decision to temporarily or permanently relocate or resettle communities has been always criticized, as the reasons behind these decisions may not necessarily consider the well-being of the affected people. These reasons could be politically biased like in the case of the Philippines (Hirano 2012; Carrasco et al. 2016), to facilitate the logistics of humanitarian assistance (Johnson 2010), or even because of commercial interests, as Shumaker and Conti (1985) argue 'the poor and powerless are sometimes relocated when their homes lands become valuable to others' (p. 240).

In general, the forced removal or involuntary relocation of a community from their environment involves a disruption of their attachment to their physical environment (Brown and Perkins 1992) and a fractured relationship between space and place (Zetter and Boano 2010). People coexisting within a community develop strong interconnections (Shumaker and Conti 1985) with these bonds providing a framework for a collective sense of identity, stability and confidence. This collectivity is threatened when there is a disruption due to a sudden event like disasters or forced relocation (Brown and Perkins 1992). Shumaker and Conti (1985) observed that 'people with strong community ties were least likely to relocate regardless of the real or perceived hazards existing in the area'. Edelstein and Wandersman (1987) noted that disaster-affected people do not perceive the effects as the mere loss of the physical community, but the loss of social ties and networks that make up their human surroundings. The quick disruption of these attachments, which developed slowly over the time, can be repaired or re-created although there might be a long-term phase of dealing with the loss (Brown and Perkins 1992).

In order to recreate these bonds, confidence in the continuity is crucial, where the big challenge is how to imagine a realized past with an idealized future in order to create a new homeworld in the present (Pader 1994). The answer could be in the observation of the behaviours of disaster-affected communities. For instance, Oliver-Smith (1986) noted that after the 1970 Yungay Avalanche in Peru, those survivors quickly allocated to camps started to give familiar street names to the paths between their tents. Similarly, in Japan, after the 2011 tsunami, residents of the town of Ofunato organized the Tanabata Festival, one of the most important summer traditions in Japan, only 4 months after their village was devastated in order to maintain their traditional cultural patterns.

4 Role of Open and Communal Spaces for Community Recovery

The nature of the relationships between the disaster-affected people as individuals and as a group is decisive for the development of senses of identity, community belonging and participation in collective activities. This results in a sense of attachment to a physical and sociocultural environment which improves the well-being of the community members (Davidson and Cotter 1991). Hence, research into the importance of a common physical space and its role in the process of recovery and the re-establishment of normality and continuity is paramount in a full understanding of disaster recovery.

Spaces that can be shared by all members of a community promote a balance between public and private activities as part of a communal life, which also provides channels for movement, the nodes of communication and the common grounds for play and relaxation (Carr 1992). In contrast, effects of the lack of shared amenities and open spaces community spaces are clearly observed as a detriment of the quality of life of the people which also include negative effects in the physical as well as the mental health of the people (Naimi-Gasser 2012), loss of cultural values of the society (Chitrakar 2016), and compromise the safety of a community due to the lack of neighbourhood cohesion and fear of crime resulting from the loss of community attachment (Conklin 1971; Rohe and Burby 1988; Brown et al. 2003).

This paper demonstrates the issues highlighted above with special reference to two case studies – one in Japan and another in the Philippines. The research looks at the links between the spatial organization of the settlements with the social organization of the communities to demonstrate linkages and commonalities. The relationship of the construction of a sense of community is confronted with the resident-initiated use of the social spaces, which are understood as active actors in the development of place and community attachment. The creation of these key ties evidences the recovery and re-creation of the lost social capital and the capacity of adaptation of displaced people by disasters.

4.1 Case Study 1: Creation of Spaces for Interaction in Temporary Housing in Japan

The 2011 Great East Japan Earthquake and Tsunami was one of the most devastating natural disasters to strike Japan in modern times and caused more than 23,000 deaths, 121,806 destroyed homes and over 470,000 displaced persons (Reconstruction Agency 2011). In the 2–6 months of the disaster, temporary housing was completed (built by the central government, coordinated with prefectural and local governments and directly supervised by local governments), and residents were allocated into 39,032 temporary shelters, built in 614 complexes in seven prefectures along the region of Tohoku.

The progress of housing reconstruction and resettlement projects, led by the government, facilitated the reduction of evacuees to 120,000 by February 2017. Approximately 40,000 of these residents of temporary housing were then allocated to either on-site rebuilt houses, new houses in resettlement sites or public housing complexes (Reconstruction Agency 2017). The evidence reveals that after 6 years disaster victims are still inhabiting houses intended to last for 2 years and brings into question the temporality of temporary housing within particular settlements.

Temporary housing is not initiated as a permanent solution, and the decision to build these is often criticized as socially undesirable, economically unsustainable and culturally inadequate (Bolin and Stanford 1991; Barakat 2003; Johnson 2010). Moreover, the condition of secondary displacement that the affected communities face once they move to a permanent house further fragments the relationship between the people and their physical and sociocultural environments (Zetter and Boano 2010). However, the pressure to provide an environment for the victims to develop their normal activities along with the complexity to provide permanent houses on the short-term results in the urgent need to provide temporary settlements. Unfortunately, in practice, the temporality of these houses is often confusing for the residents. Thus, it is necessary to consider the long-term spectrum and the implications in the residents' lives and communal relationships.

The process of post-disaster recovery and reconstruction after the 1995 Hanshin-Awaji Earthquake has important lessons for the subsequent planning of the post-disaster recovery plans for the 2011 Great East Japan Earthquake and Tsunami. In case of the 1995 earthquake, it was reported that only after 5 years the last residents of the temporary settlements moved to their permanent home (Sakamoto 2014), and it is expected that for the Tohoku victims, the permanence in temporary housing can be extended up to 10 years (The Japan Times 2016). Therefore, the recovery path is often undertaken over the longer-term and extends beyond the recovery of the physical environment into the called 'life recovery' stage that connects with the recreation of identity, community security and the establishment of a sustainable society (Sakamoto 2014).

The temporary houses at Tohoku were built on available land including sports grounds and parks, diminishing spaces where people could informally gather. There are small sites in remote areas like the settlement in Oshika Peninsula (see Fig. 1, left) where most of the residents are elderly, with limited accessibility to the city and

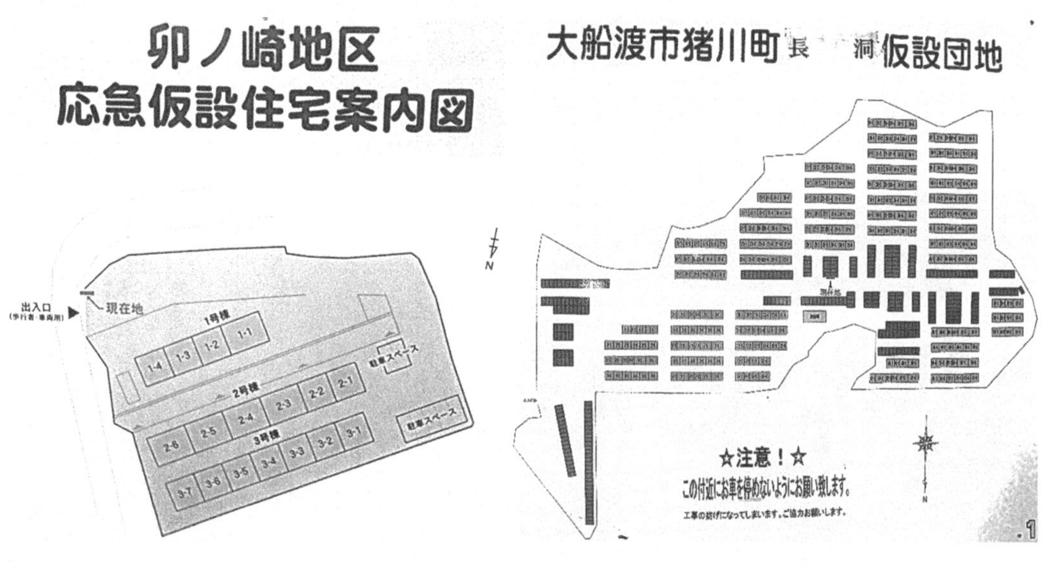

Fig. 1 Different settlement layouts of temporary housing in Tohoku

limited provision of communal spaces. In this settlement, the local residents requested one NGO to construct a gazebo to be used as a social space for the women of the settlement (see Fig. 1, right). This space was built by the NGO volunteers on privately owned land after the residents formally asked permission to the owners for their temporary use.

The possibilities for community interaction and their complexities were completely different in the larger settlements. For instance, in the settlement in the city of Ofunato (see Fig. 2, left), there were hundreds of residents in housing blocks or rows. There were also parking spaces and indoor communal spaces provided (see Fig. 2, right). These spaces were used for official activities or events of the community such as the distribution of relief supplies within the settlement with the support of the residents and volunteers as seen in the photo. Although the use of these indoor spaces are intended to promote the communication in the settlement, they do not consider the casual and spontaneous social interaction among the residents.

The development of the sense of place (Massey 2005) and the public life defines the role of a space as social (Carmona et al. 2003), where formal and informal communal activities take place and gather different groups of residents and facilitate the progress of integration and development of neighbourhood and community roots (Rivlin 1987). The photos in Fig. 3 show community festive/ceremonial activities (Fig. 3, left) in one of the temporary settlements built in a school sports ground. Elderly, children, women and disabled residents participate actively in the event. In the other photo (Fig. 3, right), it is shown a group of female residents gathered at the

Fig. 2 Different types of communal spaces for residents of temporary housing

Fig. 3 Spontaneous communal activities in open spaces perform by residents of temporary housing settlements

entrance of the housing rows. These are spontaneous and voluntary expressions of residents' interaction, communication and community collaboration and group attachment.

The lack of recreation spaces is another issue which limits the possibilities for a quick recovery of the stresses caused by disasters. Children who experience the disruption effects to lose their home, family and friends, change of schools and the resulting anxieties might result in long-term negative consequences (Woolley and Kinoshita 2015). The context in temporary housing settlements was lack of playgrounds for children; they also suffer due to the lack of interaction spaces. Woolley and Kinoshita (2015) observed in the temporary housing in Tohoku that children would play anywhere; they also observed the risks of children playing in parking lots or other dangerous locations. Volunteers and NGOs focus their community support activities to the provision of playgrounds and other recreational spaces for residents.

Shopping spaces are signs of post-disaster recovery, but they also mean places for passive social contact and casual interactions which are linked with an increased sense of community (Francis et al. 2012). In temporary settlements one of the local amenities that were crucial for independence of the residents was the local shops (Fig. 4, left). The location of these sites is generally in peri-urban areas, where there is limited access to commercial areas. Community gardening (Fig. 4, right) is another local activity that promoted the social interaction among the residents of the temporary housing. Elderly women, volunteers and children actively participated in these activities, which also add an aesthetic value to the open spaces in the settlements. Both stores and community gardens are spaces that are not limited by their utilitarian role; they actually promote the natural interaction of the residents, combining social enjoyable and recreational purposes.

4.2 Case Study 2: Social Spaces and 'Fiestas' in Permanent Housing in the Philippines

The Philippines is one of the most vulnerable countries to natural disasters, where regular climatic events such as tropical storms are becoming a permanent threat to coastal cities. The change in the patterns of frequency and intensity of the storms has been connected to the effects of climate change, but also the massive destruction caused by them is the result of fast population growth and unplanned city growth. Cities in the country have received influx from internal migrants; many of them settled in highly vulnerable areas and established marginal settlements or squatters.

In the city of Cagayan de Oro in the southern Philippines, the Typhoon Washi hit the city between December 16 and 17, 2011, destroyed urban areas along river banks and near the coastal

Fig. 4 Temporary community initiated public open spaces and community gardening

Fig. 5 Communal spaces in resettlement sites

line, which were inundated and the informal and precarious settlements were the most affected. The government approach for recovery and reconstruction was to resettle the disaster victims in safer areas which were located from 5 to 20 km away from the city centre and former settlements. The local government also declared the pre-disaster settlements as 'no-built zones', preventing the returning of former settlers to these highly vulnerable areas (Carrasco et al. 2015).

The post-disaster recovery plan drafted by the local government preferred the rapid construction of permanent houses in resettlement sites and opposed to the construction of temporary housing. Consequently, in late April 2012, the first group of beneficiaries moved into the permanent houses, less than 5 months after the disaster. The decision-making and the management of the resettlement projects were mainly top-down through a partnership between the local government and NGOs, where any possible participation of resettled communities was simply dismissed.

Permanent housing was built in available public land purchased by the local government prior the disaster. The process of planning of the new settlements was done simultaneously with the land development, construction of housing and the infrastructure works. The pressure for quick results leads to the omission of crucial issues like the appropriate land classification regarding the safety and suitability for housing, lack of technical supervision during the construction of houses and inadequate provision of facilities and communal amenities. Spaces for community interaction were provided in most of the resettlement sites; however, they were not always available for the daily use of the residents (see Fig. 5, left). Therefore, residents built alternative meeting spaces which are preferred for the unplanned community events and casual meetings (Fig. 5, right).

Streets are components of urban communication, transport of vehicles and people and exchange of information and simultaneously enable planned and occasional contacts of people. Streets in the resettlement sites in Cagayan de Oro became centres for community interaction (Fig. 6, right), together with the shops that are located along the access roads to the settlements. These spaces became the stops for shopping but also for socialization (Fig. 6, left). In occasions, streets became the extension of the housing, where people dry the firewood collected from the surroundings for cooking (Fig. 6, right). Unfortunately, the role of these social spaces has positive and negative impacts, since they also become insecure spaces in night-time due to the high crime rates and the fear to the drug smugglers that operate in these areas.

Carr (1992) points out that the involvement of the people in the design and management of the spaces promote the connections between people and places. Also the most frequent use of the spaces reinforces the sense of community. Residents of resettlement sites organize their

Fig. 6 Alternative use of streets as communal meeting areas

Fig. 7 'Fiestas' decoration and use of open public spaces during local festive events

'fiestas' which are community festivals associated with religious festivities, which is part of the local culture in the Philippines. The 'fiestas' are the most important events for a community, in which all the members actively participate, mobilizing the people for the decoration of their streets and communal spaces (Fig. 7). The community identity is reinforced during these events which evidence community leadership at different levels in the neighbourhood.

5 Discussion

Disasters are sudden events which cause involuntary displacement producing a situation of 'placelessness' whose causes are beyond the forced change of communities' built environment. Krug et al. (1998) observed that together with the loss of property and bereavement, the disruption of social networks is linked to mental health problems. The consequent effects involve uncertainty about individual and collective continuity of the lifestyles, identities, social relationships and attachment to a community and physical environment. The road to recovery and resilience of disaster-affected communities are not limited to the reconstruction of housing. Restoration of the social fabric and sense of community result crucial components of reconstruction of social structures, where diversity and inclusion of the most vulnerable must be prioritized in order to minimize the extreme psychological effects of disasters.

Although disaster-affected communities might start to rebuild their social bonds after in the short term, as it was evidenced in the Philippines and Indonesia (Navarra et al. 2012), the traumatic effects of the disaster challenge the recovery of victims, and the community disconnection can be observed in the middle or even long terms. Ohto et al. (2015) observed that the suicide rates among the victims of the 2011 tsunami in Japan increased 4 years after the disaster. Another study on the victims of the 1995 earthquake in Japan showed that after 2 years of the disaster, there was an increase in the rates of the death of elderly residents living isolated from their neighbours (Tanaka et al. 2010).

This study observed how communities spontaneously used the available spaces to develop multiple activities, such as the community gardens, and casual meetings in the circulation paths between housing blocks in Japan and the alternative use of roads as spaces for social interaction in the Philippines. The positive outcomes of big events in the community promote inclusion and leadership in the community as we observed the festivals in Japan and the religious festivities in the Philippines. In the cases analysed, residents do not show a passive attitude; remarkably, displaced residents actively took the initiative to change the use of available open spaces and even arrange the construction of social spaces by themselves or directly coordinate to receive support from external organizations. A clear example was the project led by elderly women for the construction of a social space in a remote temporary settlement in Japan. Similarly, to the construction of alternative meeting spaces in the Philippines, which despite the precariously of the construction quality, fulfils its purposes more effectively than the community spaces built by the NGOs.

The understanding of both the role of the physical environment and the capacities to influence in the social behaviour of the people results crucial for the planning, construction and further development of human-centred settlements. Multiple public and community activities are acknowledged as instruments for the restoration and creation of ties between people that recently meet and expand their social network in the new communities. Both regular and occasional use of public and communal spaces, in particular open spaces, promote leadership in the communities at the time of organization and participation in festivals which mobilize the members of the community and enhance the sense of cultural identity.

6 Conclusions

This study demonstrates that the recovery of the physical environment is not limited to housing, but to the structures that encourage the restoration and creation of social ties. They reinforce the sense of continuity and capacity of adaptation to the new conditions and support the reduction of levels of post-traumatic stresses caused by disasters.

The Japanese case analysed evidenced the ambiguity in the definition of temporality of temporary settlements and the preferred temporary spaces of casual interaction instead of the less accessible communal centres built by the government. In the Philippines, the observed precariousness of the communal spaces built by the residents defined spaces that might be considered non-permanent. Thus, for communities in transition, it is necessary to analyse the criteria of appropriateness of social spaces and consider permanent and temporary structures which may be adapted to the peoples' needs.

This paper discussed the urgency to recover or create social, cultural and spatial attachments for involuntary displaced people and their positive effects in the residents. Sociocultural connections provide security and inclusion of different groups in the community. Social spaces promote gradual and sustainable community integration. Displaced communities proved to be independent and proactive. Hence, the role of external stakeholders such as governments, NGOs and volunteers should be oriented to facilitate and support the creation of these spaces regardless of the permanent or temporary nature of the settlements.

These issues are fundamental for the recovery of displaced communities and must be considered in the planning of recovery plans, understanding the potentials of the uses of public and other social spaces for community cohesion as a starting point to promote recovery, hazard mitigation and resilience.

References

Allan, P. & Bryant, M. (2010, March). *The critical role of open space in earthquake recovery: A case study*. In: Proceedings of the 2010 NZSEE Conference (2010, Nueva Zealand) (pp. 1–10).

Barakat, S. (2003). Housing reconstruction after conflict and disaster. Humanitarian Policy Group, Network Papers, 43, 1–40.

Bolin, R., & Stanford, L. (1991). Shelter, housing and recovery: A comparison of US disasters. *Disasters, 15*(1), 24–34.

Brown, B. B., & Perkins, D. D. (1992). Disruptions in place attachment. In I. Altman & S. Low (Eds.), *Place attachment* (pp. 279–304). New York: Springer.

Brown, B., Perkins, D. D., & Brown, G. (2003). Place attachment in a revitalizing neighborhood: Individual and block levels of analysis. *Journal of Environmental Psychology, 23*(3), 259–271.

Carmona, M., Heath, T., Oc, T., & Tiesdell, S. (2003). *Public spaces-urban spaces: The dimensions of Urban Design*. Oxford: Architectural Press.

Carr, S. (1992). *Public space*. New York: Cambridge University Press.

Carrasco, S., Ochiai, C. & Okazaki, K., (2015). *Disaster Induced Resettlement: Multi-Stakeholder interactions and decision making following Tropical Storm Washi in Cagayan de Oro*. In Philippines, paper presented at the 11th International Conference of The International Institute for Infrastructure Resilience and Reconstruction (I3R2): Complex Disasters and Disaster Risk Management, Seoul, South Korea.

Carrasco, S., Ochiai, C., & Okazaki, K. (2016). Impacts of resident-initiated housing modifications in resettlement sites in Cagayan de Oro. *Philippines. International Journal of Disaster Risk Reduction, 17*, 100–113.

Chitrakar, R. M. (2016). Meaning of public space and sense of community: The case of new neighbourhoods in the Kathmandu Valley. *International Journal of Architectural Research: ArchNet-IJAR, 10*(1), 213–227.

Conklin, J. E. (1971). Dimensions of community response to crime problem. *Social Problems, 18*, 373.

da Silva, J. (2010). *Lessons from Aceh, key considerations in post-disaster reconstruction*. Rugby, Warwickshire: Practical Action Publishing Ltd.

Davidson, C. (2010). Multi-actor arrangements and project management. In G. Lizarralde, C. Johnson, & C. Davidson (Eds.), *Rebuilding after disasters: From emergency to sustainability*, Spon Press, London and New York, NY, pp. 88–109.

Davidson, W. B., & Cotter, P. R. (1991). The relationship between sense of community and subjective Well-being: A first look. *Journal of Community Psychology, 19*(3), 246–253.

Edelstein, M. R., & Wandersman, A. (1987). Community dynamics in coping with toxic contaminants. In I. Altman & A. Wandersman (Eds.), *Neighborhood and community environments* (pp. 69–112). New York: Springer.

Félix, D., Branco, J. M., & Feio, A. (2013). Temporary housing after disasters: A state of the art survey. Habitat International, 40, 136–141.

Francis, J., Giles-Corti, B., Wood, L., & Knuiman, M. (2012). Creating sense of community: The role of public space. *Journal of Environmental Psychology, 32*(4), 401–409.

Hirano, S. (2012). *Learning from urban transitional settlement response in the Philippines: Housing. Land and property issues*. Baltimore: Catholic Relief Services (CRS).

Jha, A. K., Jha, A. K., Barenstein, J. D., Phelps, P. M., Pittet, D., & Sena, S. (2010). *Safer homes, stronger communities: A handbook for reconstructing after natural disasters*. International Bank for Reconstruction and Development, The World Bank, Washington, DC.

Johnson, C. (2010). Planning for temporary housing. In G. Lizarralde, C. Johnson, & C. Davidson (Eds.), *Rebuilding after disasters: From emergency to sustainability*, Spon Press, London and New York, NY, pp 70–87.

Kondo, T., & Karatani, Y. (2016). Spatial planning for housing recovery after the great East Japan earthquake. In S. Greiving, M. Ubaura, & J. Tesliar (Eds.), *Spatial planning and resilience following disasters: International and comparative perspectives*, Polocy Press, Great Britain. pp 41–54.

Krug, E. G., Kresnow, M. J., Peddicord, J. P., Dahlberg, L. L., Powell, K. E., Crosby, A. E., & Annest, J. L. (1998). Suicide after natural disasters. *New England Journal of Medicine, 338*(6), 373–378.

Maly, E. (2014). *Towards a people-centered housing recovery after the triple disaster. Human security and Japan's triple disaster: Responding to the 2011 earthquake, tsunami and Fukushima nuclear crisis*, Routledge, NY pp.108–126.

Massey, D. (2005). *For space*. London: Sage.

Naimi-Gasser, J. (2012). The remembered trees. In J. D. Barenstein & E. Leemann (Eds.), *Post-disaster reconstruction and change: Communities' perspectives* (pp. 137–156). CRC Press. Taylor and Francis Group. Boca Raton, Fl. U.S.

Navarra, M. Q., Niehof, A., Van Der Vaart, W., van der Horst, H. M., & Suliyanto, S. E. (2012). The disruption and rebuilding of social capital in involuntary resettlement in the Philippines and Indonesia. *International Journal of Social Sciences and Humanity Studies, 4*(2), 307–324.

O'Brien, D., & Ahmed, I. (2011). Donor-driven housing, owner-driven needs: Post-tsunami reconstruction in Aceh, Indonesia. In E. Kaufman (Ed.), *Pressures and distortions, City dwellers as builders and critics: Four views* (pp. 237–337). New York: Rafael Viñoly Architects PC.

Ohto, H., Maeda, M., Yabe, H., Yasumura, S., & Bromet, E. E. (2015). Suicide rates in the aftermath of the 2011 earthquake in Japan. *The Lancet, 385*(9979), 1727.

Oliver-Smith, A. (1991). Successes and failures in post-disaster resettlement. *Disasters, 15*(1), 12–23.
Oliver-Smith, A. (1986). The martyred city: Death and rebirth in the Andes. University of New Mexico Press.
Pader, E. J. (1994). Sociospatial relations of change. In I. Altman & A. Churchman (Eds.), *Women and the environment* (pp. 73–103). New York: Springer.
Partridge, W. (1989). Involuntary resettlement in development projects. *Journal of Refugee Studies, 2*, 373–384.
Quarantelli, E. L. (1995). Patterns of sheltering and housing in US disasters. Disaster prevention and management: an international journal, 4(3), 43–53.
Reconstruction Agency. (2011). Towards Reconstruction "Hope beyond the Disaster" 25 June 2011 Report to the Prime Minister of the Reconstruction Design Council in response to the Great East Japan Earthquake. Tokyo, Japan.
Reconstruction Agency. (2017). Current Status of Reconstruction and Challenges, March 2017. Tokyo, Japan.
Rivlin, L. G. (1987). The neighborhood, personal identity, and group affiliations. In I. Altman & A. Wandersman (Eds.), *Neighborhood and community environments* (pp. 1–34). New York: Springer.
Rohe, W. M., & Burby, R. J. (1988). Fear of crime in public-housing. *Environment and Behavior, 20*, 700–720.
Sakamoto, M. (2014). *6 human security and life recovery. Human security and Japan's triple disaster: Responding to the 2011 earthquake, tsunami and Fukushima nuclear crisis* (p. 89).
Shumaker, S. A., & Conti, G. J. (1985). Understanding mobility in America. In I. Altman & C. Werner (Eds.), *Home environments* (pp. 237–253). Springer. Boston, MA.
Tanaka, M., Takahashi, C & Ueno, Y. (2010). The conditions of isolated death occurrences and its background in public temporary housing, Original source in Japanese: 応急仮設住宅における「孤独死」の発生実態とその背景. 日本建築学会計画系論文集, *75*(654), 1815–1823.
The Japan Times. (2016). *Some Tohoku disaster-displaced to remain in shelters up to 10 years*. The Japan times. Retrieved from http://www.japantimes.co.jp/news/2016/03/07/national/social-issues/11-tohoku-disaster-displaced-remain-shelters-10-years-study-finds/#.WXfkX4SGNaQ
UNOCHA and IFRCC (2015). Shelter After Disaster, Second Edition. Imprimerie Chirat, Lyons, France.
Woolley, H., & Kinoshita, I. (2015). Space, people, interventions and time (SPIT): A model for understanding Children's outdoor play in post-disaster contexts based on a case study from the triple disaster area of Tohoku in north-East Japan. *Children & Society, 29*(5), 434–450.
Zetter, R., & Boano, C. (2010). Space and place after natural disasters and forced displacement. In G. Lizarralde, C. Johnson, & C. Davidson (Eds.), *Rebuilding after disasters: From emergency to sustainability*, Spon Press, London and New York, NY, pp. 206–230.

Community Participation to Build Back Better: Evidence from the 2015 Nepal Earthquakes

April Aryal, Suzanne Wilkinson, and Alice Chang-Richards

Abstract

Community-based recovery is a process where the recovery is enhanced by using social capital, placing the community at the center of the recovery process. The community's capacity to aid recovery should be instrumental to recovery processes and is thought to enhance the ability of the community to build back better. The transfer of experience and knowledge from the community could accelerate recovery, as community needs are at the forefront of the recovery. Using ancient and traditional knowledge coupled with modern technologies can enhance recovery. A better recovery is unlikely without active participation from the community. This study focuses on post-disaster recovery based on community participation following the Nepal earthquakes. It aims to understand the role of active community participation in building back better and its implications for improving recovery. This research considers practical implications from the case study of Nepal. The methods used in this research include observations after the earthquakes in Nepal on April and May of 2015. Active community participation was observed after the Nepal earthquakes (25 April and 12 May 2015). Volunteers from the community were active in the immediate rescue, and locally available materials were used to construct temporary shelters. The community was active in preparing food and in the distribution of water supply. Findings suggest the importance of social relationships, trust, respect, and value for each other enhanced recovery. Building better community cohesiveness and assisting community-based recovery help withstand future disasters. Social cohesion, social bonds, relationships, communications, neighborhood, and festivals are important for enhanced recovery. This paper reinforces the need for active community participation in disaster recovery.

Keywords

Disaster · Community · Build back better · Social capital · 2015 Nepal earthquake

1 Introduction

The concentration of population and assets and the embedded conditions of socioeconomic and spatial vulnerabilities generate disaster risk in urban areas affected by natural hazards (Gencer 2013). When rapid urbanization and frequent natural disasters combined, they substantially

A. Aryal (✉) · S. Wilkinson · A. Chang-Richards
University of Auckland, Auckland, New Zealand
e-mail: aary953@aucklanduni.ac.nz;
s.wilkinson@auckland.ac.nz;
yan.chang@aucklanduni.ac.nz

A. Asgary (ed.), *Resettlement Challenges for Displaced Populations and Refugees*,
Sustainable Development Goals Series, https://doi.org/10.1007/978-3-319-92498-4_13

increase the risk to which people are exposed, especially in developing countries (Correa 2011). Nepal is in a high-risk seismic zone, and Kathmandu Valley is vulnerable to frequent earthquakes. From 1900 to 2005, over 13,000 disaster events were recorded with a total of 7,400,000 human casualties in Nepal (Aryal 2012). Every 75 years, there is a high probability of a devastating earthquake in Nepal (Marahatta 2007). The earthquakes of Nepal on 25 April and 12 May 2015 caused more than 9000 deaths and approximately \$10b (half of the total GDP) as per the Nepal Government. Out of 75 districts, 15 districts were severely damaged including the capital city Kathmandu.

This study focuses on how community participation aided post-disaster recovery to build back better. The paper aims to understand the role of active community participation and its value in improving recovery. The paper examines how community resilience builds during recovery especially in the vulnerable areas which appear to lack capacity.

Displacement and dislocation of social networks are common following a disaster. The affected people should be consulted at the community level, as it is essential to try to preserve existing social networks, which tend to form the basis of support and mutual aid among the affected households (PDNA 2005). The aim is to assist the community response in allowing them to return to pre-disaster phase, more resilient and at the same time less vulnerable. Once the reconstruction is to be done, if we fail to acknowledge the preexisting vulnerabilities, the restoration will be exactly as vulnerable as prior to disaster. When the restoration is of same codes as before the hazard, the post-disaster communities will confront the similar difficulties when exposed to future disasters (Lyons 2009). A common way of conceptualizing the recovery is to think in terms of build back better (BBB). Recovery can be enhanced with BBB, which leads to the formation of disaster-resilient communities in the future. The concept of BBB is to encompass the overall process to enhance a community's all aspects, physical, social, environmental, and economic, in the recovery phase, and ultimately to design a disaster-resilient community (Mannakkara and Wilkinson 2013). BBB is a pathway to utilize the reconstruction process as an opportunity for a community's development of all circumstances of physical, social, environmental, and economic to construct a better disaster-resilient community (Clinton 2006). Mannakkara and Wilkinson (2013) developed and tested a BBB framework in different disaster-affected communities, eventuating in the following framework (Fig. 1).

In order to build back better, focus should be on three main categories: risk reduction, community recovery, and effective implementation.

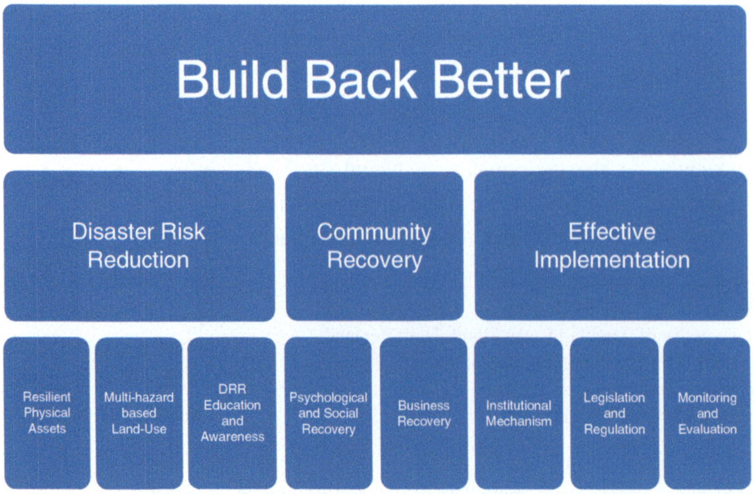

Fig. 1 Revised BBB framework Mannakkara and Wilkinson (2013)

The category risk reduction relates to improvement of resilience to natural disaster response to physical environment, while community recovery concerns enhancing the social and economic conditions of the community. Effective implementation refers to the means by which risk reduction and community recovery should take place. Using this framework, this study can illuminate the recovery progress of communities in Nepal. Build back better after the disaster is the opportunity of disaster response for society's improvement. Build back better provides a tool to assist and creates a process for strengthening the community, so they can cope with, adapt to, and recover from the loss and disruption encountered through the experience of the disaster. How effectively this is done is a function of how well people, communities, and societies can work together and use their resources to deal with the problems encountered (Paton et al. 2014). This study is focused on the importance of build back better in light of community recovery.

While evidence of physical loss associated with disaster is extensive, the existence of beneficial consequences has often been overlooked by research that focused primarily on the immediate aftermath of adverse (Gow and Paton 2008). The traditional method of recovery based on technological intervention means infrastructure is restored to the pre-disaster conditions positioning the community in same vulnerabilities if exposed to similar kinds of disasters. This is only building back to pre-disaster state, not build back better (Lyons 2009). Without appropriate policy guidelines, in particular, a commitment to building back better, the pace of the action could be delayed due to obstructive deliveries, inability to employ practical applications to exact recovery difficulties as a result of forfeiture of obligation to the reconstruction course, lack of ability to start ideas for risk and vulnerability mitigation, and ultimately damage of community recovery and the quality of life (Myburgh et al. 2008).

The strategies that community members tend to adopt for building back better depend on the resources (Chamlee-Wright 2010) that they can access, their expectations about their community's prospects for recovery, and, ultimately, community-level collective narratives (Chamlee-Wright and Storr 2011a). Aldrich studied the post-disaster responses from four different communities with unlike social capital embedded within – Tokyo following the 1923 earthquake, Kobe after the 1995 earthquake, Tamil Nadu after the 2004 Indian Ocean Tsunami, and New Orleans post-Katrina – and resulted the coordinate recovery ability is higher with vigorous social networks (Aldrich 2012).

In disasters, family ties are central to resilience because kin commonly serve as the first providers of assistance (Beggs et al. 1996; Drabek and Boggs 1968; Aldrich 2010; Garrison and Sasser 2009). Individuals assume family members, especially an immediate family of parents/stepparents, children, and siblings, will support each other in disasters, with 85% identifying at least one family member and 36% identifying only family members among their social capital networks for disaster assistance (Meyer 2013). Disaster scholars have used social capital to understand the trajectory of individuals (based on what resources are accessed through social networks) as well as communities (based on levels of trust, collective action, and other public goods). Social networks (Szreter & Woolcock 2004) provide financially (e.g., loans and gifts for property repair) and nonfinancial resources (e.g., search and rescue, debris removal, child care during recovery, emotional support, sheltering, and information) (Aldrich and Meyer 2014). The social bonding serves as social insurance, which can be called informal insurance. Social ties can serve as an informal recovery structure, providing victims with information, financial help, and physical aid. Bonding social capital can reduce individuals' likelihood of seeking formal aid from organizations (Small 2010) during disasters (Beggs et al. 1996). Moreover, to be taken seriously, community efforts to craft their own redevelopment plans had to meet professional standards, requiring further diversion of social capital away from the provision of direct support within the community (Chamlee-Wright and Storr 2011b). Community is put at the heart of the BBB framework by Mannakkara and Wilkinson (2013), where psychosocial elements of community are at the core of recovery and economic outcomes are enhanced.

2 Methodology

This study uses qualitative research based on the observations and in-field experience after the Nepal earthquakes as well as the existing secondary sources to look at community recovery in light of the BBB framework. The existing frameworks of build back better are studied to understand recovery in light of the BBB concept, and the focus is on the community aspects of building back better. The measurement of recovery is based on the observations of different disaster-affected places, and the analysis is based on the site visits to different places in terms of recovery. The researcher is from the disaster-affected region, so the comprehension of recovery is also based on local understanding. A part of the analysis is also from the experts working in the disaster recovery. The secondary data including scientific articles were collected to understand current community recovery and the role of social capital in post-disaster situations. The secondary data collection and questionnaire survey took place prior to approval from University of Auckland Human Participants Ethics Committee (UAHPEC) referenced 019444. The various social components like festivals, carnivals, and other ceremonies are considered as the tools for recovery. The traditional knowledge of the community is also considered as enhancing recovery.

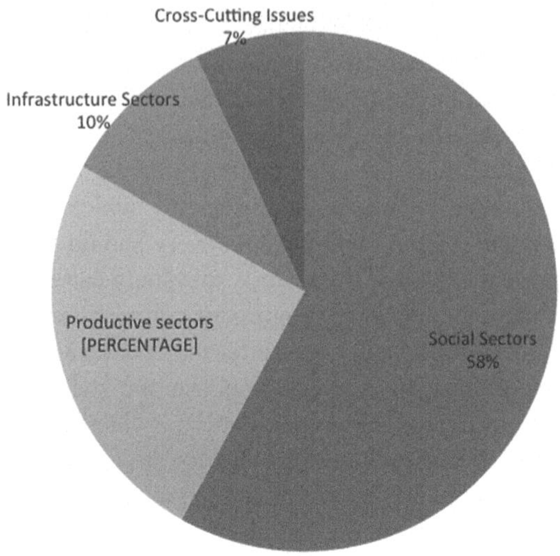

Fig. 2 Share of disaster effects across sectors as per the PDNA report (2015)

3 Build Back Better During Recovery of Kathmandu: A Community Focus

Following the earthquakes, the damage assessments found that World Heritage sites were more affected than non-heritage sites. Most of the construction of the World Heritage sites dated back to the twelfth century. The lack of proper retrofitting of these buildings meant they were vulnerable to collapse. PDNA (2015) demonstrated the effects of the disaster across sectors, finding that the social sector was the most affected (Fig. 2).

Damage showed that most of the ground floors of the buildings were used for business purposes and were served without any internal walls, which ultimately acted as soft-story effect and caused building collapse. In addition, the drinking water supply was interrupted along with other infrastructural damages. Due to the complications in infrastructure damage, local communities were not reachable for several weeks.

Communities mobilized to make sure that they could recover. The observation after the disaster shows although the disaster was devastating, it seems to have at least some positives to social bonding and coordination. The local people started forming small groups and helping to the nearest locality, as immediate rescues were needed. People provided assistance to their neighbors. In the BBB framework, community recovery breaks down into various propositions, and it is at this level that the community, and agencies supporting the community, can be seen to be either improving or worsening the recovery process.

Urban development of Kathmandu works alongside the importance of community and social life. This community strength is a social capital and is one of the foundations for build back better. The members of Nepalese communities, the individual, and the family form community networks with the internal bonding among the neighbors. The identity, nationalism, friendship, and neighborhood are the factors of social bonding. The intercommunity networks incorporate various social relationships, e.g., marriages, festival invitations, etc., forming a bridge between different communities. There is a strong relationship formed after marriages between different communities, and the marriage brings different communities together. Bridging activities and associations hold together individuals from different locations, identities, and language groups (Schuller et al. 2000). The communities of Nepal are composed of different ethnicities and castes, but they share the same culture, in many cases. The layers of the community are composed of different economic groups, different ethnic groups, different language-speaking people, and different cultural groups, but they are bound by some common religious and social values. The city was segregated into heterogeneous occupation classes or caste groups; however, there was social harmony among inhabitants (Bhandari 2014). The existing social capital is very important in the context of recovery and build back better. The BBB framework sets out the key propositions that contribute to peoples' vulnerability: exposure to hazards and stresses, fragile livelihoods, future uncertainty, and weak governance. The existing frameworks and their implications for post-disaster recovery of Nepal. The propositions of the BBB framework are ambitious, and the assessment below shows how far these can be achieved and how reasonable the propositions are in terms of community recovery in Nepal.

Key propositions for building back better – community	Observations from recovery in Nepal
Personnel support to each family	The individual families are receiving grants from government (USD 500 per family)
Organization of specialized assistance to vulnerable group	Since the earthquake, and there could have been more achievement toward the recovery if there has been specific development toward the identification and assistance to vulnerable groups
Psychological support and counseling	This is not coordinated, but some NGOs and INGOs are working on post-disaster trauma and providing the psychological assistance required
Building social cohesion	During the emergency period, there was a huge social cohesion with a lot of communities working together, but currently social cohesion has started to diminish
Rebuilding public facilities	There has been huge efforts toward the strengthening of school and other public facilities
Community involvement in designing recovery plan and implementation	The current recovery method of supply-driven recovery is currently practiced in Nepal rather than community-led approach
Transparency with affected communities	The community is not always aware of the recovery plan and the execution of the recovery plan
Local governments must be empowered to manage recovery efforts	The local governance is weak, the country is without elected local body from more than 2 decades, and this could be seen to slow recovery

Build back better propositions adapted from Mannakkara and Wilkinson (2013)

3.1 Personnel Support to Each Family

There are no specific case managers appointed for families, and it is very unlikely to happen in the case of Nepal due to the shortage of such managers and lack of past experience. Most of the families received financial assistance either from government or from nongovernment. As per the steering committee of the National Reconstruction Authority (NRA), the housing grant for families who lost their homes during the earthquakes of 2015 is Rs. 300,000 (US$ 3000), which was previously US$ 2000. The NRA began distributing the first installment of the housing grants in March 2016 and had until the end of September 2016 to complete distribution to 407,004 families.

3.2 Organization of Specialized Assistance to Vulnerable Group

There is a strategic recovery objective in the Post-Disaster Recovery Framework (PDRF) published by NRA, which postulates "Strengthen the capacity of people and communities to reduce their risk and vulnerability, and to enhance social cohesion." The framework addressed the implementation of effective measurements for gender equity, social inclusion, closing the gender gap, the participation of women and marginalized groups, and raising awareness and capacity of women and vulnerable and marginalized groups. To action this, there was the identification of vulnerable communities. Vulnerable communities needed to have specific focus, and their needs taken into account, in order to achieve building back better.

3.3 Psychological Support and Counseling

Although there is no specific strategic framework from the government and NRA for the support and counseling toward mental health, some NGOs and INGOs are working for the psychological support and mental health prior to the disaster and were able to assist post-disaster.

3.4 Building Social Cohesion

The focus of the vision 2 of the PDRF was social cohesion, but the study and various observations showed diminishing social cohesion and connectivity over the duration of recovery to date. There was an increment in the social cohesion and strength immediately after the disaster. This social cohesion needs to be facilitated for build back better.

3.5 Rebuilding Public Facilities

The community facilities that were in specific focus for rebuilding were schools. School rebuilding took priority. The Post-Disaster Needs Assessment (PDNA)) claimed 21,169 classrooms had been fully destroyed and 27,452 classrooms were partially damaged at 7923 schools. The Department of Education began reconstruction of classrooms, with the support of different agencies, and was making rapid progress. Rebuilding public facilities creates hope in the community and has a healing affect and assists with building back better.

3.6 Community Involvement in Designing Recovery Plan and Implementation

The disaster recovery practice in Nepal is supply driven rather than participatory and is top to down approach (Kadel 2011). This reflects less community participation in the formulation of disaster recovery plans and recovery process. The community members are unaware of the plans and strategies developed for the recovery, and even in other development and planning projects, they have minimal input. If the inclusion of community during formulation and implementation could have been adjusted, the better recovery of the community could be achieved. Inclusion of community is at the heart of build back better.

3.7 Transparency with Affected Communities

The transparency has always been the issue for Nepal. It is ranked 116th among 177 countries around the world with the score of 31 on the scale of 0–100 (Transparency International 2014). It could be very efficient for the recovery if the transparency with the communities is developed and maintained. This would then facilitate building back better.

3.8 Local Governments Must Be Empowered to Manage Recovery Efforts

The country is without elected local body from more than 2 decades. All of the government authorities, including reconstruction authority, were without local elected governments. Hence the lack of governance with transparency and the lack of local input into recovery planning could slow recovery and does not facilitate build back better approaches.

4 Nepal Post-disaster Recovery Framework

The Nepalese government proposed a framework for recovery. The framework had elements of build back better. Below is the framework and comments on how the framework worked in practice.

Nepal Post-Disaster Recovery Framework	Observations from recovery in Nepal
Restore and reconstruct disaster-resilient housing, government buildings, and cultural heritage in rural areas and cities	Restoration and reconstruction of resilient housing especially in rural areas was slow, but the restoration of cultural heritage is progressing well. More focus required for rural housing reconstruction
Strengthen the capacity of people and communities to reduce their risk and vulnerability and to enhance social cohesion	The recovery could have been accelerated, if the active community participation was encouraged and facilitated
Restore and improve access to services and improve environmental resilience	Programs to action this were needed to improve the environment resilience and access to services
Develop and restore economic opportunities and livelihoods and reestablish productive sectors	There are some government and nongovernment programs for the training, including construction, which is generating the livelihood and reestablishment opportunities

Post-Disaster Recovery Framework (NRA 2016)

The Nepal Government framework for recovery incorporates elements of build back better for community recovery but is too generic to provide focus points for action. The framework could have been better if the community actions were more comprehensive and there had been community participation in the framework formulation.

5 Discussion

The build back better framework is seen in the context of the Nepal recovery framework. The framework used by Mannakkara and Wilkinson (2013) provides more opportunities for community engagement and also addresses more complex social recovery situations. The BBB framework seems more directive compared to the one proposed by the government. Community participation is at the heart of BBB, and the reflection of existing Nepal approach to recovery is not fully based on sound BBB principles. Strengthening of the capacity is not part of the recovery process in practice but is fundamental to BBB. However, when observing implementation, local people in Nepal are receiving training to build safer houses and to develop infrastructures in the earthquake-resilient method. The construction skills being developed are important in the promotion of the livelihood recovery of the local people. The necessity of providing personal-level support, encouragement, and information regarding development and recovery is the important part of build back better. Resources to the community for activities such as festival celebrations, and weekly participatory programs, could increase the social bonding and could improve recovery through active participation. The participatory programs within and among communities are important to identify the available resources and to develop trustworthiness among the community members and to enhance the intercommunity relationship to the sharing of resources. Building better community cohesiveness and assisting community-based recovery help withstand future disasters. Social cohesion, social bonds, relationships, communications, neighborhood, and festivals are important for the recovery. The BBB framework advocates for this in a community-based approach to recovery.

6 Conclusion

This study is based on observations and a case study approach to understand how to enhance recovery through build back better. Community participation is the core to recovery, especially in developing countries. How to encourage and promote social cohesiveness is one of the key challenges post-disaster. The BBB framework advocates community-led recovery with government oversight through implementation programs. Creative decision-making should hold community-based approaches to make applicable policies and strategies, which create more resilient societies. The concept of community resilience is important in building tools to reduce future hazard vulnerability. Building better community cohesiveness and assisting community-based recovery help withstand future disasters. Social cohesion, social bonds, relationships, communications, and healthy neighborhoods create the foundations for building back better.

References

Aldrich, D. P. (2010, June). Fixing recovery: Social Capital in Post-Crisis Resilience. *Journal of Homeland Security, 6*, 1–10.

Aldrich, D. P. (2012). *Building resilience: Social capital in post-disaster recovery*. Chicago/London: The University of Chicago Press.

Aldrich, D. P., & Meyer, M. A. (2014). Social capital and community resilience. *American Behavioral Scientist, 59*(2), 254–269.

Aryal, K. R. (2012). The history of disaster incidents and impacts in Nepal. *International Journal of Disaster Risk Science, 3*(3), 147–154. https://doi.org/10.1007/s13753-012-0015-1.

Asian Development Bank, Japan Bank for International Corporation and World Bank .(2005). *Preliminary damage and needs assessment. Sri Lanka 2005 post-tsunami recovery programme.* Colombo: Asian Development Bank, Japan Bank for International Corporation and World Bank.

Beggs, J. J., Haines, V. A., & Hurlbert, J. S. (1996). Situational contingencies. *Social Forces, 75*(1), 201–222. https://doi.org/10.2307/2580762.

Bhandari, R. B. (2014). Social capital in disaster risk management; a case study of social capital mobilization following the 1934 Kathmandu Valley earthquake in Nepal. *Disaster Prevention and Management, 23*(4), 314–328.

Chamlee-Wright, E. (2010). Community resilience in New Orleans east: Deploying the cultural toolkit. In *Community disaster recovery and resiliency* (pp. 99–122). London: Taylor & Francis Group.

Chamlee-Wright, E., & Storr, V. H. (2011a). Social capital as collective narratives and post-disaster community recovery. *The Sociological Review, 59*(2), 266–282.

Chamlee-Wright, E., & Storr, V. H. (2011b). Social capital, lobbying and community-based interest groups. *The Sociological Review, 149*(1–2), 167–185.

Clinton, W. J. (2006). *Lessons learned from tsunami recovery; key propositions for building back better.* New York: Office of the UN Secretary-General's Special Envoy for Tsunami Recovery.

Correa, E. (2011). *Populations at risk of disaster: A resettlement guide*. Washington, DC: Global Facility for Disaster Reduction and Recovery, World Bank.

Drabek, T. E., & Boggs, K. S. (1968). Families in disaster: Reactions and relatives. *Journal of Marriage and Family, 30*(3), 443–451.

Garrison, M. E. B., & Sasser, D. D. (2009). Families and disasters: Making meaning out of adversity. In *Lifespan perspectives on natural disasters: Coping with Katrina, Rita, and other storms* (pp. 113–130). New York: Springer.

Gencer, E. A. (2013). *The interplay between urban development, vulnerability, and risk management: A case study of Istanbul metropolitan area* (Vol. 7). New York: Springer.

Gow, Kathryn & Paton, Douglas (2008). In The phoenix of natural disasters: Community resilience. New York: Nova Science Publishers, Inc.

Kadel, M. (2011). *Community participation in disaster preparedness planning: A comparative study of Nepal and Japan*. Tokyo: Asian Disaster Reduction Center.

Lyons, M. (2009). Building back better: The large-scale impact of small-scale approaches to reconstruction. *World Development, 37*, 385–398.

Mannakkara, S., & Wilkinson, S. (2013). Supporting post-disaster social recovery to build back better. *International Journal of Disaster Resilience in the Built Environment, 6*(2), 126–139. https://doi.org/10.1108/IJDRBE-06-2013-0019.

Marahatta, P. S. (2007). *Cultural capital for earthquake vulnerability reduction: The case of Kathmandu Valley.* Master's Thesis, Erasmus University, Rotterdam, Netherlands.

Meyer, M. A. (2013). *Social capital and collective efficacy for disaster resilience: Connecting individuals with communities and vulnerability with resilience in hurricane-prone communities in Florida.* Doctor of Philosophy PhD, Colorado State University, Colorado.

Myburgh, D., Seville, E., & Wilkinson, S. (2008). *Summary of learnings from the 4th international i-rec conference, 'Building Resilience – Achieving effective post-disaster reconstruction'*. Paper presented at the Resilient Organizations, Christchurch, New Zealand.

NRA. (2016). *Post-disaster recovery framework*. Kathmandu: Nepal Government.

Paton, D., Johal, S., & Johnston, D. (2014). Community recovery following earthquake disasters. In; Beer M., Kougioumtzoglou I., Patelli E., (eds) *Encyclopedia of Earthquake Engineering*, 1–8. Springer, Berlin, Heidelberg.

PDNA, 2005, *Preliminary damage and needs assessment. Sri Lanka 2005 post-tsunami recovery programme*. Colombo: Asian Development Bank, Japan Bank for International Corporation and World Bank.

PDNA, N. P. C. (2015). *Post disaster needs assessment*. Government of Nepal Kathmandu.

Schuller, T., Baron, S., & Field, J. (2000). Social capital: A review and a critique. In T. Schuller, S. Baron, & J. Field (Eds.), *Social capital: Critical perspectives* (pp. 1–38). Oxford: Oxford University Press.

Small, M. L. (2010). *Unanticipated gains: Origins of network inequality in everyday life*. New York: Oxford University Press.

Szreter, S., & Woolcock, M. (2004). Health by association? Social capital, social theory, and the political economy of public health. *International Journal of Epidemiology, 33*(4), 650–667.

Transparency International. (2014). *Transparency International Corruption Perceptions Index 2013*. Transparency International.

Effects of Disasters on Displaced Workers

Alice Chang-Richards, Erica Seville, Suzanne Wilkinson, and Bernard Walker

Abstract

Natural disasters can have significant impacts on the workforce in affected regions. There are often widespread disruptions to labour supply due to displacement of people from their jobs, either by disrupting their place of work or by disrupting a worker's ability to attend work. This research aims to investigate the patterns of impact that disasters have on the workforce and the employment and livelihood issues that emerge during post-disaster recovery. By using comparative case study approach, this research compares recent disaster events, including the June 2013 Southern Alberta floods in Canada, the 2010 and 2011 Queensland floods in Australia, the 2010 and 2011 Canterbury earthquakes in New Zealand, the 2011 Great East Japan earthquake and tsunami and the 2008 Wenchuan earthquake in China. It was found that common disaster effects on displaced workers included job and worker displacement, loss of income, disruptions to workers' livelihoods and creation of additional participation barriers, particularly for females, youth and individuals with lower skill sets. Comparison of different disaster events also revealed insights into how disasters can change the local labour market structure post-disaster. General economic conditions, sectoral structure as well as business and individual coping mechanisms all influence livelihood outcomes for the affected workers. As the post-disaster recovery progresses in Queensland (Australia), Canterbury (New Zealand) and Tohoku (Japan), coordination of employment and livelihood initiatives with housing and other welfare policies is critical for ensuring that job opportunities are available to everyone, especially those with disadvantage.

Keywords

Natural hazard · Displacement · Workforce · Livelihood · Recovery

This chapter is based on a Resilient Organisations report prepared for APEC Natural Disasters Workforce Project "Building natural disaster response capacity: Sound workforce strategies for recovery and reconstruction in APEC economies" published in November 2013. The APEC Natural Disasters Workforce Project Office has granted consent for using materials in the report for publications.

A. Chang-Richards (✉) · S. Wilkinson
University of Auckland, Auckland, New Zealand
e-mail: yan.chang@auckland.ac.nz;
s.wilkinson@auckland.ac.nz

E. Seville
Resilient Organisations, Christchurch, New Zealand
e-mail: erica.seville@resorgs.org.nz

B. Walker
University of Canterbury, Christchurch, New Zealand
e-mail: bernard.walker@canterbury.ac.nz

A. Asgary (ed.), *Resettlement Challenges for Displaced Populations and Refugees*,
Sustainable Development Goals Series, https://doi.org/10.1007/978-3-319-92498-4_14

1 Introduction

Natural disasters can have significant impacts on the labour market in affected regions. Impacts usually involve temporary or longer-term disruptions to the local economy through direct and indirect effects on infrastructure, business and the workforce (Venn 2012). Current understanding of the impact of disasters on labour markets largely relies on post-disaster economic assessment in which indicators of labour market outcomes, such as job losses, labour force participation and unemployment, are measured.

The functioning of labour markets is reliant on a network where many determinants of a worker's social and economic wellbeing come to play, particularly in a crisis situation. These determinants relate to workplaces, communities and livelihoods (Skoufias 2003). The way natural hazards disrupt labour supply is by displacing people from their normal lives and jobs. Damage to properties and infrastructure often forces people to evacuate to areas outside the disaster zone. According to Venn (2012), mass displacement can lead to severe labour market disruptions, making it difficult for displaced people to retain their pre-disaster jobs and putting a strain on local labour markets in the areas to which people have been evacuated.

Workers in some countries can absorb the effects of disasters well, quickly returning to pre-disaster jobs and retaining livelihoods. However, in other countries, disasters can be major shocks for the workforce, with the events resulting in prolonged unemployment, loss of income and continued displacement. Therefore, comparison of different disaster events can reveal insights into how disasters can change the workforce structure post-disaster. This research aims to investigate the patterns of impact that disasters have on the workforce and the employment and livelihood issues that emerge during post-disaster recovery.

By using comparative case study approach, this research compares recent disaster events, including the June 2013 Southern Alberta floods in Canada, the 2010 and 2011 Queensland floods in Australia, the 2010 and 2011 Canterbury earthquakes in New Zealand, the 2011 Great East Japan earthquake and tsunami and the 2008 Wenchuan earthquake in China. There is great value in learning from past disaster experiences for achieving an improved understanding of effects of disasters on displaced workers. Such learnings will assist decision makers in developing policy settings and measures to support employment and social cohesion in these circumstances, thereby sustaining the capacity for livelihood recovery.

2 Literature Review

2.1 Job Displacement as a Result of Disasters

Over the past 5 years, approximately 80% of disaster-induced displacement has occurred in Asia (IDMC 2013). In 2012 alone, countries located on the Pacific Ring of Fire, such as China, Japan, Peru, the Philippines and the USA, experienced large-scale displacement with over 100,000 people in each event. Disasters displaced people not only from homes but also jobs. For example, research on the impact of floods on employment in the US found that floods decrease employment in an affected region on average by 3% (Camilo 2007).

In the case of Hurricane Katrina, an estimated 1.1 million people were evacuated from their homes. There were sharp variations in labour market outcomes for evacuees who returned to their pre-hurricane addresses compared with those that did not (BLS 2006; Groen and Polivka 2008; Zissimpoulos and Karoly 2010). Ten months after the hurricane, the unemployment rate for evacuees who had not returned to their homes was 26%, much higher than for those who had returned (BLS 2006). Even those who returned had higher rates of unemployment than those who were in the Katrina-affected states but did not evacuate (Zissimpoulos and Karoly 2010).

When displacement happens in a densely populated region, more detrimental labour market outcomes are likely. Hurricane Sandy in late

October 2012 struck the most densely populated region of the USA, displacing more than 775,000 people (BLS 2012a). In the following month, the US Bureau of Labor Statistics (2012b) found that employers took 1759 mass layoff actions involving 173,558 workers, measured by new filings for unemployment insurance benefits.

Some studies have found that people who relocate to a new area need additional assistance and protection to find a job due to losses of their social capital (Giglio and Wiseman 2010; IFRC 2012). Even for those who are not evacuated or who return to their homes relatively quickly after a disaster, disruptions to social and physical infrastructure as well as disruptions to their livelihood can create a barrier to ongoing labour force participation (Venn 2012).

The need to replace lost homes, infrastructure and commercial properties generates a significant demand surge for construction workers. As a result of this change of employment patterns, female workers tended to be badly affected as they were less likely to be able to take advantage of new job opportunities during the reconstruction phase (Craigie et al. 2012). Time needed for construction training means young people, particularly school leavers, have less chance to take up immediate employment in construction (Chang-Richards et al. 2012). Surge demand signalled in the construction market was unable to provide a panacea for workforce issues in the aftermath of a disaster. There were often skill shortages coexisting with relatively high numbers of unemployment and low rates of labour force participation.

Impacts can be felt across the geography and the demography of a region. In Chile, the most affected industries varied by region. In terms of economic sectors, three regions, namely, Bio Bio, Maule and O'Higgins, played a vital role before the disaster. The three regions represented a majority of job losses following the 2010 earthquake and tsunami. Across the three regions, service industries were most affected (United Nations 2010). An OECD study shows that around 46% of jobs lost were borne by women, but 85% of jobs created went to men. Overall, women were most affected, accounting for 60% of net job losses (Venn 2012).

2.2 Workforce Policy Strategies in Development

Set within a broad evolutionary framework, this section outlines the pathways through which institutional changes and renewed disaster experiences influence managing workforce issues following a crisis. It is found that the features and processes of different social development systems (UN, ILO, World Bank, OECD and APEC) shape their workforce strategies (Table 1).

Table 1 Comparison of workforce policy priorities in international developments

Agency	Priorities of workforce strategies	Context
World Bank	Social safety nets Social insurance Labour market programmes	Crisis
UN	Job creation Reintegration of workforce into livelihood	Post-conflict
ILO	Quality of jobs (decent work) Promotion of skills (training and education, capital assistance to SMEs and self-employed) Social protection	General labour employment support and post-disaster
OECD	Social protection – especially for the most vulnerable (youth, low-skilled immigrants) Upskilling and labour productivity	General labour employment policy and new projects looking at effects of disasters on displaced workers
APEC	Skills promotion Job creation Social safety nets	General labour employment policy and new projects looking at sound workforce strategies in disaster settings

Note: *UN* United Nations, *ILO* the International Labour Organisation, *OECD* the Organisation for Economic Co-operation and Development, *APEC* Asia-Pacific Economic Cooperation

A review of different policy frameworks in relation to workforce strategies in a crisis setting shows three key focus areas: (1) social protection, (2) skills promotion through job training and skills development and (3) job creation. Policy assistance in different systems also targeted different populations. For instance, while extensive work has been done by the World Bank and the UN with respect to the role of social safety nets in responses to crises, much of their work is focused on developing countries and/or within the context of poverty alleviation. Nevertheless, the World Bank and the UN's policies provide a strong foundation for agencies like OECD and APEC which have a narrower focus to look at good practice in workforce strategies for post-disaster recovery and reconstruction.

In APEC countries, for example, the ILO has advocated natural disaster responses to be more closely linked to social protection policies, with more dialogue between stakeholders to ensure the value of partnerships with businesses and the private sector are rallied. The new *Japan Earthquake Project*[1] launched by ILO's Japan office aims to collect and compile lessons and good practices from the recovery efforts in Japan and to share them with a wide circle of the international community.

There is recognition that when employment-related human impacts are ill-handled, pre-existing conditions in the labour market are most likely to combine to produce more vulnerabilities for those who are displaced from their jobs (Zedlewski 2006). A "one-size-fits-all" approach may not address some structural problems rooted in a labour market. There is a need to shift from post-disaster labour market response to developing improved processes pre-event to better cope with disaster shocks on the workforce.

3 Research Methodology

The issue of livelihood and employment is one of the least researched areas in disaster settings (Alexander et al. 2006; Budidarsono et al. 2007). However, it holds the key to unlock the solutions to addressing the United Nations' Sustainable Development Goals (SDGs). In particular, livelihood was considered to be at the heart of measures to combat poverty and to promote sustained economic growth and decent work (Wilkinson et al. 2016). There are some crucial gaps in the literature when discussing livelihood and employment in a disaster setting. Firstly, there is a lack of information on the private sector, particularly its partnership with public employment services. Efforts by private sector actors deserve attention due to their unconventional and innovative approaches to workforce strategies (Giglio and Wiseman 2010). Another gap is related to thematic focuses and methodology. There is no systematic case study analysis of the employment issues after a disaster. Only a few studies have been done with a focus on different aspects of employment issues (Holzer and Lerman 2006; Venn 2012).

To fill these gaps, it is necessary to examine employment and labour-related measures in the context of disaster displacement by using a case study approach. The creation of a "learning culture" in order to build upon past disaster experiences and apply their lessons can help reduce future risks (Lorch 2005). Case studies and the cross-case comparison will complement the work of other international organisations (e.g. ILO, World Bank, OECD) by providing more concretely the evidence on the pattern of effects of disasters on displaced workers. As part of the APEC Natural Disasters Workforce Project, authors have worked with the government officials within the Labour and Employment Department of countries including Australia, New Zealand, Canada, Japan and China to collect case study data. Data for the case studies below have been gathered and analysed by using the content analysis method (Krippendorff 1980).

International comparisons are made more complex by differences in political, social, economic and cultural environments between countries. In order to incorporate these differences, a conceptual framework was developed. In line with the components of individual case studies, the comparative analysis reported in this paper focuses on three broad themes including:

[1] http://www.ilo.org/asia/whatwedo/projects/WCMS_209150/lang--en/index.htm

(a) Pre-disaster labour market conditions
(b) Characteristics of the natural disaster event
(c) Impacts of natural event on the labour market

3.1 Case Study 1: The 2010 and 2011 Queensland Floods in Australia

From November 2010 to early January 2011, significant flooding occurred in Queensland with three quarters of the state declared a disaster zone. The widespread flooding events, followed by Severe Tropical Cyclone Yasi, resulted in the tragic death of 36 people (Queensland Reconstruction Authority 2012). Over 2700 houses and 3500 businesses were affected by floods, forcing about 15,500 people to evacuate.

As of March 2012, 131,000 insurance claims were reported with an estimated value of AUD 3.8 billion (Insurance Council of Australia 2012). There were estimated losses of AUD 875 million in primary industries, primarily the sugar, fruit and vegetable subsectors. The estimated cost of the disaster in terms of damage and economic impacts was around 15.7 billion or around 1% of the economy's GDP (World Bank & Queensland Reconstruction Authority 2011).

3.2 Case Study 2: The 2010 and 2011 Canterbury Earthquakes in New Zealand

A series of large earthquakes struck the Canterbury region of New Zealand in late 2010 and 2011. The two major ones were a 7.1 magnitude earthquake that hit west of Christchurch on 4 September 2010 and the deadly 6.3 magnitude earthquake on 22 February 2011. The second quake took the toll of 185 lives and caused substantial destruction of buildings, widespread land damage and rock falls. The damage was largely centred in the city of Christchurch.

More than 60% of Christchurch's central business district (CBD) buildings were severely damaged (CERA 2012). Another 60% of the 5000 businesses in the CBD and 50,000 employees were displaced. More than one third of central city businesses were unable to operate, with another third relocating to makeshift premises. Over 150,000 homes (around three quarters of Christchurch's housing stock) sustained some damage from the earthquakes. The total number of individual building, land and contents claims received exceeds 600,000 (Earthquake Commission 2011). The latest figures released on 28 April 2013 by the Government suggest that the rebuild with improvements could reach NZD 40 billion with high levels of uncertainty remaining. This damage is about 19% of New Zealand's GDP.

3.3 Case Study 3: The 2011 Great East Japan Earthquake and Tsunami

The Great East Japan earthquake with a magnitude of 9.0 on the Richter scale occurred on 11 March 2011 about 160 kilometres off the east coast of the island of Honshu, Japan. It was the biggest recorded earthquake ever to hit Japan and the world's fifth largest since 1900 (Japan's Reconstruction Agency 2012). As of 28 May 2012, Japan's Extreme Disaster Management Headquarters listed 15,883 fatalities with an additional 2676 people still missing.

The US Congressional Research Service estimated that the overall cost of the earthquake and tsunami could be between USD 195 billion and 305 billion, making it the most expensive natural disaster on record (Nanto et al. 2011). A further assessment undertaken by the Japanese Cabinet Office (2011), however, shows that the economic impact of this event was 16.9 trillion yen (approximately USD 207 billion) or around 4% of Japan's GDP.

3.4 Case Study 4: The June 2013 Southern Alberta Floods in Canada

In June 2013, large areas of Southern Alberta experienced a deluge of heavy rainfall that caused disastrous levels of flooding. The floods devastated areas spanning 55,000 square kilometres and affected more than 100,000 people and approximately 10,000 homes (Flood Recovery Task Force 2013). The scale and impacts of this event resulted in an unprecedented state of provincial emergency to be declared in Alberta. During the first few days after 20 June 2013, there were 29 local states of emergency declared in the province.

The Alberta Government estimated that approximately 5.1 million hours of work were lost during the flooding, resulting in CAD 485 million of lost economic output by the private sector. Factoring in the public sector working losses, the total economic impact of the floods reached approximately CAD 500 million. It represents about 0.2% of Alberta's annual GDP or about 2.2% of June's GDP (Parkatti 2013). The professional, scientific and technical service industry experienced about one in four employees missing work due to the flooding (losing CAD 61 million), followed by finance, insurance and real estate (losing CAD 31 million) and information, culture and recreation (losing CAD 25 million).

3.5 Case Study 5: The 2008 Wenchuan Earthquake in China

On 12 May 2008, a devastating earthquake measuring 8.0 on the Richter scale hit Wenchuan County in China's Sichuan Province. The earthquake was followed by over 30,000 aftershocks affecting 4667 towns, and large landslides occurred in some places (Ministry of Finance et al. 2012). A population of 46 million in Sichuan, Gansu and Shaanxi provinces was affected by this event (International Recovery Platform 2010). Official statistics reported that 69,227 people were killed in this disaster and 17,923 were missing (Ministry of Finance et al. 2012).

Around 7,789,000 housing units collapsed, and 24,590,000 were damaged during the earthquake (Paterson et al. 2008). As of 11 September 2008, the State Planning Group (2008) reported that approximately 34,125 kilometres of roads, 1263 reservoirs, 7444 schools and 11,028 hospitals were damaged in the earthquake. The total economic loss in the quake-affected areas was estimated at RMB 845 billion (approximately USD 150 billion) which was around between 1% and 3% of China's GDP. A number of cities and towns which functioned as major industrial and agricultural bases in the region were among the hardest hit. More than 1,150,000 farmers lost farmlands and production assets in the earthquake. In total, this disaster had left 372,000 people unemployed, and more than 15 million people were forced to evacuate (Ministry of Finance et al. 2012).

4 Results and Discussion

This section will report the findings with regard to what a disaster actually does to the displaced workforce. What are the similarities between the studied events? From these comparisons, we discuss the practical and theoretical implications.

Loss of lives and destruction of buildings and infrastructure are the inevitable consequences of natural hazards that take place where people have settled. In terms of lives lost or population affected, the recent Southern Alberta floods were moderate compared to other four international counterparts studied in this paper. By comparing event severity, however, the magnitude of these disasters was overwhelming. Each has been rated the strongest and most expensive natural disaster in the economy's/region's history. Table 2 below compares the magnitude and impact of the five examined disaster events. With the exception of the Wenchuan earthquake, damage estimates of the other four events are preliminary as the recovery evolves.

In a comparison of the five natural disasters by their overall losses, the Great East Japan earthquake and tsunami and the Wenchuan earthquake caused many more deaths, destroyed substantially more buildings and together represented a financial loss of more than USD 300 billion. In the case of the Canterbury earthquakes, despite the fact that the event had a lower number of deaths, a smaller number of units damaged and lower dollar value attributed to total loss, it was

Table 2 Impact of the five examined disaster events

Country	Australia	New Zealand	Japan	China	Canada
When	November 2010– January 2011	4 September 2010 and 22 February 2011	11 March 2011	12 May 2008	June 2013
Event	Floods	Earthquake	Earthquake and tsunami	Earthquake	Floods
Population affected	200,000	460,000	400,000 most directly affected	46 million	100,000 in Southern Alberta
% of regional population	4.4%	81.3%	4.3%	52.5%	2.5%
Human losses	36	185	15,883	69,227	4
Damage estimates	AUD 15.7 billion	NZD 40 billion	JPY 16.9 trillion	RMB 845 billion	CAD 500 million

Source: Parker and Steenkamp (2012), Parkatti (2013) and case study data

the earthquake event that caused serious damage to the national economy.

If the concentration of losses (e.g. the percentage of national GDP) in a region and the intensity of population affected (e.g. the ratio of people affected to the total population in the area) are significant, the loss can certainly overwhelm the local capacity. This is most evident in Canterbury where more than 80% of the population were affected by the earthquakes, and the estimated repair and rebuild costs reached up to 19% of the entire nation's GDP. Similarly, although with a lower proportion of GDP damage, the Wenchuan earthquake occurred in places of high people density, affected half of the population and caused a significant challenge to disaster relief, response and recovery services.

The disaster events have all disproportionately affected the major industries in the region. After the Queensland floods and the Southern Alberta floods, the majority of businesses in resources sectors ceased operations temporarily. In Canterbury, tourism as the major revenue-generating sector had experienced the biggest blow from the earthquakes. While the total damage was relatively modest compared to the size of Japan's economy, disruption to nuclear electricity generation from the 2001 earthquake meant severe short-term disruption to industrial production and economic activity across the economy.

The Alberta economy, by comparison, has been relatively resilient, and the wider Canada economy appears to have been little affected. The immediate impact on output appears to be much more muted than that following the other four events. However, disasters happen against a background context of other pressures on the economy. Apart from the Wenchuan earthquake which occurred prior to the 2008/2009 global financial crisis, all the other studied disasters happened in countries while they were coping with economic disturbances, even though they were faring relatively well in this crisis.

During the financial crisis, with few exceptions the APEC countries experienced much higher unemployment in 2009 compared to 2008 (APEC HRDWG 2010). This is particularly the case in countries studied in this paper. Figure 1 below summarises the trend of unemployment in these countries prior to their respective disaster event. Unemployment was rising in 2009 in Australia, New Zealand, Japan and Canada. From 2009 onwards, with one exception (New Zealand), unemployment fell in the other three countries to the point of the studied event.

As alluded to earlier, it is difficult to separate the impact of disasters on labour market indicators from those of the economic recession. Disasters do not completely change pre-disaster economic conditions; instead they simply magnify trends or conditions in place before disaster strikes (Comerio 1998). It is useful to look beyond the impacts to local circumstances at the time of the disaster: to implications of the losses

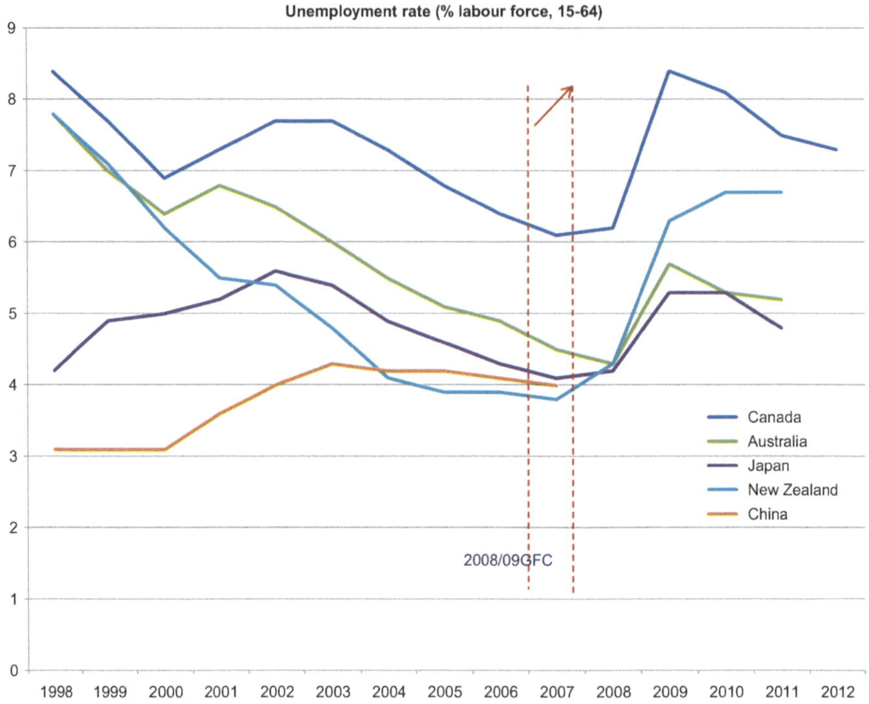

Fig. 1 Trends of unemployment rate of case study countries up to the disaster. (Source: Data for Canada, Australia, Japan and New Zealand are from the OECD Employment Outlook database; data for China is from the World Bank Jobs Data. The disaster event for each of these was at the end point of the line on the graph)

in order to determine the effects of a disaster on the labour market.

Disasters often displace workers through the destruction of productive assets and infrastructure. Cross-case comparison described above concludes that a labour market can be fundamentally changed due to the restructuring of economic activities as a consequence of the disaster itself or of the reconstruction. By comparing case study information, the following section summarises the implications of a natural disaster for workforce, including:

- *Physical damage and disruptions caused by disasters can cause fluctuations in the economic cycle which disproportionately affect less-skilled or vulnerable workers.* As demonstrated in the case studies, natural events would directly impact on labour-intensive sectors, such as manufacturing, services and agriculture industries. Loss of production assets, facilities and closure of workplaces can lead to significant job displacement of employees in these sectors. Similar to an economic cycle, a disaster can bring about fluctuations in the economic cycle and cause structural adjustment in certain industries. This sudden structural adjustment would particularly harm the labour force that is less skilled or those who are vulnerable to job loss or other age- or health-related problems.
- *Disasters tend to create additional participation barriers through the disruptions to individual workers' livelihood.* Disasters can have direct personal impact. Some household-related issues, as highlighted in the case studies, such as displacement from their homes, housing repairs, dealing with insurance payouts, family issues and childcare arrangements, may hinder continued workforce participation in the labour market. When other factors such as psychological wellbeing and

health are added into the mix, the combination of issues has the potential to continue to influence labour participation over time.

- *Compounding the workforce displacement problem are the difficulties facing individuals with lower skills.* Disasters in case studies had a marked impact on the patterns of labour demand. Many of the displaced workers have skills that are not required for newly available jobs. Young people and female workers of lower skills in the retailing and services sectors, for instance, were less likely to transfer into other sorts of jobs in booming industries like construction. Workers with single skills or lower skills, such as people with disability or with low language, literacy and numeracy skills, can be left behind from this structural adjustment.
- *Displacement of workers, especially those evacuees who moved to elsewhere outside the disaster zone, is associated with more detrimental labour market outcomes.* The case study of the Canterbury earthquakes has highlighted the employment effects of evacuees. Those who moved out of the region and who are employed elsewhere were more likely to have lower rates of labour force participation and employment and higher rates of unemployment than those who did not move. This finding, however, is consistent with prior studies of labour market effects following Hurricane Katrina (e.g. Giglio and Wiseman 2010; Holzer and Lerman 2006).

5 Conclusions

Building on the previous literature review, this study sheds light on how a disaster differs from other shocks and disturbances to a local economy and workforce. In general, aggregate patterns in case studies show short-lived negative workforce effects. These findings are consistent with previous studies of aggregate labour market effects following hurricanes and other disasters such as earthquakes and floods (e.g. (Belasen and Polachek 2008; Sawada et al. 2011)).

Across examined case studies, weak economic conditions may compound the disaster impacts and widespread damage, worsen employment in certain sectors and put a brake on further recovery in the livelihood. While the occurrence of a given disaster is largely exogenous, the impact that it has on the population is not (Baez and Santos 2008). Post-event labour market indicators show that many of the people who were temporarily displaced from jobs have already or will soon find new jobs on their own, either in the same area or elsewhere. But others need help to manage their transitions back to the labour market.

Compared to other demographic groups, higher job displacement results from Canterbury and Tohoku in relation to female workers and youth have confirmed the patterns identified in the literature review that female workers and youth groups may face more severe and sustained consequences. With major events, the implications of displacement are especially important to consider.

Comparison shows that the occurrence of natural disasters tends to intensify deficiencies in the labour market through physical and economic destruction and subsequent disruption. However, disentangling the causal impacts of disasters on the workforce is a complex empirical task. The lack of consistent data makes it difficult to undertake robust comparative analysis. Although informative, the aggregate findings in this paper do not illustrate how different groups fare in the labour market over the longer-term recovery period. There might be some pervasive effects of disasters on the workforce of a particular group in the studied countries.

The multiple aspects of disaster impacts on different socio-demographics make it a complex undertaking for policy makers, firstly to identify these groups and individuals and secondly, to provide the support that is needed to address the barriers to their labour force participation. As the recovery progresses in Queensland, Canterbury and Tohoku, coordination of employment initiatives with housing and other welfare policies is critical for ensuring that employment opportunities are available to everyone, especially those with disadvantage.

Acknowledgement This research was funded by the Asia-Pacific Economic Cooperation Secretariat (Project: Building Natural Disaster Response Capacity – Sound Workforce Strategies for Recovery and Reconstruction; Project Number, HRD 01/2012A). Authors would like to thank the APEC Natural Disasters Workforce Project Office and Nick Mowbray and William La of the Australian Government Department of Employment for providing support and many officials from the Governments in Australia, New Zealand, Japan, Canada and China for assisting with the collection of case study data.

References

Alexander, B., Chan-Halbrendt, C., & Salim, W. (2006). Sustainable livelihood considerations for disaster risk management: Implications for implementation of the government of Indonesia tsunami recovery plan. *Disaster Prevention and Management, 15*(1), 31–50.

APEC HRDWG. (2010). *Human resources and the global economic crisis: Effective social safety nets and labour market policies in APEC member economies*. Retrieved from Singapore.

Baez, J. E., & Santos, I. V. (2008). On shaky ground: The effects of earthquakes on household income and poverty. In *RPP LAC-MDGs and poverty – 02/2008*. New York: RBLAC-UNDP.

Belasen, A.R., & Polachek, S.W. (2008). How hurricanes affect employment and wages in local labor markets. *The Institute for the Study of Labor (IZA) discussion paper no. 3407, March 2008*.

BLS (Bureau of Labor Statistics). (2006). The labour market impact of hurricane Katrina: An overview. *Monthly Labour Review, 129*(8), 3–10.

BLS (Bureau of Labor Statistics). (2012a). Hurricane Sandy: A pre-storm look at affected areas. *Spotlight on statistics, November 2012*.

BLS (Bureau of Labor Statistics). (2012b). Mass Layoffs – November 2012. *BLS news release, For release 10:00 a.m. (EST) Friday, december 21, 2012*.

Budidarsono, S., Wulan, Y.C., Budi, Joshi, L., Hendratno, S. (2007). *Livelihoods and forest resources in Aceh and Nias for A sustainable forest resource management and economic progress*. Working paper. World Agroforestry Centre.

Cabinet Office. (2011). Disaster Situation, June 24, 2011. http://www.bousai.go.jp/2011daishinsai/pdf/110624-1kisya.pdf.

Camilo, S. (2007). The impact of flood hazards on local employment. *Applied Economics Letters, 14*(15), 1123–1126.

CERA. (2012). *Christchurch economic recovery plan, presented by S. Wakefield*. Paper presented at the logistics of natural disaster recovery 2012 international conference, 5–7 March 2012, Wellington.

Chang-Richards, Y., Wilkinson, S., Seville, E. (2012). *Resourcing of the Canterbury rebuild: Changes and emerging themes*. Retrieved from Auckland.

Comerio, M. C. (1998). *Disaster hits home: New policy for urban housing recovery*. Berkeley/Los Angeles/California: University of California Press.

Craigie, R., Gillmore, D., & Groshenny, N. (2012). Matching workers with jobs: How well is the New Zealand labour market doing? *Reserve Bank of New Zealand Bulletin, 75*(4), 3–12.

Earthquake Commission. (2011). *Briefing to the incoming minister*. Retrieved from.

Flood Recovery Task Force. (2013). Southern Alberta 2013 Floods: The provincial recovery framework. http://alberta.ca/albertacode/images/Flood-Recovery-Framework.pdf: The Government of Alberta.

Giglio, K., & Wiseman, S. (2010). *The workforce and economic recovery: effects of Hurricane Katrina*. Retrieved from Santa Monica, California.

Groen, J., & Polivka, A. (2008). The effects of Hurricane Katrina on the labour market outcomes of evacuees. *American Economic Review, 98*(2), 43–48.

Holzer, H. J., & Lerman, R. I. (2006). Employment issues and challenges in post-Katrina New Orleans. In *Rebuilding opportunity and equity into the new New Orleans*. Washington, D.C.: The Urban Institute.

IDMC. (2013). *Global estimates 2012: People displaced by disasters*. Retrieved from Geneva.

IFRC. (2012). *World disasters report 2012: Focus on forced migration and displacement*. Retrieved from Geneva.

Insurance Council of Australia. (2012). Disaster statistics. http://www.insurancecouncil.com.au/statistics.

International Recovery Platform. (2010). *Wenchuan earthquake 2008: Recovery and reconstruction in Sichuan Province*. Retrieved from Kobe.

Japan's Reconstruction Agency. (2012). Current situation and measures for reconstruction on 21 may 2012. www.reconstruction.gp.jp/topics/120521genjototorikumi.pdf.

Krippendorff, K. (1980). *Content analysis: An introduction to its methodology*. Beverly Hills: Sage Publications.

Lorch, R. (2005). What lessons must be learned from the tsunami? *Building Research & Information, 33*(3), 209–211.

Ministry of Finance, GFDRR (Global Facility for Disaster Reduction and Recovery), & World Bank. (2012). *Supporting sustainable post-earthquake recovery in China*. Retrieved from Beijing.

Nanto, D.K., Cooper, W.H., Donnelly, J.M., & Johnson, R. (2011). *Japan's 2011 Earthquake and Tsunami: Economic effects and implications for the United States*. Retrieved from Washington, D.C.

Parkatti, M. (2013). Impact of southern Alberta flooding on hours worked and GDP. *Economic commentary, Alberta Government*.

Parker, M., & Steenkamp, D. (2012). The economic impact of the Canterbury earthquakes. *Reserve Bank of New Zealand Bulletin, September 2012, 75*(3), 13–25.

Paterson, E., Re, D.D., & Wang, Z. (2008). The 2008 Wenchuan earthquake: Risk management lessons and implications. *Risk management solutions (RMS) publications*. Nework, CA, USA.

Queensland Reconstruction Authority. (2012). *Rebuilding a stronger, more resilient Queensland: The capacity to prepare for, withstand, respond to and recover from disasters*. Retrieved from Brisbane.

Sawada, Y., Bhattacharyay, R., & Kotera, T. (2011). Aggregate impacts of natural and man-made disasters: A quantitative comparison. *The research institute for economy, trade, and industry, discussion paper series 11-E-023*. Tokyo, Japan.

Skoufias, E. (2003). Economic crises and natural disasters: Coping strategies and policy implications. *World Development, 31*(7), 1087–1102.

State Planning Group of Post-Wenchuan Earthquake Restoration and Reconstruction. (2008). *The state overall planning for post-Wenchuan earthquake restoration and reconstruction*. Retrieved from Beijing.

United Nations. (2010). *The chilean earthquake of 27 february 2010: an overview*. Retrieved from Santiago, Chile.

Venn, D. (2012). Helping displaced workers back Into jobs after a natural disaster: Recent experiences in OECD Countries. *OECD social, employment and migration working papers, no. 142, OECD publishing*. Paris, France.

Wilkinson, E., Kirbyshire, A., Mayhew, L., Batra, P., Milan, A. (2016). *Climate-induced migration and displacement: closing the policy gap*. Retrieved from London.

World Bank, & Queensland Reconstruction Authority. (2011). *Queensland recovery and reconstruction in the aftermath of the 2010/2011 flood events and cyclone Yasi*. Retrieved from Brisbane, Australia.

Zedlewski, S.R. (2006). Building a better safety net for the new New Orleans. *Building opportunity and equity into the new New Orleans.*

Zissimpoulos, J. M., & Karoly, L. A. (2010). Employment and self-employment in the wake of Hurricane Katrina. *Demography, 47*(2), 345–367.

Analyzing Flood Fatalities in Vietnam Using Statistical Learning Approach and National Disaster Database

Chinh Luu and Jason von Meding

Abstract

Floods and storms have had a severe impact on the people of Vietnam over many years, particularly regarding an unacceptably high death toll. However, it still lacks studies on flood-related fatalities in Vietnam. This research aims to explore flood fatalities on a national scale and analyze damage-influencing attributes related to flood fatalities using the national disaster database of Vietnam and statistical learning approach. Records covering 27 years from 1989 to 2015 indicate at least 14,927 flood mortalities in Vietnam. The analysis results of statistical learning methods show that housing impact factor has the most considerable influence on flood fatalities. The results can provide implications for housing policies for the poor in flood-prone areas. The objective of reduction in mortality in disasters is under Goal 11 of Sustainable Development Goals.

Keywords

Flood fatalities · Flood damage · Multiple linear regression · Random forest

C. Luu (✉) · J. von Meding
University of Newcastle, Newcastle, Australia
e-mail: ThiDieuChinh.Luu@uon.edu.au;
Jason.VonMeding@newcastle.edu.au

1 Introduction

A flood is a very complex phenomenon, that is, the links of natural environment, people, and social system (Slobodan 2012). Global flood exposure and flood frequency are projected to increase especially in many low-latitude regions in Asia and Africa (Hirabayashi et al. 2013). The global disaster database, EM-DAT (Fig. 1), shows that the frequency of flood and storm events is correlated with flood fatalities.

Disaster data collection and analysis are increasingly prominent (UNISDR 2015). The analysis could provide necessary information for policy-setting and decision-making process in disaster risk reduction (IRDR 2014). Various disaster databases are available on national and global scales, which are summarized and listed in by Grasso and Dilley (2013) and Simpson et al. (2014). The exploration of these databases could contribute to better understanding of disaster risk, which is the first priority in the Sendai Framework (UNISDR 2015).

The Centre for Research on the Epidemiology of Disasters supported Vietnam and other Asian countries to develop their national disaster databases (Below et al. 2010). The Vietnamese national disaster database is Damage Assessment and Needs Analysis or DANA. The Central Committee for Flood and Storm Control of

A. Asgary (ed.), *Resettlement Challenges for Displaced Populations and Refugees*, Sustainable Development Goals Series, https://doi.org/10.1007/978-3-319-92498-4_15

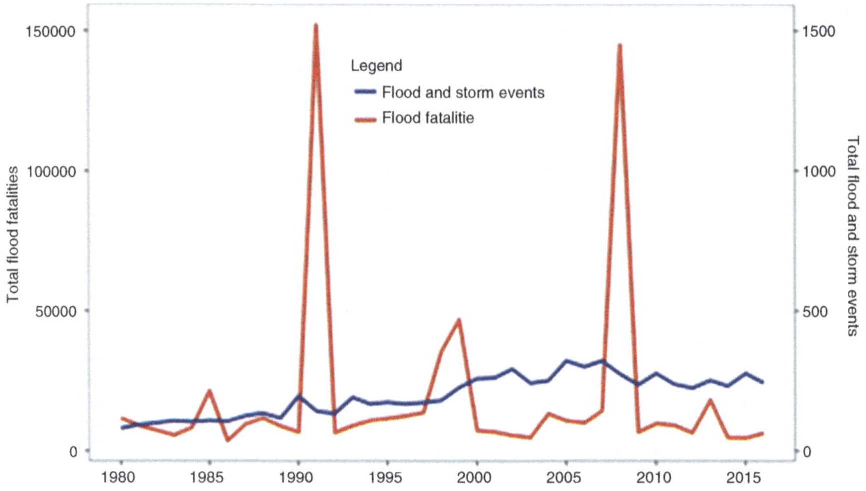

Fig. 1 Flood fatalities and flood events between 1980 and 2016 at global scale (compiled from EM-DAT database http://emdat.be)

Vietnam had been jointly developing the DANA database (MARD 2006; Hughey et al. 2011).

Analysis of flood damage data could provide crucial information to decision-makers in the field of flood risk management and adaptation planning (Merz et al. 2010). Although previous research investigated DANA, such as Hughey et al. (2011) and Nhu et al. (2011), there has not been any investigation related to flood fatalities.

There are many studies on flood-related deaths in developed countries (e.g., Coates 1999; Ashley and Ashley 2008; FitzGerald et al. 2010; Di Mauro and de Bruijn 2012; Sharif et al. 2015) while only a few studies in developing countries (e.g., Paul and Mahmood 2016). The presence of studies on flood-related mortalities in developing economies is minimal despite the fact that the mortality rate in these countries is significant. There is no systematical research on flood-related fatalities in Vietnam.

Studies on flood fatalities often have two main approaches: (1) developing predictive models (Di Mauro et al. 2012; Jonkman and Vrijling 2008; Zhai et al. 2006; Di Mauro and de Bruijn 2012; Jonkman et al. 2002) and (2) analyzing the causes of fatalities (Ashley and Ashley 2008; Paul and Mahmood 2016; Sharif et al. 2015; Coates 1999; Jonkman et al. 2009; Jonkman and Kelman 2005).

Statistical learning is the combination of classical statistics and computer science (James et al. 2013a). The applications of statistical learning are increasingly applied in genetics, medical science, business, and flood risk management field (Merz et al. 2013; Hasanzadeh Nafari et al. 2016). However, there is a lack of research on the application of statistical learning techniques to analyze the relationship between flood damage attributes on flood fatalities.

The present study aims to explore flood fatalities on a national scale and analyze damage-influencing attributes related to flood fatalities using the national disaster database of Vietnam and statistical learning approach.

2 Disaster Database

The Central Committee for Flood and Storm Control of Vietnam has developed the national disaster database since 1989 through DANA database. Flood damage data is collected via one template including 12 categories with many flood impact indicators, for example, fatalities, agriculture impacts, housing damages, infrastructure damages, and economic loss. The database provides hydrometeorological disaster damage information at the national level (Hughey et al.

2011). It only stores direct losses on monetary of reconstruction and recovery of damaged property and infrastructure, and does not report indirect losses, for instance, business disruption and production interruption (Wang et al. 2010).

3 Flood Fatalities in Vietnam

EM-DAT defined death or fatality including person confirmed dead and missing person presumed dead (Below et al. 2010). The loss of human life is considered one of the most critical indicators in assessing flood risk (Maaskant et al. 2009). The flood mortalities are very low in developed countries such as Australia (Mojtahedi and Oo 2016) and Scotland (Crichton 2004). Meanwhile, the flood fatalities in Vietnam are unacceptably high, with at least 14,972 flood mortalities between 1989 and 2015 (Fig. 2).

Spatial patterns of flood fatalities by provinces in Vietnam are generated using the compiled flood damage data from DANA database and ArcGIS 10.1 software in Fig. 2. Quang Nam, Ca Mau, and Quang Ngai provinces have the highest death toll of more than 800 people during the observation period. The second highest flood fatality level is from 401 to 800 people in nine provinces including Nghe An, Thanh Hoa, Da Nang, Thua Thien Hue, Binh Dinh, Khanh Hoa, Kien Giang, Dong Thap, and An Giang.

4 Statistical Learning Methods

Based on the DANA database, this study uses the two following statistical learning methods, multiple linear regression model and random forest, to measure the relative influence of flood damage attributes on fatalities. We present the detail underpinning methodology of these methods in this section.

4.1 Multiple Linear Regression Model

Multiple linear regression technique is employed to analyze the independent factors relating to the flood fatalities, which is set as a dependent variable or an outcome. After that, LMG method is used to obtain the relative importance of flood fatality determinants based on the regression model.

Multiple linear regression model aims to find an equation to describe the relationship between X (independent variables) and Y (dependent variable). A multiple linear regression model (population) with p independent variables is described as in Eq. (1) to examine the linear relationship between one dependent (Y) and two or more dependent variables (x_i).

$$Y_i = \beta_0 + \beta_1 x_{1i} + \beta_2 x_{2i} + \ldots + \beta_p x_{pi} + \varepsilon \quad (1)$$

where

β_0: intercept
$\beta_1, \beta_2, \ldots, \beta_p$: regression coefficients
Y: dependent variable
x_{pi}: independent variables
ε: residuals

Weights of independent variables can be generated based on relative importance of variables (Hair Jr. et al. 2014). Bi (2012) reviewed new methods for generating relative importance from regression methods and recommended the LMG method for raw data. LMG indicator, which was proposed by Lindeman, Merend, and Gold in 1980 (Lindeman et al. 1980), is applied to assess the relative importance of variables of the multiple linear regression model in this study. The explanation of LMG is as follows (Lindeman et al. 1980; Bi 2012):

$$\text{LMG}(x_k) = \frac{1}{p!} \sum_{r\text{permutation}} \text{seq}R^2\left(\{x_k\} | r\right) \quad (2)$$

where $r = 1, 2, \ldots, p!$ and $seqR^2(\{x_k\} | r)$ denotes sequential sum of squares for the regressor x_k in the ordering of regressors in r-th permutation.

4.2 Random Forest

Another method, random forest, for regression is applied to the same database. Random forest algorithm, for both regression and classification,

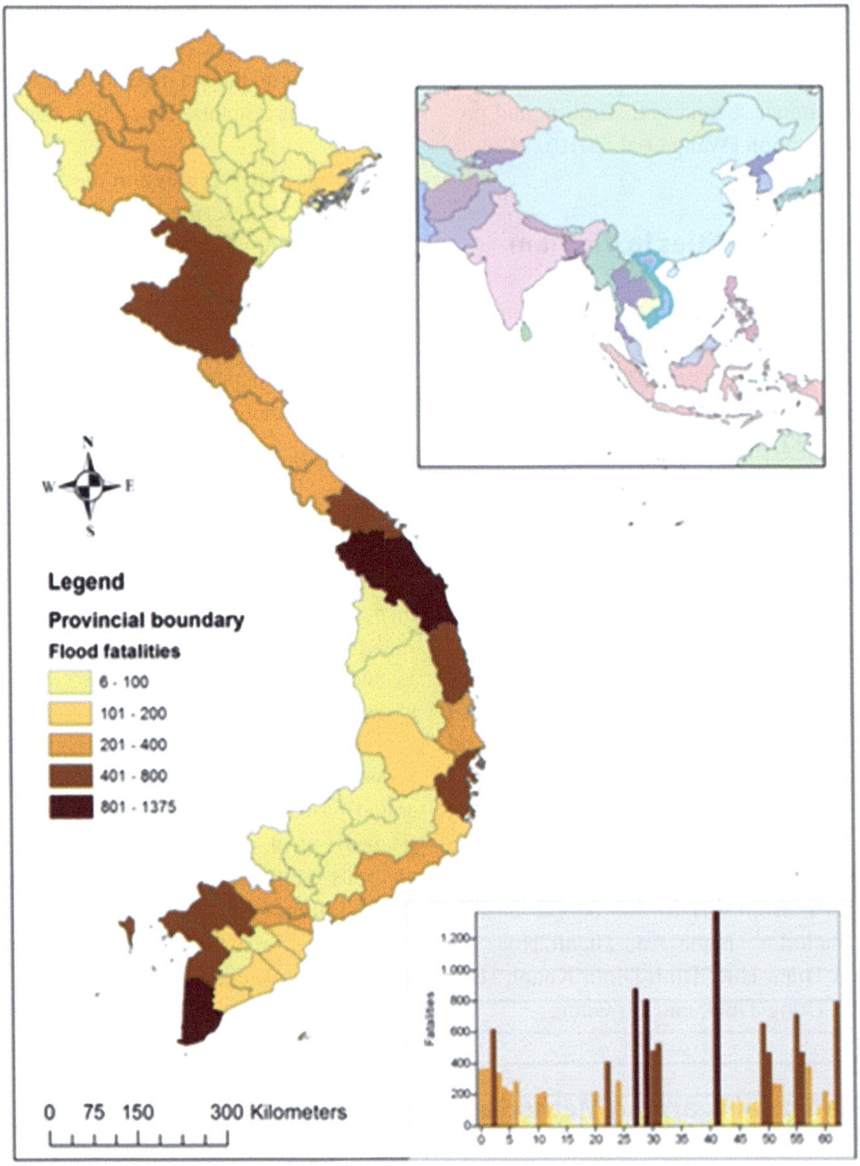

Fig. 2 Spatial patterns of flood fatalities by provinces in Vietnam from 1989 to 2015

is a panacea for all data science problems. This method constructs multitude of decision trees and selects the best as the final result which can be used to build predictive models. The random forest algorithm can be presented in the following steps (Liaw and Wiener 2002):

1. Draw n_{tree} bootstrap sample from the original dataset. A sample of these n_{tree} is taken at random with replacement.
2. For each of these samples, develop an unpruned regression tree: randomly sampling m_{try} of predictors and selecting the best split from these variables at each node.
3. Predict new dataset by aggregating the predictions of the trees (average for regression).

Random forest for regression is constructed by growing trees depended on a random vector such that the predicted tree takes numerical values as opposed to class labels. C. Strobl et al. (2008) suggested that the conditional importance for random forest should be applied for the highly correlated predictor variables when this tool reflects the actual impact of each variable. The conditional importance technique is chosen for the dataset of this study.

5 Application

5.1 Multiple Linear Regression Analysis

Multiple linear regression model is applied to flood damage data of DANA. The collected dataset includes 27 samples in 63 provinces (27 years from 1989 to 2015). One year is considered an observation, so there are 1701 observations. Flood fatality is set as a dependent variable. Flood damage attributes are set as independent variables from X1 to X10. Data transformation with logarithm function is applied to both outcome and independent variables for better fitting the normal distribution due to large variation in the flood damage data (Zhou et al. 2017).

Cross-validation methods are used to validate the multiple linear regression model. Cross-validation aims to evaluate whether a model has the good predictive ability for a new dataset. K-fold cross-validation and leave-one-out cross-validation (LOOCV) techniques can provide the best cross-validation estimate (James et al. 2013b). The k-fold cross-validation and LOOCV results are generated using "boot" package (Canty and Ripley 2016) in R statistical software (R Core Team 2016). The results in Fig. 3 show that mean square errors of both LOOCV and k-fold CV models with a degree of polynomial from 2 to 10 are low values and approximately the same. Therefore, the model is validated.

We use R statistical software to run the multiple linear regression model with the transformed data. The model has adjusted R-squared of 0.601 and residual standard error of 0.82. After that, we run "*relaimpo*" package (Grömping 2006) in R (R Core Team 2016) to generate the weights or relative importance of attributes. The weights of attributes or independent variables are generated based on LMG indicator as in Eq. (2). The result is shown in Table 1.

5.2 Conditional Importance for Random Forest

Random forest algorithm aimed to find the relative influence of independent factors (flood impacts from X1 to X10) to a dependent outcome (flood fatalities). The conditional importance for random forests was conducted using '*party*' package (Hothorn et al. 2006; Strobl et al. 2007, 2008) in R statistical software. The result is shown in Fig. 4. Housing impact factor (variable X1) has the highest influence on fatalities of this model.

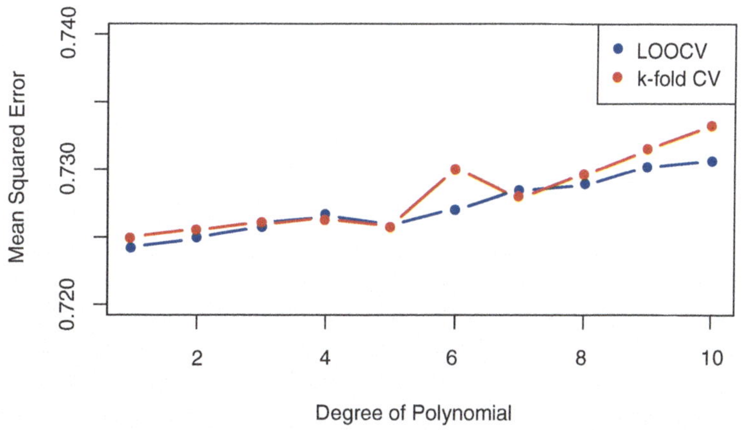

Fig. 3 Error curves of tenfold CV and LOOCV with different random splits and ten times repeated of the two cross-validation methods

Table 1 The relative importance of flood fatality determinants based on multiple linear regression model and LMG indicator after normalization

Determinants	Explanatory	Unit	Symbol	Relative importance	P value
Housing	Number of houses damaged		X1	0.226	<0.0001
Education	Number of classrooms damaged		X2	0.141	<0.0001
Healthcare	Number of clinics damaged		X3	0.085	0.0013
Agriculture	Area of paddy inundated and farm produce damaged	ha	X4	0.118	0.007
Irrigation	Volume of earth and rock eroded, washed away, and redeposited (of dikes, canals, and reservoirs)	m^3	X5	0.104	0.0122
Transportation	Volume of earth and rock eroded, washed away, and redeposited (of roads and highways)	m^3	X6	0.114	<0.0001
Fisheries	Area of fish and shrimp feeding damaged	ha	X7	0.099	<0.0001
Telecommunication	Number of telephone poles broken		X8	0.027	0.6374
Electricity	Number of high voltage electrical towers and electrical distribution poles collapsed		X9	0.055	0.7024
Materials	Volume of cement damaged, salt lost, clinker wetted, coal drifted	ton	X10	0.03	0.0059

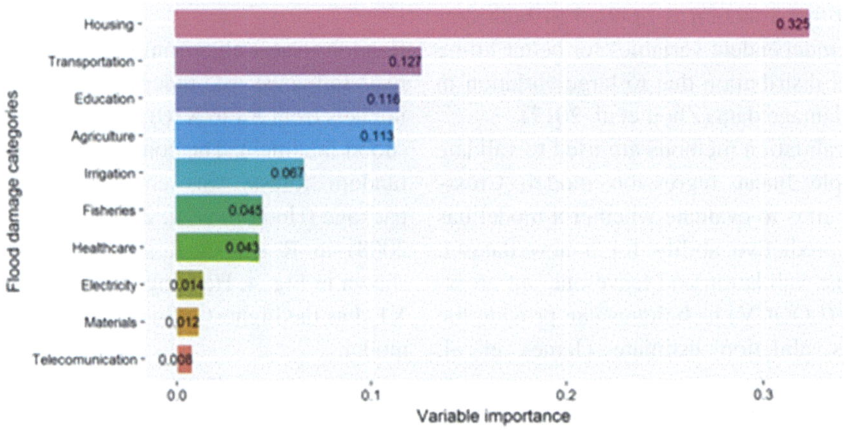

Fig. 4 Variable importance scores for the ten predictors in regression tree model by conditional variable importance for random forest

6 Discussion and Conclusion

The interrelation between sustainable development and disaster risk reduction is recognized in Sendai Framework (UNISDR 2015). Part of that, the objective of reduction in mortality in disasters is stated under Goal 11 of Sustainable Development Goals (UN 2015). A summary of flood fatalities in Vietnam and the investigation of damage-influencing attributes on flood fatalities in this study can support future efforts to mitigate fatality in flood disaster and have implications for flood risk management activities.

The recorded damage data suggests that flood hazards have had severe impacts on Vietnamese people, with over 14,927 dead and missing or at least 553 flood mortalities per year between 1989 and 2015 (Fig. 2). The susceptibility has been continuing with 264 flood fatalities in 2016 (UN Country Team in Vietnam 2016). The flood fatalities are unexpectedly high due to the passive response activities of local authorities and residents to flood and storm events. Therefore, flood risk management needs focusing on a proactive approach or mitigation and preparedness activities.

The analysis results from multiple linear regression analysis (Table 1) and conditional importance for random forest (Fig. 4) show that housing damage factor has the most significant influence on flood mortalities. The more houses damaged and flooded, the more fatalities are. The people who lived in a house that is destroyed or flooded are more likely to be killed. The houses of farmers (over 70% of Vietnamese are farmers) are mostly one-story and in poor conditions. They are not strong enough to withstand storms or floods, and they provide no room to escape in high flood depth cases.

Floods and storms affected low-income communities disproportionately, especially in riverine and coastal areas in Vietnam. The rural poor are particularly vulnerable to flood events. The affected households often received extremely limited financial support from the local government, which was stipulated in Decree 67/2007/ND-CP of the government. The poverty rate, as is common on a global scale, is invariably linked to disaster impact.

The results may provide information on community awareness and safety regulations. In particular, our results can be used to recommend government policies that focus on supporting the poor in upgrading their houses in flood-prone areas to mitigate flood fatalities.

This study used the data available in the national disaster database for the analysis. It is, therefore, limited to analyzing the relative influence of damage attributes on fatalities in Vietnam. We call for more detailed studies on flood fatalities such as predictive models and the causes of flood fatalities. However, it is required the disaster loss database documented details on the causes, age, and gender of flood fatalities.

In conclusion, the present study proposes an approach to investigate the damage-influencing attributes related to flood fatalities using statistical learning techniques and a national disaster database. The results show that housing damage factor has the most significant influence on flood mortalities in Vietnam. Our research provides a better understanding of flood fatalities in Vietnam by analyzing and reporting on flood mortalities using statistical learning approach and a national disaster database. The output can produce a reference for the decision-making process in flood risk management in Vietnam.

References

Ashley, S. T., & Ashley, W. S. (2008). Flood fatalities in the United States. *Journal of Applied Meteorology and Climatology, 47*(3), 805–818. https://doi.org/10.1175/2007jamc1611.1.

Below, R., Vos, F., & Guha-Sapir, D. (2010). *Moving towards harmonization of disaster data: A study of six Asian databases*. Brussels: Centre for Research on the Epidemiology of Disasters.

Bi, J. (2012). A review of statistical methods for determination of relative importance of correlated predictors and identification of drivers of consumer liking. *Journal of Sensory Studies, 27*(2), 87–101. https://doi.org/10.1111/j.1745-459X.2012.00370.x.

Canty, A., & Ripley, B. (2016). Boot: Bootstrap R (S-plus) functions. *R package version, 1*, 3–18.

Coates, L. (1999). Flood fatalities in Australia, 1788-1996. *Australian Geographer, 30*(3), 391–408. https://doi.org/10.1080/00049189993657.

Crichton, D. (2004). *Flood risks in the former Grampian Region since Devolution* (A Research Report for WWF by David Crichton, Vol. 16 November 2015): WWF Scotland.

Di Mauro, M., & de Bruijn, K. M. (2012). Application and validation of mortality functions to assess the consequences of flooding to people. *Journal of Flood Risk Management, 5*(2), 92–110. https://doi.org/10.1111/j.1753-318X.2011.01131.x.

Di Mauro, M., De Bruijn, K. M., & Meloni, M. (2012). Quantitative methods for estimating flood fatalities: Towards the introduction of loss-of-life estimation in the assessment of flood risk. *Natural Hazards, 63*(2), 1083–1113. https://doi.org/10.1007/s11069-012-0207-4.

FitzGerald, G., Du, W., Jamal, A., Clark, M., & Hou, X. Y. (2010). Flood fatalities in contemporary Australia (1997-2008). *Emergency Medicine Australasia, 22*(2), 180–186. https://doi.org/10.1111/j.1742-6723.2010.01284.x.

Grasso, V. F., & Dilley, M. (2013). *A comparative review of country-level and regional disaster loss and damage databases*. New York: United Nations Development Programme Bureau for Crisis Prevention and Recovery.

Grömping, U. (2006). Relative importance for linear regression in R: The package relaimpo. *Journal of Statistical Software, 17*(1), 1–27. https://doi.org/10.18637/jss.v017.i01.

Hair, J. F., Jr., Black, W. C., Babin, B. J., & Anderson, R. E. (2014). *Multivariate data analysis* (7th ed.). Harlow: Pearson Education Limited.

Hasanzadeh Nafari, R., Ngo, T., & Mendis, P. (2016). An assessment of the effectiveness of tree-based models for multi-variate flood damage assessment in Australia. *Water, 8*(7), 282. https://doi.org/10.3390/w8070282.

Hirabayashi, Y., Mahendran, R., Koirala, S., Konoshima, L., Yamazaki, D., Watanabe, S., et al. (2013). Global flood risk under climate change. *Nature Climate Change, 3*(9), 816–821. https://doi.org/10.1038/nclimate1911.

Hothorn, T., Bühlmann, P., Dudoit, S., Molinaro, A., & Van Der Laan, M. J. (2006). Survival ensembles. *Biostatistics, 7*(3), 355–373.

Hughey, E., Bell, H., & Chatman, M. (2011). Who needs what? A case study of post-disaster damage and needs assessment (DANA) in Vietnam. *Risk, Hazards & Crisis in Public Policy, 2*(4), 1–24. https://doi.org/10.2202/1944-4079.1097.

IRDR (2014). *IRDR Peril Classification and Hazard Glossary* (DATA Project Report No. 1). Beijing: Integrated Research on Disaster Risk (IRDR).

James, G., Witten, D., Hastie, T., & Tibshirani, R. (2013a). Introduction. In *An introduction to statistical learning: With applications in R* (pp. 1–14). New York: Springer New York.

James, G., Witten, D., Hastie, T., & Tibshirani, R. (2013b). Resampling methods. In *An introduction to statistical learning: With applications in R* (pp. 175–201). New York: Springer.

Jonkman, S. N., & Kelman, I. (2005). An analysis of the causes and circumstances of flood disaster deaths. *Disasters, 29*(1), 75–97. https://doi.org/10.1111/j.0361-3666.2005.00275.x.

Jonkman, S. N., Maaskant, B., Boyd, E., & Levitan, M. L. (2009). Loss of life caused by the flooding of New Orleans after hurricane Katrina: Analysis of the relationship between flood characteristics and mortality. *Risk Analysis, 29*(5), 676–698. https://doi.org/10.1111/j.1539-6924.2008.01190.x.

Jonkman, S. N., van Gelder, P. H. A. J. M., & Vrijling, J. K. (2002). Loss of life models for sea and river floods. *Flood defence, 1*, 196–206.

Jonkman, S. N., & Vrijling, J. K. (2008). Loss of life due to floods. *Journal of Flood Risk Management, 1*(1), 43–56. https://doi.org/10.1111/j.1753-318X.2008.00006.x.

Liaw, A., & Wiener, M. (2002). Classification and regression by randomForest. *R news, 2*(3), 18–22.

Lindeman, R. H., Merenda, P. F., & Gold, R. Z. (1980). *Introduction to bivariate and multivariate analysis*. Glenview: Scott, Foresman and Company.

Maaskant, B., Jonkman, S. N., & Bouwer, L. M. (2009). Future risk of flooding: An analysis of changes in potential loss of life in South Holland (the Netherlands). *Environmental Science & Policy, 12*(2), 157–169. https://doi.org/10.1016/j.envsci.2008.11.004.

MARD (2006). Guideline on natural disaster damage and needs assessment. http://www.ngocentre.org.vn/webfm_send/1533. Accessed 14 Oct 2016.

Merz, B., Kreibich, H., & Lall, U. (2013). Multi-variate flood damage assessment: A tree-based data-mining approach. *Natural Hazards and Earth System Science, 13*(1), 53–64. https://doi.org/10.5194/nhess-13-53-2013.

Merz, B., Kreibich, H., Schwarze, R., & Thieken, A. (2010). Review article "assessment of economic flood damage". *Natural Hazards and Earth System Science, 10*(8), 1697–1724. https://doi.org/10.5194/nhess-10-1697-2010.

Mojtahedi, S. M. H., & Oo, B. L. (2016). Coastal buildings and infrastructure flood risk analysis using multi-attribute decision-making. *Journal of Flood Risk Management, 9*(1), 87–96. https://doi.org/10.1111/jfr3.12120.

Nhu, O. L., Thuy, N. T. T., Wilderspin, I., & Coulier, M. (2011). *A preliminary analysis of flood and storm disaster data in Vietnam* (Global Assessment Report on Disaster Risk Reduction, Vol. 30 Sept 2016): United Nations Development Programme Vietnam.

Paul, B. K., & Mahmood, S. (2016). Selected physical parameters as determinants of flood fatalities in Bangladesh, 1972–2013. *Natural Hazards*. https://doi.org/10.1007/s11069-016-2384-z.

R Core Team. (2016). *R: A language and environment for statistical computing*. Vienna: R Foundation for Statistical Computing.

Sharif, H. O., Jackson, T. L., Hossain, M. M., & Zane, D. (2015). Analysis of flood fatalities in Texas. *Natural Hazards Review, 16*(1), 04014016. https://doi.org/10.1061/(asce)nh.1527-6996.0000145.

Simpson, A., Murnane, R., Saito, K., Phillips, E., Reid, R., & Himmelfarb, A. (2014). *Understanding risk in an evolving world: Emerging best practices in natural disaster risk assessment*. Washington DC: Global Facility for Disaster Reduction and Recovery, the World Bank.

Slobodan, P. S. C. (2012). *Floods in a changing climate: Risk management (part of international hydrology series)*. Cambridge, UK: Cambridge University Press.

Strobl, C., Boulesteix, A.-L., Zeileis, A., & Hothorn, T. (2007). Bias in random forest variable importance measures: Illustrations, sources and a solution. *BMC Bioinformatics, 8*(1), 25.

Strobl, C., Boulesteix, A. L., Kneib, T., Augustin, T., & Zeileis, A. (2008). Conditional variable importance for random forests. *BMC Bioinformatics, 9*, 307. https://doi.org/10.1186/1471-2105-9-307.

UN. (2015). *Transforming our world: The 2030 Agenda for Sustainable Development (a/RES/70/1)*. New York: United Nations.

UN Country Team in Vietnam. (2016). Vietnam: Situation reports. https://reliefweb.int/organization/unct-vietnam. Accessed 16 Oct 2017.

UNISDR. (2015). *Sendai framework for disaster risk reduction 2015–2030*. Geneva: United Nations Office for Disaster Risk Reduction (UNISDR).

Wang, X., Mahul, O., & Stutley, C. (2010). *Weathering the storm: Options for disaster risk financing in Vietnam*. Washington, DC: Global Facility for Disaster Reduction and Recovery, the World Bank.

Zhai, G., Fukuzono, T., & Ikeda, S. (2006). An empirical model of fatalities and injuries due to floods in Japan. *JAWRA Journal of the American Water Resources Association, 42*(4), 863–875. https://doi.org/10.1111/j.1752-1688.2006.tb04500.x.

Zhou, Q., Leng, G., & Feng, L. (2017). Predictability of state-level flood damage in the conterminous United States: The role of hazard, exposure and vulnerability. *Scientific Reports, 7*(1), 1–11. https://doi.org/10.1038/s41598-017-05773-4.

Suitability of Different Decision-Making Methods Applied for Analysing Sustainable Post-disaster Temporary Housing

S. M. Amin Hosseini, Oriol Pons, and Albert de la Fuente

Abstract

Increasing population of urban areas by 2050, especially in areas highly prone to natural hazards, could potentially threaten vulnerable cities in the face of future disaster events. Alternative temporary housing (TH) needs to be assessed as an important element of local resiliency. In this regard, there are some interrelated factors that can have antithetical impacts on the different cases of TH. Thus, proper decision-making models can be highly useful to deal with this complex multifaceted process. These models should be flexible and adaptable to existing local and boundary conditions since a TH strategy could lead to different satisfactions and perceptions. This chapter aims to present a multi-criteria decision-making (MCDM) method for choosing the most sustainable temporary housing alternative. To this end, general requirements of the decision-making process for dealing with temporary housing selection are specified and organized based on the assessed cases. Finally, all possible decision-making methods are assessed to determine the most suitable one based on correspondence between the methods and the requirements. As a result, this study concludes that a robust MCDM method should incorporate concept method to objectively measure the satisfaction degree of every stakeholder involved. In this sense, the Integrated Value Model for Sustainable Assessment (MIVES) is a suitable method for assessing the sustainability of temporary housing since this includes the value function method (based on the utility theory).

Keywords

Sustainable post-disaster temporary housing · Natural disasters · Emergency management · MCDM · MIVES

1 Introduction

Regardless of the prosperity level, almost all affected areas are struggling with post-disaster housing (PDH) aftermath of natural disasters. Temporary housing (TH) is the first priority after the emergency response phase (Hidayat 2010), because TH offers security and safety to displaced population (DP) so they can return to pre-disaster conditions (Collins et al. 2010; Johnson 2007a). However, most temporary housing units (THUs) that have been used in previous recovery programs have been criticized by most experts (Johnson 2009). In general, THUs usually do not satisfy stakeholders due to numerous weaknesses

S. M. A. Hosseini (✉) · O. Pons · A. de la Fuente
Universitat Politècnica de Catalunya (UPC), Barcelona, Spain
e-mail: amin.hosseini@upc.edu; oriol.pons@upc.edu; albert.de.la.fuente@upc.edu

A. Asgary (ed.), *Resettlement Challenges for Displaced Populations and Refugees*, Sustainable Development Goals Series, https://doi.org/10.1007/978-3-319-92498-4_16

and shortcomings. According to experts (Barakat 2003; Chandler 2007; El-Anwar et al. 2009a, b; Hadafi and Fallahi 2010; Johnson 2002; Coffey and Trigunarsyah 2012), these units have economic, social, and environmental problems.

According to Lizarralde and Davidson (2006), PDH strategies often fail to address the DP expectations. In this regard, Simon (1996) stated that dealing with complex emergency situations cannot rely only on decision-makers due to the bounded rationality (cited by (Kapucu and Garayev 2011)). Additionally, decision-making processes are usually implemented after natural disasters under high pressure and stressful conditions in extremely tight timeframes. Meanwhile, it is necessary to consider long-term planning (Kennedy et al. 2008) and all stakeholders' participation in decision-making to achieve suitable outcomes. Furthermore, Davidson (2009) stated that even for building construction in normal situations, it is necessary to consider stakeholders' characteristics, such as culture, in order to achieve appropriate organizational forms. Additionally, it should be emphasized that the organizational strategy has great impact on the supervisors' roles, which is one of the key issues for PDH provision (Gharaati and Davidson 2008).

Additionally, according to the United Nations Disaster Relief Organization (UNDRO) (1982), each affected area has individual conditions that lead to choose its particular strategy. Furthermore, different natural disasters have diverse impacts (Lindell and Prater 2003), which need to be considered individually. Therefore, decision-makers need to choose a suitable strategy to deal with PDH issue, which embraces intertwined interior and exterior factors that could have antithetical impacts on each particular case (Hall 1962) (cited by (Johnson 2007a)). Thus, if decision-makers do not apply previous recovery strategies, there is no platform for decision-making process. Moreover, when previous strategies are used, there is no guarantee to achieve similar outcomes. In this regard, Kapucu and Garayev (2011) stated that traditional decision-making approaches cannot be used in emergencies, which need flexible tools. Therefore, it is necessary to have a model that could cover human errors and consider the correspondences and interconnections between previous cases and new cases.

To this end, the main *objective* of this chapter is to present a suitable decision-making tool to deal with PDH by defining the features of decision-making of PDH and considering the tools, which have been applied in this case. Therefore, this study defines requirements (characteristics) of decision-making of recovery programs. Then, an appropriate tool, which embraces exclusively the main TH sustainability requirements, is defined. Additionally, a number of case studies are analysed in order to determine emergency management requirements, outcomes, and sustainability indicators. This research aims to provide some answers to the following two questions: (1) What are the main requirements for the decision-making process for PDH? (2) What is the most suitable method to assess the PDH sustainability focusing on the aforementioned main requirements?

2 Research Background

This study analyses decision-making models that have been applied to assess PDH suitability. Although most academic studies about PDH are new, the number of these studies which are outstanding is high. However, few studies have considered the sustainability and optimization of PDH (El-Anwar et al. 2009a, b; Yi and Yang 2014). Additionally, compared with other considered issues, there are few studies that have been conducted focusing on decision-making tools for PDH issues. Previous studies that have developed decision-making methods are the following:

A generic decision model for PDH (Peng et al. 2014)

A decision process for secure site location (Hale and Moberg 2005)

Selection of fixed seismic shelters by the Technique for Order Preference by Similarity

to Ideal Solution (TOPSIS) (Chua and Su 2012)
Considering earthquake evacuation capacity by Analytic Hierarchy Process (AHP) (Ma et al. 2011)
Settlement suitability by geographical information system (GIS) (Alparslan et al. 2008)
Earthquake refugee shelters by combination of GIS and entropy methods (Li et al. 2013)
Analysing the sustainable site selection and decision-making methods by GIS and multi-attribute decision-making (MADM) (Omidvar et al. 2013)
Urban shelter locations based on covering models (Wei et al. 2012)
TH technology aftermath of Bam earthquake using MIVES (Hosseini et al. 2016a)
Selecting suitable site location of TH using MIVES (Hosseini et al. 2016b)
Hierarchical location models for earthquake-shelter planning (Chen et al. 2013)
Selecting site of temporary sheltering using fuzzy algorithms (Nojavan and Omidvar 2013)
Optimizing PDH allocation (El-Anwar et al. 2009a)
Optimizing TH assignments to minimize displacement distance (El-Anwar and Chen 2012)

3 Decision-Making Process for Post-disaster Housing

Decision-making process of post-disaster housing could be accomplished mainly by two approaches: (1) choosing suitable options among limited alternatives. For instance, decision-making models can be applied to find out an appropriate approach of TH site location among camp and yard of DP's pre-disaster housing. For another example, it could be referred to the process of choosing a suitable THUs site location among determined alternative sites (see (Hosseini et al. 2016b) and (Omidvar et al. 2013)). (2) determining suitable possible alternatives without having initial alternatives is the second approach. For instance, a model is used for choosing a proper settlement by considering all areas (see (Alparslan et al. 2008)). Indeed, the main difference among these approaches is related to the number of available alternatives. In the first approach, the number of alternative is limited, while in the second category, numerous alternatives are considered.

3.1 Case Studies

This research analyses five different cases from the management point of view. These cases are (1) Marmara and Düzce earthquake in Turkey (1999); (2) Bam earthquake in Iran (2003), (3) L'Aquila earthquake in Italy (2009), (4) earthquake and tsunami in Indonesia (2004), and (5) Hurricane Katrina in the USA (2005). These cases are shown in Table 1. The assessment of these cases demonstrates that decision-making processes can be one of the elements that has major impacts on success or failure of PDH programs. Additionally, different indicators with diverse interactions were involved in the studied cases. Furthermore, the importance of indicators can vary from case to case based on disasters' types and scales, local characteristics, and resiliency. In line with this, it is difficult to guarantee that the PDH program that has been useful for one case will be suitable for another case with different conditions. Thus, in the aftermath of a disaster event, authorities are confronted with the complicated decision-making process of finding the best PDH alternative, in which specific local conditions determine the suitability of previous PDH solutions.

The recovery program of five cases varied based on the local characteristics and resiliency. Almost all of these cases struggled with recovery program and, consequently, faced difficulties, as shown in Table 1. Moreover, the main common problems of the five cases regardless of these cases' differences are the following: (1) late delivery, (2) mismatching with local culture, (3) inappropriate organization strategy, and (4) strategy deficiency in dealing with tenants.

Table 1 Information of the five case studies

Case study	Hazard	Intensity	Problem issue	Reference
Turkey 1999	Earthquake	Mw = 7.4 and Mw = 7.2	(a) Site location, (b) long-term plan, (c) facilities, and (d) environment	Arslan and Unlu (2006), Tas et al. (2007), Johnson (2007b), McConnan (1998)
Iran 2003	Earthquake	Ms = 6.5	(a) Site location, (b) material, and (c) emigrant	Amini Hosseini et al. (2013), Ghafory-Ashtiany and Hosseini (2008), Khazai & Hausler (2005)
Indonesia 2004	Earthquake tsunami	Mw = 9.2	(a) Complexity, (b) site location, and (c) quality	Da Silva (2010), Doocy et al. (2006), Steinberg (2007)
USA 2005	Hurricane flood	Category 3	(a) Dispersal of DP, (b) utilities, and (c) environment	Chandler (2007), McCarthy (2008), Nigg et al. (2006)
Italy 2009	Earthquake	Mw = 6.3	(a) Site location, (b) delay in reconstruction, and (c) cost	Alexander (2010), Özerdem and Rufini (2013), Rossetto et al. (2014)

4 Analysis of Cases

In the aftermath of two earthquakes in Turkey in 1999, the amount of damaged buildings caused losses of around $5 billion (Erdik 2000). The recovery program utilized THUs because of the DP numbers and the harsh climate conditions that forced the Turkish government to erect THUs for DP. Nevertheless, THUs have been criticized by researchers due to their negative impacts. On the contrary, decision-makers in Indonesia's recovery program due to the earthquake and tsunami in 2004, which caused approximately 220,000 deaths and 10,000 injured people (Steinberg 2007), decided to apply a self-built or community-built program by considering the local potentials and DP characteristics. Indeed, a top-down approach was applied in the case of Turkey and a bottom-up approach for the Indonesia (Dikmen et al. 2012). Since the bottom-up approach considers all beneficiaries' requirements by empowering DP (Dikmen et al. 2012), this approach has proven to be more successful in terms of adaptation to culture, local skills, and climate conditions (El-Anwar et al. 2009a, b; Johnson 2007b). Additionally, regardless of the recovery program approaches, the recovery program of Aceh, Indonesia, led to long delivery, higher expenses, and poor quality because the strategy was changed several times (Da Silva 2010).

According to (Fois and Forino 2014; Ghafory-Ashtiany and Hosseini 2008; Johnson 2007b), in the aftermath of the natural disasters in Turkey, 1999; Bam, 2003; and Italy, 2009, some of the THUs were vacant because of the DP's rejections. These situations, which occurred due to unsuitable decision-making methods for covering all requirements and indicators, led to waste of time and investment, as well as DP's dissatisfaction. For instance, Johnson (2007b) stated that from THUs, which had been provided after the 1999 earthquakes in Turkey, 2.5 per cent were vacant. Although in a post-analysis this percentage could be considered low, any costs dedicated to unused THU could had been an invaluable budget in the aftermath of the disaster. On the other hand, if the TH strategy provided by the authorities is inappropriate, then DP could look for unofficial self-built THU. This was the case of the Colombian recovery program after the Armenia earthquake, 1999 (Johnson et al. 2006). Therefore, weaknesses of decision-making process somehow result in unacceptable and sometimes unexpected outcomes.

Alexander (2004) stated that without the consideration of the magnitude of losses, the local financial statue of the affected area has considerable impacts on its resilience to disaster. In this regard, the assessment of different aforementioned cases completely demonstrates how diverse local financials' impact on final decision-making is. Furthermore, other local potentials such as building construction industry with different technology and speed also lead to diverse decisions. For instance, the local potentials such as building construction industry of Italy in the wake of L'Aquila earthquake, which led to 308 deaths and 67,500 DP (Alexander 2010), were able to provide these accommodations with great speed based on the required standards. However, the recovery strategy aftermath of L'Aquila earthquake designed by the government had sustainability problems (Fois and Forino 2014). Additionally, different local conditions, such as climate condition, forced authorities to choose individual PDH strategy. As an example, the Turkey recovery program utilized THUs because of DP numbers and the harsh climate conditions. In this sense, the Indonesian decision-makers in 2004 had to change the initial strategy because tents declined under the tropical conditions (Da Silva 2010; Steinberg 2007).

Additionally, some people of non-affected areas migrated to the disaster-affected areas after the 1999 earthquake in Turkey and the 2003 Bam earthquake. Meanwhile, the trailers of the Katrina recovery program were not a perfect solution for returning DP (McCarthy et al. 2006) despite the fact that the trailers' quality can be considerably higher than THUs of another case. Furthermore, in the wake of the Hurricane Katrina, which caused 1570 deaths of Louisiana residents, $40–50 billion economic loss (Kates et al. 2006), and 770,000 DP (Weiss 2006), emergency managers were aware of the hurricane. This awareness did not exist for authorities of the Bam earthquake. Meanwhile, the decision-makers had the ability to erect tents and THUs on the yard of DP's pre-disaster housing. This possibility was not applicable to the recovery program aftermath of the Hurricane Katrina because according to Chandler (2007), the area was pumped dry by September 20th.

5 Findings

The assessment of the above-mentioned cases demonstrates that there are some similarities and dissimilarities between different recovery programs. Additionally, all considered decision-making process aspects of the study cases could be organized in three factors shown in Fig. 1. Indeed, final decisions are derived from these three main vertexes. *Properties* group embraces all material and immaterial things that have formed characteristics of the affected area, such as financial powers, technology, facilities, population characteristic, climate conditions, etc.

In general, the Properties factor can be broken down into four components: belongings, attributes, environmental, and technical aspects. The *Belongings* category considers substantial components of the area, such as buildings including public and private accommodations and services, utilities, facilities, infrastructures, available areas, etc. and their qualities. The *Attributes* category takes into account the wide immaterial complex of economic, social, and political characteristics in the area based on numerous factors, such as life standards, livelihood, welfares, cooperative spirit, abilities, etc. The *Environmental* aspect embraces climate conditions, geographical aspects, and potential threats. The *Technical* aspect considers local abilities to deal with providing temporary and permanent housing related to technical capacities, including construction methods, skilled/expert human resources, material availability, construction firms and companies, reuse systems, transportation quality, etc.

The *Requirements* vertex, which consists of many diverse physical and psychological aspects (Hosseini et al. 2016c), takes into account all essentials for returning the post-disaster situations to the pre-disaster or better situations, especially in terms of DP. The last vertex is *Limitations* group, which embraces all factors that cause difficulties and restrict to arrive at the solutions and achieve suitable requirements, such as timing, number of DP, natural hazard types and effects, etc. Some of the properties could become limitations such as climate conditions, especially when natural hazard happens during a season with unsuitable climate conditions.

Therefore, PDH strategy is the complicated multifaceted process that includes many different factors with diverse interconnections and sometimes antithetical impacts on different cases. A decision-making tool needs to be a comprehensive method, and it should be customizable. Decision-makers should be able to

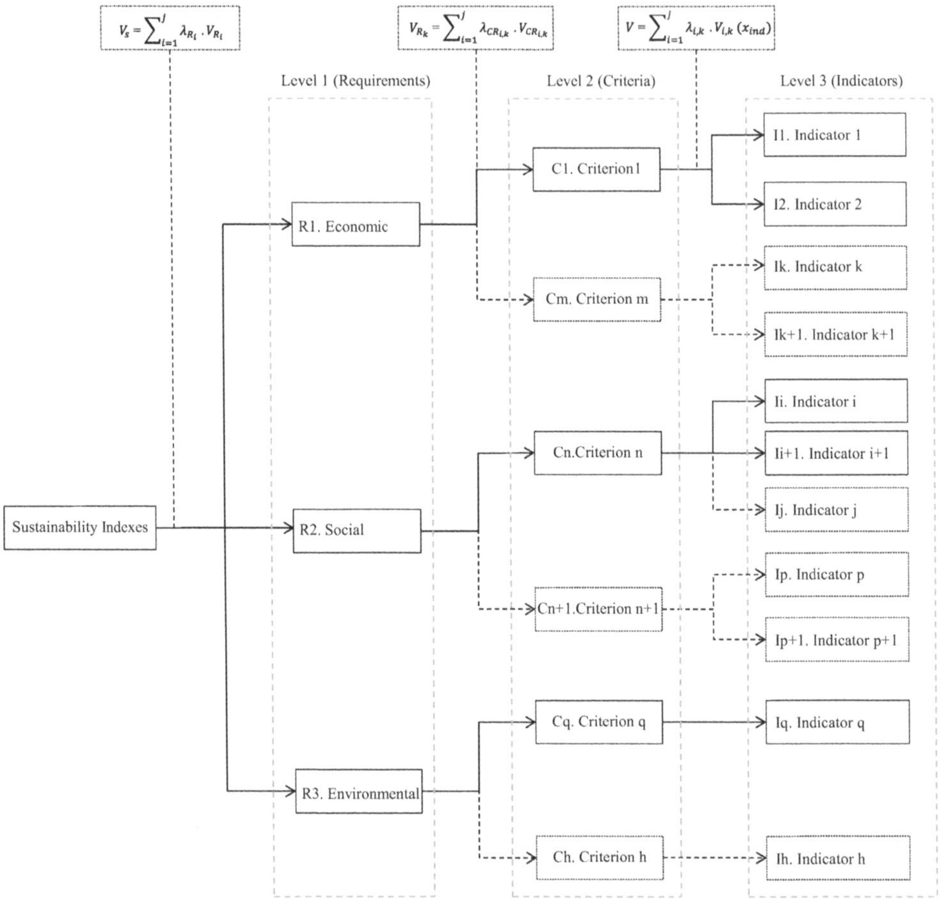

Fig. 1 Three main factors of decision-making process

improve this tool by determining priorities of indicators or/and adding new specific indicators. This addition would be based on individual characteristics of each case, which embraces features of population and area.

6 Required Characteristics for Post-disaster Decision-Making Models

In general, to deal with recovery programs in the aftermath of natural disasters, decision-makers need to determine specific characteristics of these programs in order to design or choose a suitable decision-making model. According to (Hayles 2010; Pearce 2003; von Meding et al. 2016; Zavadskas et al. 2016) and their case studies, the decision-making processes of PDH embrace the following factors: (1) many diverse beneficiaries with different expertise are involved in decision-making process, (2) numbers of organizations participate in parallel, (3) there are distinct short- and long-term requirements, (4) its indicators have diverse interconnections with linear and non-linear functions, (5) an individual strategy for each recovery program is needed, and (6) although it is essential to have a pre-plan for dealing with natural disasters, there is uncertain initial information, and final requirements are determined in the post-disaster period.

Therefore, decision-making models that are applied to determine suitable PDH and TH alternatives based on the individual characteristics of each case should be (1) easy to understand, (2) customizable, (3) quick enforceable, (4) able to satisfy all beneficiaries' concerns (5) able to consider diverse quantitative or qualitative indicators with different units, (6) able to incorporate the utility theory, and (7) flexible to incorporate changes.

7 Considering Suitability of Models and Tools

Multi-criteria decision-making (MCDM) techniques that have been applied for PDH assessments in previous research projects are shown in Table 2. This study assesses these methods based on the required characteristics of decision-making methods to select the most suitable methods. However, Ozernoy (1989) stated that choosing a suitable MCDM is an MCDM problem that needs to apply a decision-making process. In this regard, Zanakis et al. (1998) stated that it is very difficult to answer this question: which method is more suitable for a specific problem? (Zanakis et al. 1998). Nevertheless, the present study determines the best decision-making for the study cases taking into account the strengths and weaknesses of the assessed tools.

Almost all compensation methods could be risky for decision-making on recovery program issues because it is possible to choose an alternative with ineligible features. For instance, distance from source of danger can be one of the most important indicators for site location of THUs. Thus, in order to choose the most suitable site among two alternatives, it is difficult to select which one would be better than the other when adding the assessment of numerous other important indicators as well. In such cases, the indicator-weighting system could help to avoid choosing unsuitable alternatives. Additionally, the utility theory, which is the base of the MIVES method, also prevents from unsuitable solutions. Furthermore, satisfactions of stakeholders for most indicators are not based on linear functions. Thus, utility theory needs to be applied to address a suitable strategy for PDH. For example, distance from source of danger indicator cannot be assessed with a linear function by considering only distances from site alternative to source of danger. Therefore, with regard to Table 2 and the aforementioned PDH essentials, the authors conclude that all methods have advantages and disadvantages but the MIVES appears to be the most suitable method.

Table 2 Main characteristics of the assessed decision-making tools

Method	Main characteristic	Reference
AHP	Experts' knowledge, priority theory, hierarchical structure analysis, flexible, ranking irregularities, pairwise comparison, rank reversal	Aruldoss et al. (2013), Triantaphyllou (2013), Velasquez and Hester (2013)
TOPSIS	Tendency of monotonically increasing or decreasing utility, shortest distance from the positive ideal and farthest from negative one, alternative ranking method, widely applied method, difficult to weight	Aruldoss et al. (2013), Stanujkic et al. (2013), Triantaphyllou (2013), Velasquez and Hester (2013)
MIVES	Value function based on the utility theory, experts' knowledge, alternative ranking and selection method, sustainability assessment tool, easy to understand, a combination of techniques, requirements tree	Cuadrado et al. (2015a, b), del Caño et al. (2012), del Caño et al. (2015)
ELECTRE	Alternative selection method, time consuming, outranking relations, coordination indexes, alternative pairwise comparison, different outputs from other methods	Aruldoss et al. (2013), Triantaphyllou (2013)
SAW	Almost simplest and oldest method, popular to practitioners, intuitive, sometimes illogical results	Stanujkic et al. (2013), Triantaphyllou (2013), Velasquez and Hester (2013)
Fuzzy theory	Widely applied method, ability of imprecise input and insufficient information, difficult, time consuming	Hwang and Yoon (2012), Velasquez and Hester (2013)

According to Aruldoss et al. (2013), MCDM, such as the methods presented in Table 2, can reach strong decisions for considerable complex issues involving multi-criteria. These experts also state that each method has special uniqueness. In regard to their simplicity, according to Stanujkic et al. (2013), SAW is one of the simplest tools. In this sense, AHP is a simple method as well. On the other hand, most methods have the ability to deal with PDH. Additionally, the assessed methods embrace most required characteristics of decision-making techniques with different qualities, as shown in Table 2. Nevertheless, although all assessed tools cover all essentials, MIVES is the only method that takes into account value functions based on the utility theory. However, Hwang and Yoon (2012) indicated that it is possible to replace the simple additive weighting function of SAW method by additive utility function.

8 MIVES Methodology

The Integrated Value Model for Sustainable Assessment (MIVES) from the Spanish institutions consists of a multi-criteria decision-making method that incorporates the concept of value function (Alarcon et al. 2011). This model considers the main sustainability requirements (economic, environmental, and social). In addition, by means of the value functions, the satisfaction degree of the involved indicators, which might have different units, can be assessed. According to (Alarcon et al. 2011) and (San-José Lombera and Garrucho Aprea 2010), MIVES presents rate satisfaction on a scale from 0 to 1, where 0 indicates minimum satisfaction (S_{min}) and 1 indicates maximum satisfaction (S_{max}). MIVES was developed by three different Spanish institutions (UPC, UPV, and Labein-Tecnalia), and this work is the initial application in the field for industrial buildings (Alarcon et al. 2011).

MIVES has been used more recently to assess the sustainability and to make decisions in the fields of:

1. University professors (Viñolas et al. 2009)
2. Economic decisions (Ormazabal et al. 2008; Pujadas et al. 2017)
3. Industrial buildings (San-José Lombera and Rojo 2010)
4. The Spanish Structural Concrete Code (Aguado et al. 2012)
5. Sewerage concrete pipes (Viñolas 2011)
6. School edifices (Pons and Aguado 2012)
7. Developing the probabilistic method MIVES–EHEm–Mcarlo for large and complex edifices (del Caño et al. 2012)
8. Structural concrete columns (Pons and de la Fuente 2013)
9. Wind turbine supports (de la Fuente et al. 2014)
10. TH (Hosseini et al. 2016a, b, 2017)

According to MIVES, a specific tree that includes requirements, criteria, and indicators which is shown in Fig. 2 is developed to assess the sustainability of alternatives. The designed tree must contain minimum indicators, which are independent from each other and calculable to be assigned in formula. The MIVES tree has three different hierarchical levels. The first level of the tree includes the economic, environmental, and social requirements, the second hierarchical level has the criteria, and the third and last level has the indicators. Unlike requirements and criteria, indicators are measurable variables to quantify each alternative site.

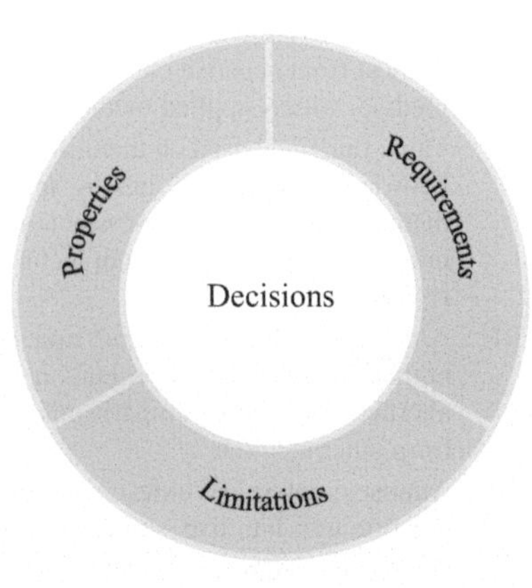

Fig. 2 MIVES tree including requirements, criteria, and indicators

By determining a value function for each indicator according to MIVES equations, it is possible to quantify each attribute. According to (Alarcon et al. 2011) and (San-José Lombera and Garrucho Aprea 2010), to determine the satisfaction value, there are four stages as follow: (1) determine the tendency (increase or decrease) of the value function, (2) determine the points in order to find minimum (S_{min}, value 0) and maximum (S_{max}, value 1) satisfaction, (3) determine the shape of the value function (linear, concave, convex, S-shaped), and (4) determine the mathematical expression of the value function.

According to Alarcon et al. (2011), in a concave curve when the value of the indicator starts to increase, satisfaction rapidly increases. A concave curve is chosen when most alternatives are close to the minimum satisfaction, as shown in Fig. 3. In a convex function, when the value of the indicator starts to increase, the satisfaction slightly increases. Unlike the previous case, the convex function is selected when approaching the maximum satisfaction point is more important than moving away from the minimum satisfaction point. In this last case, most alternatives are close to the maximum satisfaction point, as shown in Fig. 3.

A linear function presents a steady increase in satisfaction. An S-shaped function is a combination of concave and convex functions, as shown in Fig. 3. In an S-shaped function, a considerable increase in satisfaction is obtained in the middle range of values. This S-shape is chosen when most alternatives are centralized into a middle range, as shown in Fig. 3.

The parameters, tendency, and shape of the value function for each indicator are determined from international guidelines, scientific literature, National Building Regulations, and the background of experts that participated in the seminars. In the next step, the value function is obtained based upon the general exponential in MIVES Eq. (1).

$$V_i = A + B.\left[1 - e^{-k_i \cdot \left(\frac{|X_{ind} - X_{min}|}{C_i}\right)^{P_i}}\right] \quad (1)$$

A:	The response value X_{min} (indicator's abscissa), generally $A = 0$
X_{ind}:	The considered indicator abscissa which generates a value V_i
P_i:	A shape factor that determines if the curve is concave or convex or is linear or shaped as an "S"
C_i:	Factor that establishes, in curves with $P_i > 1$, abscissa's value for the inflexion point
K_i:	Factor that defines the response value to C_i
B:	The factor that prevents the function from getting out of the range (0.00, 1.00) is obtained by Eq. (2)

The sets of indicator values ($V_i(x_i)$) that are between 0 and 1, according to the satisfaction range, are generated by Eq. (1). [1]

$$B = \left[1 - e^{k_i \cdot \left(\frac{|X_{max} - X_{min}|}{C_i}\right)^{P_i}}\right]^{-1} \quad (2)$$

After the assessment of indicator value for each alternative, the formula that is presented in Eq. (3) needs to be applied. In this equation, the indicator value ($V_i(x_i)$) has previously been determined, and the weights (λ_i) are assigned to determine the sustainability value of each branch. For the multi-criteria case, the additive formula corresponding to Eq. (3) is applied to determine the sustainability value of each level including indicators, criteria, and requirements.

$$V = \sum \lambda_i . V_i(x_i) \quad (3)$$

$V_i(x_i)$:	The value function of each indicator and each criterion
λ_i:	The weight of considered indicator or criterion

In this step, the weights of the requirements, criteria, and indicators (λ_i) are assigned by using the Analytic Hierarchy Process (AHP) based on previous studies, local characteristics, and the knowledge of the experts involved in seminars.

9 Discussion

MIVES creates an opportunity for decision-makers to assess different indicators with diverse characteristics as do most decision-making

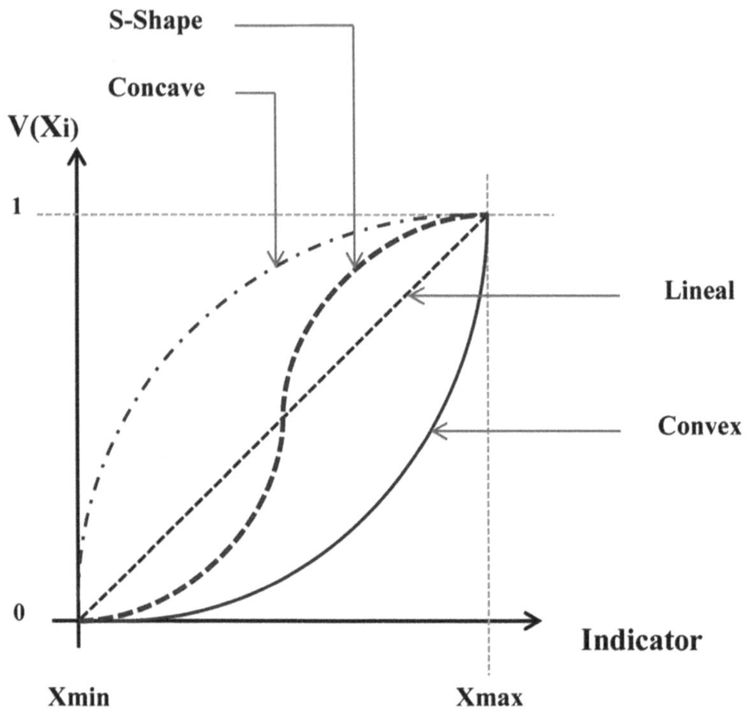

Fig. 3 Value function types

models. However, as MIVES includes the value function concept based on the utility theory concept, indicator values of each alternative are obtained more accurately. Meanwhile, for most issues, especially for housing and PDH, relations between parameters and satisfaction degree of the stakeholders are not based on linear functions.

One of MIVES advantages is the consideration of indicators independently although this method is a compensation decision-making model. In order to define a MIVES model for PDH, which contains intertwined factors, decision-makers who are experts in this area are needed. These decision-makers should have the ability to define independent indicators that embrace all required aspects. In this regard, the MIVES model can be improved by considering interconnections of indicators. In this sense AHP, which is the weighting assignment system that is applied in MIVES method, makes indirect connections between indicators. Furthermore, this study suggests applying the Analytic Network Process (ANP) to use advantages of AHP and in order to strengthen the interconnections of indicators.

Additionally, as different cities have various local living standards and characteristics, the weight of indicators, criteria, and requirements would be different from one area to another (Davis 1982; Johnson 2007a). Therefore, the MIVES methodology, which can be used for different locations with diverse characteristics without being limited by the present conjuncture, has more suitability for this issue. Furthermore, this model is capable of engaging local specialist and authorities from divers' departments in decision-making processes.

Indeed, in MIVES methodology, indexes' weights are determined by experts during several seminars and meeting using AHP. It is an appropriate idea to use AHP method, which helps to organize the process efficiently, to reduce the model complexity and subjectivity and decrease possible disagreements between the team members (del Caño et al. 2015). To this end, the participation of all stakeholders in management processes is needed (Kapucu and Garayev 2011). Moreover, these stakeholders should have different expertise, or their assignation of weights could not reach the most suitable outcomes. In this sense, MIVES uses seminars of experts to determine the weights that assist to find out suitable results and eliminate outliers.

Additionally, in MIVES weights assignation process, other approaches such as the Shannon's entropy can be applied (can see (Hosseini et al. 2016b)). Furthermore, decision-makers who apply MIVES can change weights easily and quickly in order to analyse different scenarios and results (can see (Hosseini et al. 2016a, b)). In this regard, when weights of each requirement, which includes more criteria and indicators, are changed, sustainability value is changed extremely compared to change of weights of other requirements. However, this is a common characteristic of all requirements of weighting-based decision-making models.

10 Conclusion

This chapter explored and examined the suitability of several decision-making models for post-disaster housing by considering the techniques that had been applied by previous researchers. In order to determine their suitability, this study defines the essential requirements of decision-making methods for PDH and analyses five different case studies. The assessment of the cases with diverse characteristics in terms of financial power, social levels, and natural hazards confirms that almost all decision processes had considerable problems to arrive the solutions. In consequence, in order to define a suitable PDH, it is required to replace any old problematic decision process with the new more successful decision-making model. Additionally, this study finds that all the factors, which are involved in decision-making process, can be organized into the three main vertexes: *Properties*, *Requirements*, and *Limitations*.

This study demonstrates that the MIVES method has more appropriate characteristics for dealing with PDH compared to the other assessed methods. In general, decision-makers could obtain suitable alternatives easily and quickly by applying MIVES in an emergency situation after disaster events requiring temporary housing. Meanwhile, MIVES like other decision-making models has advantages and disadvantages, which need to be considered. However, MIVES has been selected as the most suitable model because of its positive features:

- It permits all stakeholders to participate in the decision-making process.
- It incorporates value functions based on the utility theory concept that leads to achieve more accurate results. This prevents to choose an alternative with ineligible features that are vital for all decision process, especially PDH.
- It specifies the best alternative(s), it ranks all alternatives, and it identifies the major characteristics and the appropriate area of each alternative during operation.

This study provides a platform that is going to be an opportunity for future studies in order to assess decision-making methods. This platform is based on the main requirements of the case studies' decision-making processes. However, this platform could assess other previous case studies after being properly adapted.

References

Aguado, A., Caño, A. d., Cruz, M. P., Gómez, D., & Josa, A. (2012). Sustainability assessment of concrete structures within the Spanish structural concrete code. *CONSTRUCTION Engineering and Management, 138*(2), 268–276. https://doi.org/10.1061/(ASCE)CO.1943-7862.0000419.

Alarcon, B., Aguado, A., Manga, R., & Josa, A. (2011). A value function for assessing sustainability: Application to industrial buildings. *Sustainability, 3*, 35–50. https://doi.org/10.3390/su3010035.

Alexander, D. E. (2004). Planning for post-disaster reconstruction. I-Rec 2004 International Conference Improving Post-Disaster Reconstruction in Developing Countries.

Alexander, D. E. (2010). The L'Aquila earthquake of 6 April 2009 and Italian government policy on disaster response. *Journal of Natural Resources Policy Research, 2*(4), 325–342. https://doi.org/10.1080/19390459.2010.511450.

Alparslan, E., Ince, F., Erkan, B., Aydöner, C., Özen, H., & Dönertaş, A. (2008). A GIS model for settlement suitability regarding disaster mitigation, a case study in Bolu Turkey. *Engineering Geology, 96*(3), 126–140. https://doi.org/10.1016/j.enggeo.2007.10.006 .

Amini Hosseini, K., Hosseinioon, S., & Pooyan, Z. (2013). An investigation into the socioeconomic aspects of two major earthquakes in Iran. *Disasters, 37*(3), 516–535. https://doi.org/10.1111/disa.12001.

Arslan, H. and Unlu, A. (2006), "The evaluation of community participation in housing reconstruction projects after Duzce earthquake", 2006 International Conference and Student Competition on Post-disaster Reconstruction "Meeting Stakeholder Interests", p 13, available at: https://www.grif.umontreal.ca/pages/ARSLAN_%20 Hakan.pdf (accessed October 15, 2015).

Aruldoss, M., Lakshm, T. M., & Venkatesan, V. P. (2013). A survey on multi criteria decision making methods and its applications. *American Journal of Information Systems, 1*(1), 31–43. https://doi.org/10.12691/ajis-1-1-5.

Barakat, S. (2003). *Housing reconstruction after conflict and disaster*. London: Overseas Development Institute.

Chandler, P. J. (2007). Environmental factors influencing the siting of temporary housing. Master thesis, Louisiana State University, Department of Environmental Studies , Louisiana State University.

Chen, Z., Chen, X., Li, Q., & Chen, J. (2013). The temporal hierarchy of shelters: A hierarchical location model for earthquake-shelter planning. *International Journal of Geographical Information Science, 27*(8). https://doi.org/10.1080/13658816.2013.763944.

Chua, J., & Su, Y. (2012). The application of TOPSIS method in selecting fixed seismic shelter for evacuation in cities. *Systems Engineering Procedia, 3*, 391–397. https://doi.org/10.1016/j.sepro.2011.10.061.

Collins, S., Corsellis, T., & Vitale, A. (2010). Transitional shelter: understanding shelter from the emergency through reconstruction and beyond. ALNAP.

Cuadrado, J., Zubizarreta, M., Pelaz, B., & Marcos, I. (2015a). Methodology to assess the environmental sustainability of timber structures. *Construction and Building Materials, 86*, 149–158. https://doi.org/10.1016/j.conbuildmat.2015.03.109.

Cuadrado, J., Zubizarreta, M., Rojí, E., García, H., & Larrauri, M. (2015b). Sustainability-related decision making in industrial buildings: An AHP analysis. *Mathematical Problems in Engineering, 2015*, 13. https://doi.org/10.1155/2015/157129.

Coffey, V., & Trigunarsyah, B. (2012). Rebuilding housing after a disaster: factors for failure. In Proceedings of 8th Annual International Conference of the International Institute for Infrastructure, Renewal and Reconstruction (IIIRR) (pp. 292–300). Kumamoto University.

de la Fuente, A., Armengou, J., Pons, O., & Aguado, A. (2014). New precast concrete tower system for wind – Turbine support and tool to asses its sustainability index. Civil Engineering and Management.

del Caño, A., Gomez, D., & de la Cruz, M. P. (2012). Uncertainty analysis in the sustainable design of concrete structures: A probabilistic method. *Construction and Building Materials, 37*, 865–873. https://doi.org/10.1016/j.conbuildmat.2012.04.020.

del Caño, A., Cruz, M. P., Cartelle, J. J., & Lara, M. (2015). Conceptual framework for an integrated method to optimize sustainability of engineering systems. *Energy and Power Engineering, 9*, 608–615. https://doi.org/10.17265/1934-8975/2015.07.002.

Da Silva, J. (2010). Key Considerations in Post-Disaster Reconstruction.

Davidson, C. H. (2009). The challenge of organizational design for manufactured construction. Construction innovation, 9(1), 42–57. doi:https://doi.org/10.1108/14714170910931534.

Davis, I. (1982). *Shelter after disaster*. Oxford: Oxford Polytechnic Press.

Dikmen, N., Elias-Ozkan, S. T., & Davidson, C. (2012). Comparison of post-disaster housing procurement methods in rural areas of Turkey. *Open House International, 37*(1), 28–39.

Doocy, S., Gabriel, M., Collins, S., Robinson, C., & Stevenson, P. (2006). Implementing cash for work programmes in post-tsunami Aceh: Experiences and lessons learned. *Disasters, 30*(3), 277–296. https://doi.org/10.1111/j.0361-3666.2005.00321.x.

El-Anwar, O., & Chen, L. (2012). Computing a displacement distance equivalent to optimize plans for Postdisaster temporary housing projects. *Journal of Construction Engineering and Management, 139*(2), 174–184. https://doi.org/10.1061/(ASCE)CO.1943-7862.0000601.

El-Anwar, O., El-Rayes, K., & Elnashai, A. (2009a). Optimizing large-scale temporary housing arrangements after natural disasters. *Computing in Civil Engineering, 23*(2). https://doi.org/10.1061/(ASCE)0887-3801(2009)23:2(110)).

El-Anwar, O., El-Rayes, K., & Elnashai, A. (2009b). An automated system for optimizing post-disaster temporary housing allocation. *Computing in Civil Engineering, 23*(2), 110–118. https://doi.org/10.1061/(ASCE)0887-3801(2009)23:2(110))10.1061/(ASCE)0887-3801(2009)23:2(110)).

Erdik, M. (2000). Report on 1999 Kocaeli and Düzce (Turkey) Earthquakes.

Fois, F., & Forino, G. (2014). The self-built ecovillage in L'Aquila, Italy: Community resilience as a grassroots response to environmental shock. *Disasters, 38*(4), 719–739. https://doi.org/10.1111/disa.12080.

Ghafory-Ashtiany, M., & Hosseini, M. (2008). Post-bam earthquake: Recovery and reconstruction. *Natural Hazards, 44*(2), 229–241. https://doi.org/10.1007/s11069-007-9108-3.

Gharaati, M. & Davidson, C. (2008). Who Knows Best? An Overview of Reconstruction after the Earthquake in Bam, Iran. In: Proceedings of the 4th International i-Rec Conference, University of Canterbury. Christchurch, New Zealand.

Hale, T., & Moberg, C. R. (2005). Improving supply chain disaster preparedness: A decision process for secure site location. *International Journal of Physical Distribution & Logistics Management, 35*(3), 195–207. https://doi.org/10.1108/09600030510594576.

Hall, A. D. (1962). A methodology for systems engineering. Princeton: D. Van Nostrand Company, Inc.

Hayles, C. S. (2010). An examination of decision making in post disaster housing reconstruction. *International Journal of Disaster Resilience in the Built Environment, 1*(1), 103–122. https://doi.org/10.1108/17595901111149141.

Hidayat, B. A. (2010). A literature review of the role of project management in post-disaster reconstruction. *Proceedings of 26th annual ARCOM conference* (pp. 1269–1278). Association of Researchers in Construction Management.

Hosseini, S. M. A., de la Fuente, A., & Pons, O. (2016a). Multi-criteria decision-making method for assessing the sustainability of post-disaster temporary housing units technologies: A case study in Bam, 2003. *Sustainable Cities and Society, 20*, 38–51. https://doi.org/10.1016/j.scs.2015.09.012.

Hosseini, S. M. A., de la Fuente, A., & Pons, O. (2016b). Multicriteria decision-making method for sustainable site location of post-disaster temporary housing in urban areas. *Construction Engineering and Management.* https://doi.org/10.1061/(ASCE)CO.1943-7862.0001137.

Hosseini, S. M. A., Pons, O., Arroyo, C. M., & de la Fuente, A. (2016c). Identifying temporary housing main vertexes through assessing post-disaster recovery programs. *World Academy of Science, Engineering and Technology, International Journal of Civil, Environmental, Structural, Construction and Architectural Engineering, 10*(10), 1344–1351.

Hosseini, S. M. A., Pons, O., & de la Fuente, A. (2017). A combination of the knapsack algorithm and MIVES for choosing optimal temporary housing site locations: A case study in Tehran. *International Journal of Disaster Risk Reduction.* https://doi.org/10.1016/j.ijdrr.2017.10.013.

Hwang, C.-L., & Yoon, K. (2012). *Multiple attribute decision making: Methods and applications a state-of-the-art survey* (Vol. 186). Springer Science & Business Media, Springer-Verlag Berlin Heidelberg New York.

Hadafi, F., & Fallahi, A. (2010). Temporary housing respond to disasters in developing countries-case study: Iran-Ardabil and Lorestan Province Earthquakes. *World Academy of Science, Engineering and Technology, 66*, 1536–1542.

Johnson, C. (2002). What's the big deal about temporary housing? Planning considerations for temporary accommodation after disasters: Example of the 1999 Turkish earthquakes. In 2002 TIEMS disaster management conference. Waterloo.

Johnson, C. (2007a). Impacts of prefabricated temporary housing after disasters: 1999 earthquakes in Turkey. *Habitat International, 31*(1), 36–52. https://doi.org/10.1016/j.habitatint.2006.03.002.

Johnson, C. (2007b). Strategic planning for post-disaster temporary housing. *Disasters, 31*(4), 435–458. https://doi.org/10.1111/j.0361-3666.2007.01018.x.

Johnson, C. (2009). Planning for temporary. In C. J. Gonzalo Lizarralde (Ed.), *Rebuilding after disasters: From emergency to sustainability* (pp. 70–87). New York: Taylor & Francis.

Johnson, C., Lizarralde, G., & Davidson, C. H. (2006). A systems view of temporary housing projects in post-disaster reconstruction. *Construction Management and Economics, 24*(4), 367–378. https://doi.org/10.1080/01446190600567977.

Kapucu, N., & Garayev, V. (2011). Collaborative decision-making in emergency and disaster management. *International Journal of Public Administration, 34*(6), 366–375. https://doi.org/10.1080/01900692.2011.561477.

Kates, R. W., Colten, C. E., Laska, S., & Leatherman, S. P. (2006). Reconstruction of New Orleans after hurricane Katrina: A research perspective. *Proceedings of the National Academy of Sciences, 103*(40), 14653–14660.

Kennedy, J., Ashmore, J., Babister, E., & Kelman, I. (2008). The meaning of 'build back better': Evidence from post-tsunami Aceh and Sri Lanka. *Contingencies and Crisis Management, 16*(1), 24–36. https://doi.org/10.1111/j.1468-5973.2008.00529.x.

Khazai, B., & Hausler, E. (2005). Intermediate shelters in bam and permanent shelter reconstruction in villages following the 2003 bam, Iran, earthquake. *Earthquake Spectra, 21*(S1), 487–511. https://doi.org/10.1193/1.2098907.

Li, Y., Liu, Y., & Jiao, J. (2013). A GIS-based suitability analysis of Xiamen's green space in park for earthquake disaster prevention and refuge. *Urban Planning and Design Research, 1*(1) 1–8.

Lindell, M. K., & Prater, C. S. (2003). Assessing community impacts of natural disasters. *Natural Hazards Review, 4*(4), 176–185. https://doi.org/10.1061/(ASCE)1527-6988(2003)4:4(176).

Lizarralde, G., & Davidson, C. (2006). Learning from the poor. i-Rec conference proceedings 2006: Post-disaster reconstruction: Meeting the stakeholders' interest.

McConnan, I. (1998). Humanitarian charter and minimum standards in disaster response. < The> Sphere Project.

Ma, D., Chu, J., Liu, X., & Zhao, S. (2011). Study on evaluation of earthquake evacuation capacity in village based on multi-level Grey evaluation. *Systems Engineering Procedia, 1*, 85–92. https://doi.org/10.1016/j.sepro.2011.08.015.

McCarthy, F. X. (2008). FEMA disaster housing and hurricane Katrina: Overview, analysis, and congressional options. Congressional Research Service, Library of Congress.

McCarthy, K. F., Peterson, D. J., Sastry, N., & Pollard, M. (2006). *The repopulation of New Orleans after hurricane Katrina.* Santa Monica: Rand Corporation.

Nigg, J. M., Barnshaw, J., & Torres, M. R. (2006). Hurricane Katrina and the flooding of New Orleans: Emergent issues in sheltering and temporary housing. *The Annals of the American Academy of Political and Social Science, 604*(1), 113–128. https://doi.org/10.1177/0002716205285889.

Nojavan, M., & Omidvar, B. (2013). *The Selection of Site for Temporary Sheltering using Fuzzy*

Algorithms;Case study: Tehran Metropolitan after earthquake,Municipal districts No, 1(31), 205–221.

Omidvar, B., Baradaran-Shoraka, M., & Nojavan, M. (2013). Temporary site selection and decision-making methods: A case study ofTehran, Iran. *Disasters, 37*(3), 536–553. https://doi.org/10.1111/disa.12007.

Ormazabal, G., Viñolas, B., & Aguado, A. (2008). Enhancing value in crucial decisions: Line 9 of the Barcelona Subway. *Journal of Management in Engineering, 24*(4), 265–272. https://doi.org/10.1061/(ASCE)0742-597X(2008)24:4(265).

Ozernoy, V. M. (1989). A framework for building an expert system for MCDM models selection. In *In improving decision making in Organisations* (pp. 553–562). Berlin/Heidelberg: Springer.

Özerdem, A., & Rufini, G. (2013). L'Aquila's reconstruction challenges: Has Italy learned from its previous earthquake disasters? *Disasters, 37*(1), 119–143. https://doi.org/10.1111/j.1467-7717.2012.01296.x.

Pearce, L. (2003). Disaster management and community planning, and public participations: How to achieve sustainable hazard mitigation. *Natural Hazards, 28*(2–3), 211–228. https://doi.org/10.1023/A:1022917721797.

Peng, Y., Shen, Q., Shen, L., Lu, C., & Yuan, Z. (2014). A generic decision model for developing concentrated rural settlement in post-disaster reconstruction: A China study. *Natural Hazards, 71*(1), 611–637. https://doi.org/10.1007/s11069-013-0924-3.

Pons, O., & Aguado, A. (2012). Integrated value model for sustainable assessment applied to technologies used to build schools in Catalonia, Spain. *Building and Environment, 53*, 49–58. https://doi.org/10.1016/j.buildenv.2012.01.007.

Pons, O., & de la Fuente, A. (2013). Integrated sustainability assessment method applied to structural concrete columns. *Construction and Building Materials, 49*, 882–893. https://doi.org/10.1016/j.conbuildmat.2013.09.009.

Pujadas, P., Pardo-Bosch, F., Aguado-Renter, A., & Aguado, A. (2017). MIVES multi-criteria approach for the evaluation, prioritization, and selection of public investment projects. A case study in the city of Barcelona. *Land Use Policy, 64*, 29–37. https://doi.org/10.1016/j.landusepol.2017.02.014.

Rossetto, T., D'Ayala, D., Gori, F., Persio, R., Han, J., Novelli, V., et al. (2014). The value of multiple earthquake missions: The EEFIT L'Aquila earthquake experience. *Bulletin of Earthquake Engineering, 12*(1), 277–305. https://doi.org/10.1007/s10518-014-9588-y.

San-José Lombera, J.-T., & Rojo, J. C. (2010). Industrial building design stage based on a system approach to their environmental sustainability. *Construction and Building Materials, 24*(4), 438–447. https://doi.org/10.1016/j.conbuildmat.2009.10.019 .

San-José Lombera, J. T. , & Garrucho Aprea, I. G. (2010). A system approach to the environmental analysis of industrial buildings. *Building and environment, 45*(3), 673–683. https://doi.org/10.1016/j.buildenv.2009.08.012.

Simon, H. A. (1996). The sciences of the artificial, 2nd ed., MIT Press, Cambridge, MA.

Stanujkic, D., Djordjevic, B., & Djordjevic, M. (2013). Comparative analysis of some prominent MCDM methods: A case of ranking Serbian banks. *Serbian Journal of Management, 8*(2), 213–241. https://doi.org/10.5937/sjm8-3774.

Steinberg, F. (2007). Housing reconstruction and rehabilitation in Aceh and Nias, Indonesia—Rebuilding lives. *Habitat International, 31*(1), 150–166. https://doi.org/10.1016/j.habitatint.2006.11.002.

Tas, N., Cosgun, N., & Tas, M. (2007). A qualitative evaluation of the after earthquake permanent housings in Turkey in terms of user satisfaction—Kocaeli, Gundogdu permanent housing model. *Building and Environment, 42*, 3418–3431. https://doi.org/10.1016/j.buildenv.2006.09.002.

Triantaphyllou, E. (2013). *Multi-criteria decision making methods: A comparative study* (Vol. 44). New York: Springer Science & Business Media.

von Meding, J., Wong, J., Kanjanabootra, S., & Tafti, M. T. (2016). Competence-based system development for post-disaster project management. *Disaster Prevention and Management: An International Journal, 25*(3), 375–394. https://doi.org/10.1108/DPM-07-2015-0164.

Velasquez, M., & Hester, P. T. (2013). An analysis of multi-criteria decision making methods. *International Journal of Operations Research, 10*(2), 56–66.

Viñolas, B. (2011). *Applications and advances of MIVES methodology in multi-criteria assessments*. PhD thesis, UPC, Barcelona.

Viñolas, B., Aguado, A., Josa, A., Villegas, N., & Prada, M. Á. (2009). Aplicación del análisis de valor para una evaluación integral y objetiva del profesorado universitario. *6*(1).

Wei, L., Li, W., Li, K., Liu, H., & Chenguse, A. L. (2012). Decision Support for urban shelter locations based on covering model. *Procedia Engineering, 43*, 59–64. https://doi.org/10.1016/j.proeng.2012.08.011.

Weiss, N. E. (2006). Rebuilding housing after hurricane Katrina: Lessons learned and unresolved issues. Congressional Research Service, Library of Congress.

Yi, H., & Yang, J. (2014). Research trends of post disaster reconstruction: The past and the future. *Habitat International, 42*, 21–29. https://doi.org/10.1016/j.habitatint.2013.10.005.

Zanakis, S. H., Solomon, A., Wishart, N., & Dublish, S. (1998). Multi-attribute decision making: A simulation comparison of select methods. *European Journal of Operational Research, 107*(3), 507–529. https://doi.org/10.1016/S0377-2217(97)00147-1.

Zavadskas, E. K., Mardani, A., Turskis, Z., Jusoh, A., & Nor, K. M. (2016). Development of TOPSIS method to solve complicated decision-making problems: An overview on developments from 2000 to 2015. *International Journal of Information Technology & Decision Making, 15*, 645–682. https://doi.org/10.1142/S0219622016300019.

Index

A. Asgary (ed.), *Resettlement Challenges for Displaced Populations and Refugees*,
Sustainable Development Goals Series, https://doi.org/10.1007/978-3-319-92498-4

The manufacturer's authorised representative in the EU is Springer
Nature Customer Service Centre GmbH, Europaplatz 3, 69115 Heidelberg,
Germany. If you have any concerns regarding our products, please
contact ProductSafety@springernature.com

Printed and bound by CPI Group (UK) Ltd, Croydon, CR0 4YY
05/10/2026
02228141-0020